CONTROL OF BIOLOGICAL AND DRUG-DELIVERY SYSTEMS FOR CHEMICAL, BIOMEDICAL, AND PHARMACEUTICAL ENGINEERING

CONTROL OF BIOLOGICAL AND DRUG-DELIVERY SYSTEMS FOR CHEMICAL, BIOMEDICAL, AND PHARMACEUTICAL ENGINEERING

LAURENT SIMON
Otto H. York Department of Chemical, Biological and
Pharmaceutical Engineering
New Jersey Institute of Technology

A JOHN WILEY & SONS, INC., PUBLICATION

Cover illustration: Courtesy of Laurent Simon

Published by John Wiley & Sons, Inc., Hoboken, New Jersey
Published simultaneously in Canada

For general information on our other products and services or for technical support, please contact our Customer Care Department within the United States at (800) 762-2974, outside the United States at (317) 572-3993 or fax (317) 572-4002.

Wiley also publishes its books in a variety of electronic formats. Some content that appears in print may not be available in electronic formats. For more information about Wiley products, visit our web site at www.wiley.com.

Library of Congress Cataloging-in-Publication Data:

Simon, Laurent, 1968–
    Control of biological and drug-delivery systems for chemical, biomedical, and pharmaceutical engineering / Laurent Simon.
        p. ; cm.
    Includes index.
    ISBN 978-0-470-90323-0 (cloth)
    I. Title.
    [DNLM: 1: Chemistry, Pharmaceutical–methods.   2. Models, Theoretical.   3. Technology, Pharmaceutical–methods.   QV 744]
    615.1'9–dc23

                                                                                2012011008

Printed in the United States of America

10 9 8 7 6 5 4 3 2 1

CONTENTS

PREFACE

The control of biological and drug-delivery systems is critical to providing a long and healthy life to millions of people worldwide. In living systems, maintenance of homeostasis is credited to several control mechanisms (e.g., positive and negative feedback loops). Researchers in systems biology and controlled-release devices continue to use dynamics and control theory to increase their understanding of cell behavior, to treat diseases, and to develop drug administration protocols.

As the need to develop and commercialize bio-based products becomes more prevalent, chemical engineering departments throughout the nation have begun to shift their focus from a curriculum centered on the knowledge of chemical plant operations to a program that includes biological and pharmaceutical applications. Consequently, a multidisciplinary approach is mandatory to help ensure that chemical engineering graduates secure employment in industries where expertise in bioprocess and drug delivery is needed.

This textbook combines knowledge of process dynamics and basic control theory to analyze processes in the chemical, biomedical, and pharmaceutical engineering fields. Chemical process control topics, such as external disturbances, transfer functions, and input/output models, will be covered and enhanced by examples selected in the focus areas. Armed with this information, students will be in a strong position to address issues and to solve problems that dominate both fields (i.e., biological sciences and release devices).

Because most textbooks published in these areas are written for graduate-level study, undergraduate chemical engineering students are not exposed to diversified problems in biological sciences. This book is the first of its kind to provide biological and drug-delivery applications for dynamics and control concepts taught at the undergraduate level.

An expected result of the proposed perspective is an enrichment of fundamental concepts and the development of an application-oriented environment that gives students broader career choices and a competitive edge in the job market. The new outlook is also indispensable in developing technologies and in providing effective medicine to millions of people in need of gene therapies, heart–lung bypasses and dialysis machines. Although written primarily for undergraduate chemical and biomedical engineering students, this book's focus on drug-delivery systems and its coverage of a wide range of topics in the biological sciences is expected to appeal to a large audience in pharmaceutical engineering and systems biology.

The textbook is organized so that theory is accompanied by illustrations in several areas. Chapter 1 outlines the role of process dynamics and control in a number of disciplines and a brief overview of instrumentations. Chapter 2 introduces mathematical modeling based on the physical knowledge of a system. In Chapter 3, techniques are developed to linearize process models around nominal points. The concept of deviation variables is also introduced. Stability considerations and phase diagrams are addressed in Chapter 4. The properties of the Laplace operator are described in Chapter 5. Laplace transforms of several functions and ordinary and partial differential equations are computed. Techniques for inverting Laplace transforms are provided in Chapter 6. Partial fraction expansion and the residue theorem are applied to obtain closed-form solutions for differential equations. Chapter 7 discusses derivations of transfer functions from input–output models. This approach is fundamental for controller analysis and design. Physical systems, represented by ordinary and partial differential equations, are discussed. Dynamic behaviors of open-loop systems that are introduced in Chapter 8 deal with rational and transcendental transfer functions. Strategies to derive reduced-order models are also presented. In Chapter 9, control methodologies are developed. The emphasis is placed on three widely used feedback controllers: the proportional, proportional–integral, and proportional–integral–derivative controllers. In Chapter 10, frequency response analyses are studied and methods to draw Bode and Nyquist plots are described. Techniques to analyze the stability of feedback systems are developed in Chapter 11. Examples from biological processes are provided to illustrate the implementation of these tools. In Chapter 12, tuning guidelines for feedback controllers are provided. The Smith predictor, a model-based method to help reduce the effects of dead time on closed-loop performance, is discussed in Chapter 13. Using this structure, the controller acts on a delay-free response. The fundamentals of cascade and feedforward control designs are covered in Chapter 14. Both architectures provide methods for lessening the impact of disturbances on the controlled variable. A technique for determining a relaxation time for lumped- and distributed-parameter systems is explained in Chapter 15. Based on Laplace transforms, the time to reach a steady-state value can be estimated. Examples of optimum control and design problems encountered in biomedicine are presented in Chapter 16.

This textbook is the result, in part, of my experience as an instructor of process control. I have expertise in process dynamics and control, bioprocesses, and drug-delivery systems and have written over 35 refereed articles and book chapters on biotechnology, controlled release, and mathematical modeling. My unique experiences in teaching biotransport to biomedical and chemical engineering students have exposed me to an assortment of problems that are relevant to both disciplines. My perspective on process dynamics and control has been enriched by courses such as Introduction to Biotechnology and Pharmaceutical Engineering Fundamentals. For additional information, visit my website http://www.laurentsimon.com or http://web.njit.edu/~lsimon/.

LAURENT SIMON

ACKNOWLEDGMENTS

I would like to thank Dr. Kwang Seok Kim who contributed the solutions for most of the problems for Chapters 1 through 6. In addition, I express my sincere gratitude to graduate students Ani Abraham, Aniruddha Banerjee, and Nikola Maric, who helped in the editing process and provided invaluable comments and recommendations, and to my friends and colleagues for their encouragement and support. Most of all, I am grateful to my mother, Claudie, for her unconditional love, to my wife, Lisa, for her constructive suggestions and active involvement in nearly every aspect of the book, and to my daughter, Francesca, who created in me the drive and inspiration to complete the project.

L. S.

CHAPTER 1

INTRODUCTION

Examples, pertinent to the application of process control in some areas of bioprocessing and drug delivery, are outlined below to underscore the ubiquitous nature of this technology. Concepts of disturbance variables, set points, manipulated variables, and controlled variables are introduced. Block diagrams are drawn to describe processes. A list of hardware and software required to implement control algorithms is also included.

1.1 THE ROLE OF PROCESS DYNAMICS AND CONTROL IN BRANCHES OF BIOLOGY

Biology deals with the study of living organisms and vital processes. A close examination of cellular functions reveals a sophisticated mechanism and a remarkable control system. The cells, fundamental units in all living things, are responsible for growth, maintenance, and reproduction. Branches of biology, such as biotechnology and physiology, have witnessed a substantial growth in the application of control theories to guide research and to promote discovery.

1.1.1 Applications in Biotechnology

The dynamics of bacterial growth, for example, involve *in vivo* and *in vitro* reactions (i.e., bioreactions). A microorganism, inoculated into a sterilized

Control of Biological and Drug-Delivery Systems for Chemical, Biomedical, and Pharmaceutical Engineering, First Edition. Laurent Simon.
© 2013 John Wiley & Sons, Inc. Published 2013 by John Wiley & Sons, Inc.

medium, undergoes a lag, an exponential growth phase, a stationary phase, and a death phase. Cell proliferation occurs in a bioreactor, a critical unit operation in biopharmaceutical, biochemical, and activated sludge processes, to name a few [1]. In the lag phase, there is little or no evidence of cell division as the bacteria adjust to their new environment. In microbial cell cultivation, the length of the lag phase can be attributed to the type and age of the microorganism, the size of the inoculum, the temperature of the medium, and nutrient concentration. As cells divide in a bioreactor, their number grows in an exponential fashion. An equilibrium phase (i.e., stationary phase) is achieved as the rate at which cells die is equal to the rate at which they divide. For *in vitro* processes, the lack of nutrients, pH changes, and reduced oxygen are among the factors that may explain why some cells enter the stationary phase. In the death phase, the number of viable cells decreases as nutrients deplete and lytic enzymes start to accumulate. Process dynamics and control can be applied, in biotechnology, to identify the factors that influence cell growth and help devise a procedure for maximizing the production of high-valued proteins. An efficient system needs to consider the different growth phases because of the diverse patterns and kinetics exhibited by the cells. Distinct methods are required depending on the production of (1) primary metabolites, excreted in the exponential growth phase, or (2) secondary metabolites, generated as the cells approach the stationary phase.

It is also necessary to regulate the environmental conditions (e.g., temperature, pH, dissolved oxygen [DO], and limiting nutrient) that affect the reactions occurring within the cells in order to achieve a desired outcome (e.g., product yield, cell concentration). The goal of process control is defined in these terms by Boudreau and McMillan [1]:

> Process control attempts to influence the individual sophisticated internal reactions of billions of cells by controlling their extracellular environment.

DO control is crucial in the cultivation of aerobic cells in bioreactors. Oxygen is required in aerobic respiration to produce energy, in the form of adenosine triphosphate (ATP), from glucose or another organic substrate. The energy consumed by the cells helps them to carry reactions, make products, reproduce, transport nutrients, and change locations. The control of DO in bioprocesses requires careful consideration and an understanding of process dynamics. For example, fermentations aimed at producing antibiotics can be highly viscous, which may lead to fluctuation in the DO concentration in the bioreactor [2]. Advanced algorithms, incorporating the kinetic data, were applied in real time to control DO in the production of aminoglycoside antibiotics from *Streptomyces*.

Figure 1.1 shows DO control when a mouse hybridoma cell line was used to produce an antibody against a tissue-type plasminogen activator (t-PA). Only some of the peripherals are shown in the schematic. Sampling and inoculum ports, humidifiers, and moisture traps are usually included. This product

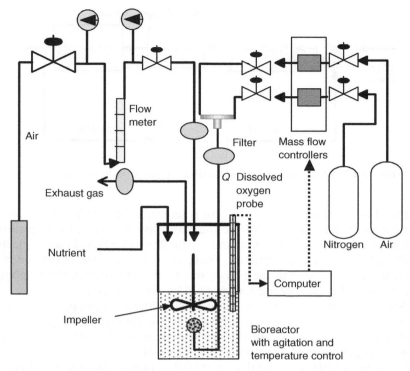

Figure 1.1. Schematic diagram for the control of dissolved oxygen.

(t-PA) has important clinical applications in heart attack research. A control strategy that depends on manipulating the airflow in the sparger, or the agitation, would not work because the hybridoma cells lack a protective cell wall and, as a result, are highly sensitive to shear forces. The basic idea is to disturb the growth environment minimally by keeping the stirring speed and gas flow rate constant. Signals from the DO probe are sent to the computer that stores a control design algorithm (i.e., control law). The computer/controller sends instructions to the mass flow controllers (MFCs) to vary the flow rates of nitrogen (F_{N2}) and air (F_{air}) while keeping a constant total gas flow rate ($Q = F_{N2} + F_{air}$). To design the control law properly, it is important to understand how the hybridoma cells respond to changes in the DO concentration and the dynamics of the DO probe. For these reasons, a fundamental knowledge of process dynamics is a critical step in control design. A *block diagram*, usually drawn to represent the system (Fig. 1.2), is a schematic representation of the interconnections or relationships among variables and processes that make up the control system. The actual DO concentration in the bioreactor is read by the DO probe, which feeds a signal to a *comparator*. The difference between a reference value, set by the operator, and the input signal is

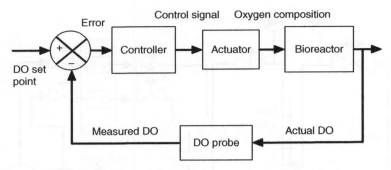

Figure 1.2. Block diagram of the feedback control of dissolved oxygen.

calculated by the comparator. This error is sent to the controller that is linked to actuators, in this case, two MFCs. The MFCs are not only able to measure the flow of gases but can also manipulate the flow rates based on the electric signals received from the computer. Then, the molar fraction of oxygen in Q is adjusted before entering the bioreactor. In this illustration, the combined comparator and controller are represented by the computer.

Also, MFCs can act as stand-alone controllers and are equipped with a mass flow sensor, a control valve, and actuators. More details will be given in the section on instrumentation. For this application, it is sufficient to say that the air and nitrogen flow rates are adjusted by the MFCs.

The *control of pH* is essential in industrial fermentation processes because of the dependence of the cell-specific growth rate and protein production on the pH. An acidic medium may be the result of depletion in the nutrient leading to the production of organic acids (e.g., acetic acids). Bioreactors are usually equipped with pH controllers. In Chinese hamster ovary (CHO) cells, CO_2 tends to accumulate in the bioreactor, especially at a high cell density [3]. The buildup of CO_2 reduces the culture pH (i.e., increase in cations) because of the following equilibrium equation:

$$CO_2 + H_2O \leftrightarrow H_2CO_3 \leftrightarrow H^+ + HCO_3^- \tag{1.1}$$

around a pH of 7.0. The addition of NaOH leads to a rise in HCO_3^- ions via

$$NaOH + H_2CO_3 \leftrightarrow Na^+ + HCO_3^- + H_2O \tag{1.2}$$

and causes the equilibrium reaction (i.e., Eq. 1.1) to shift to the left, increasing the pH. A block diagram of the process is shown in Figure 1.3. In this case, the controller manipulates the flow rates of CO_2 and NaOH in order to hold the pH at a desired value. Buffer solutions, such as bicarbonate, can also be used in the culture system to prevent significant pH changes.

The *addition of nutrient into bioreactors* often provides a way to enhance cell growth or to promote product formation. In the fed-batch mode, limiting

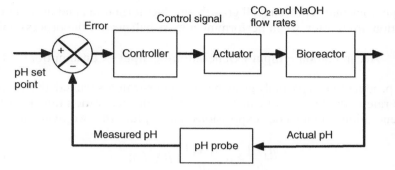

Figure 1.3. Block diagram of the feedback control of pH.

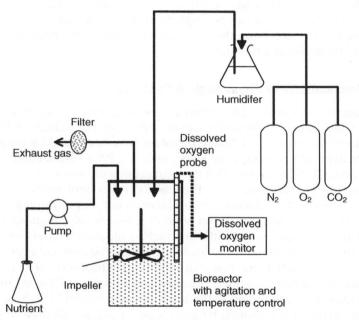

Figure 1.4. Fed-batch operation of a culture of hybridoma cells.

substrates are added to the cultivation vessel. This mode of operation is selected when the microorganism grows to a high cell density. Because the specific growth rate is often a strong function of substrate concentration, tight control of the nutrient is necessary as starvation or overfeeding can easily occur in the absence of accurate feeding protocols. A mathematical description (i.e., differential equations) of the cell culture kinetics is useful to estimate the substrate flow rate that should be added into the vessel to satisfy specified requirements. An example is shown in Figure 1.4 for hybridoma cultures [4].

Equations that represent cell growth, product formation, and nutrient consumption are written when performing a mass balance on the process unit:

$$\text{Accumulation} = \text{Input} + \text{Generation} - \text{Output} - \text{Consumption}. \quad (1.3)$$

For a product that is partially growth and non-growth-associated, the accumulation rate in the bioreactor can be set equal to the generation rate. However, the generation rate includes expressions that capture the two dynamics:

$$\frac{d[P]}{dt} = \alpha \frac{d[X_v]}{dt} + \beta X_v(t), \quad (1.4)$$

where α and β are the growth and non-growth-associated coefficients, respectively. The product P depends on the cell growth rate and the concentration of viable cells X_v. Control can be implemented to make sure that glutamine or glucose is introduced into the bioreactor at a rate that optimizes the production of monoclonal antibodies.

1.1.2 Applications in Physiological Systems

In bioreactors, cells, isolated from animal organs and tissues, are cultivated in culture media for the production of a desired product. For example, CHO cells, derived from the ovaries of Chinese hamsters, have been used for the production of protein with a therapeutic value, such as t-PA. Culture conditions are selected carefully to make sure that the extracellular environment is conducive to growth and product formation. Process control is implemented in such a context to suppress the effects of *disturbances* on key *output variables* (*regulatory control*) or to ensure that if a set point or *reference value* is changed by the operator, the bioreactor has no difficulty tracking the set point (*servo problem*). Because the environmental variables influence the biochemical events occurring within the cell, a fundamental goal of bioreactor process control is to regulate these external factors in order to preserve *homeostasis*—"the maintenance of the steady-state conditions in the internal environment" [5]. A change in metabolic activities is indicated by a decrease in pH, a reduction of glucose level, or an increase in the temperature as a result of a high viable cell concentration. In this case, external controllers are necessary to help each cell perform its metabolic function. Some pertinent questions are the following: How do cells, in their physiological surroundings, preserve metabolic equilibrium in the face of various disturbances or fluctuations? What control mechanisms are involved in the regulation of excess water by the kidneys? How do millions of interconnected neurons control the activities of muscles and allow a person to stand? How does the brain control the inner body temperature?

1.1.2.1 Pupil Light Reflex Is an Example of Physiological Control in the Human Body The pupil automatically constricts when we are exposed

to a bright light. It seems that the body is preprogrammed to let a controlled amount of light enter the eyes. The reflex is so strong that it serves to assess damage to the central nervous system or the level of human consciousness. From the way we instinctively squint when the light gets brighter, it is reasonable to deduce that there is a natural controller behaving like a thermostat that regulates a room temperature. One of the main differences between the two designs is that the inner workings of the human biological device and the control law that directs its behavior are a mystery to us, at least until we study and elucidate the underlying mechanism. In physiological systems, the engineer is being asked to characterize the plant, identify the sensor(s), and infer the type of control strategy being implemented. Goodman summarized the challenging tasks in the following terms [6]:

> These tasks are made especially difficult by the constraint that system viability cannot be threatened by probes required to make necessary interventions and measurements. Nor is one permitted arbitrarily to partition the system into discrete units or assemblies and use "stimulus–response" (input-output) techniques to characterize structure and function, i.e., procedures must be "noninvasive" or as nearly so as possible. One hypothetical analog is that of a control engineer assigned the task of improving the performance of a petrochemical plant by synthesizing an automatic process control system. Detailed design drawings and specifications, the engineer is told, are not complete. Although the plant is partially visible behind a high fence, direct access is denied. Thus, the engineer faces the challenge of determining what is inside the fence on the basis of prior knowledge of how some similar plants operate, and the freedom noninvasively to measure what goes through the wires and pipes that connect the plant to the environment. The engineer, with caution, may be permitted small variations of some input fluxes providing that plant throughput and product quality are not compromised.

There are identification methods, commonly used in control engineering, that can be applied to model how the retina responds to a change in light intensity (Fig. 1.5) [7]. A light flux of a specific intensity (L_c) falls on the retina. The

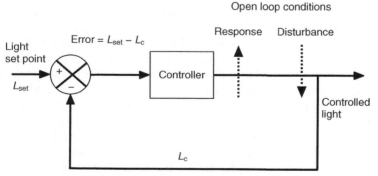

Figure 1.5. Block diagram of the pupillary light reflex system.

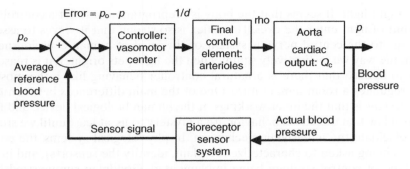

Figure 1.6. Feedback control of blood pressure using the vasomotor center.

difference between the set (or reference) point L_{set} and the controlled variable L_c is fed to the controller (i.e., the pupil neuromuscular apparatus). The latter adjusts the pupil size to drive the error to zero. Experiments were conducted to introduce disturbances to the system (i.e., change in the light flux) and to measure the response (i.e., change in the pupil area) without compromising the system.

1.1.2.2 The Control of Human Arterial Blood Pressure Provides an Illustration of Biological Control

In the block diagram shown in Figure 1.6, the aorta represents the *plant* [8]. The cardiac output (Q_c) is defined as the volumetric flow rate of blood from the heart to the aorta. Its value can be calculated by multiplying the stroke volume (i.e., volume of blood pumped with each beat) by the heart rate. The *controlled variable p* (arterial blood pressure: outward force by the blood flow against the arterial walls) is obtained by multiplying Q_c by the peripheral resistance (ρ) imposed by the arterioles. The *output p* is fed to the baroreceptor sensor system, which sends a signal to a comparator where the indicated value (p) is compared to reference p_o. Baroreceptors are pressure sensors found in the blood vessels of some mammals. These devices sense the resistance of blood flow against the vessel walls and send impulses (*action potentials*) to the brain via *glossopharyngeal* and *vagus* nerves. The error is then fed to the cardioregulatory and vasomotor centers (VMC) in the brain (i.e., *controller*). In this example, we consider the VMC as the only controller to simplify the analysis (constant Q_c). An increased activity in the VMC is accompanied by a decrease in the diameter of the arterioles (d) and an increase in the peripheral resistance since $\rho = k(1/d)^4$, where k is a proportionality constant. In reality, the VMC transmits sympathetic stimulations (Act_{VMC}) to the blood vessels. Similarly, as the activity in the VMC is decreased, d is increased, which results in a decrease in ρ [8]. From the equation $p = Q_c\rho$, a probable scenario that explains the regulation of cardiac output can be deduced. For example, if $p > p_0$, as a result of a *disturbance*, $\varepsilon < 0 \Rightarrow Act_{VMC} \downarrow \Rightarrow d \uparrow \Rightarrow \rho \downarrow \Rightarrow p \downarrow$. A similar analysis can be conducted to show how both the VMC and the cardiovascular center—which sends sympa-

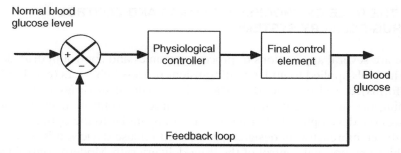

Figure 1.7. Feedback control of blood glucose by insulin in a nondiabetic person.

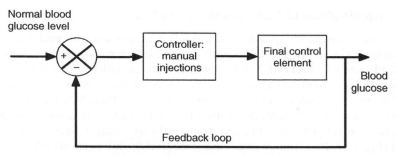

Figure 1.8. Feedback control of blood glucose by insulin in a diabetic person.

thetic stimulations to the heart affecting the heart rate and stroke volume—regulate the blood pressure.

1.1.2.3 *Blood Glucose Can Be Controlled by Exogenous Insulin Injections* Insulin-dependent diabetes mellitus (IDDM), or type 1 diabetes, is a chronic disease that may lead to blindness, heart disease, and stroke in people afflicted with the illness. In type 1 diabetes, the pancreatic beta cells, responsible for secreting the insulin hormone, have been destroyed as a result of an autoimmune process. In the absence of insulin, the way the body converts sugar (i.e., glucose) in the blood into energy, made available to all the cells in the body, is not well regulated. In other words, the cells either have too much glucose (i.e., hyperglycemia) in the blood plasma or an inadequate amount in the case of hypoglycemia. Under healthy conditions, a physiological regulator, such as the ones described above, makes sure that a constant glucose level is held. A simplified block diagram is provided in Figure 1.7. A more detailed block diagram, similar to Figure 1.6, can be drawn that includes, for example, the stimulation of the beta cells and secretion of insulin.

In type 1 diabetic patients, the control system is replaced by an external mechanism where people monitor their daily glucose concentrations and take insulin shots in order to maintain a constant blood glucose level (Fig. 1.8). In these situations, accurate and timely readings of the glucose are paramount.

1.2   THE ROLE OF PROCESS DYNAMICS AND CONTROL IN DRUG-DELIVERY SYSTEMS

There are several applications of process dynamics and control in drug delivery. The tools applied to understand transient process behaviors (e.g., Laplace transforms) can also be used to track the evolution of drugs in the body. Keeping the pharmaceutical agent at a desired concentration in the blood requires the development of programming tools similar to the optimization of a fed-batch bioreactor. In designing controlled-release devices, it is important to achieve an optimal control of the drug-delivery rate. Modern control theories are applied to optimize treatment strategies.

1.2.1   Applications in Compartmental Models

Drug delivery, using compartmental models, requires a sound understanding of pharmacokinetics, which deals with drug absorption, distribution, metabolism, and excretion (ADME). A therapeutic concentration of the active pharmaceutical ingredient (API) must be able to reach the target site. Knowledge of the transient behaviors of compartment models is therefore necessary.

The one-compartment model is used to describe uniform drug distribution in the body. Based on this representation, the body behaves like a well-stirred vessel (Fig. 1.9). This framework is implemented for intravenous (IV) boluses (i.e., instantaneous delivery) and continuous infusions. After a single dose is administered, the drug is distributed throughout the body and metabolized. If a second dose is not taken on time, the medication may be ineffective as the plasma concentration drops well below a therapeutic level. Pharmacokinetic information helps compute drug-dosing regimens. In the context of IV boluses, several injections must be administered at predetermined times in order to maintain a plasma drug concentration between the toxic and ineffective levels (Fig. 1.10). The time to reach a desired drug concentration can be estimated using dynamic analysis.

Process dynamics plays a major role in analyzing drug administration via constant-rate infusion during hospitalization. One of the main advantages of an IV infusion is the elimination of fluctuations evident in bolus IV dosing. However, the pharmacokinetics must be carefully assessed to make sure that it does not take a long time to establish a steady-state plasma drug concentration. This information can also help decide whether a combination of IV

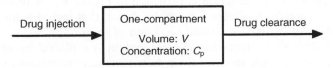

Figure 1.9. Representation of a one-compartment model.

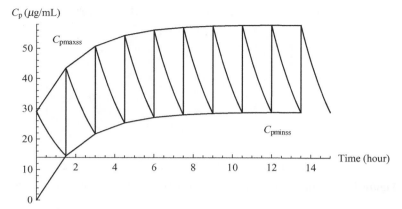

Figure 1.10. Plasma concentration profile for a multiple IV bolus regimen.

boluses and constant-rate infusions is the most appropriate way to administer the medication.

1.2.2 Applications in Controlled-Release Technology

The purpose of controlled drug release is to guarantee that the medicament is delivered at a specified rate and is sustained in the blood at a desired concentration. Unlike IV boluses and continuous infusions, needles are not used. The transdermal route enhances patient compliance, which is essential for therapy to work. A patch is placed on the skin and the medication is released continuously over a period of time. In addition, once the patch is affixed to the skin, it is no longer necessary to remember when to take the medication. *In vitro* drug permeation studies are usually conducted using Franz diffusion cells. In a typical configuration, permeation experiments are performed using excised skins (e.g., full-thickness human skin) or synthetic membranes, such as poly(ethylene-vinyl acetate) (EVA). The membrane is positioned between the donor cell and a receiver compartment, containing a continuously stirred solution (i.e., PermeGear glass diffusion cells, PermeGear, Riegelsville, PA). Drug molecules diffuse across the membrane from the donor cell and into the receiver compartment. By removing the entire receiver volume during sampling and replacing it with a fresh solution at regular intervals, the concentration in the receiver cell remains at zero (i.e., sink condition).

The system, described in Figure 1.11, can be analyzed, using tools learned in process dynamics courses (e.g., Laplace transforms). The findings can serve to inform product designers of the solvent properties and membrane thickness/composition necessary to reach a particular delivery rate or a desired plasma drug concentration, when the patch is applied to the skin. Dynamic system analysis techniques can also be implemented to describe how the transient

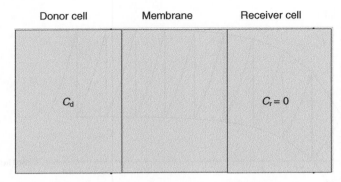

Figure 1.11. Experimental apparatus for Franz diffusion cell experiments.

response (e.g., flux, plasma drug concentration) relates to pertinent design parameters [9].

1.1.2.4   Modern Control Theory Can Be Applied to Calculate a Set of Optimum Loading Doses in a Transdermal Patch   These doses need to be administered at regular intervals in order to have a constant delivery rate over a defined period [10]. Once the percutaneous absorption kinetics is known, simulations can be conducted to estimate drug-dosage regimens appropriate for a particular treatment. The methods required in these cases (i.e., *dynamic programming*) are beyond the scope of this textbook. However, they are implemented in several processes in chemical engineering where it is necessary to compute a set of control and state histories that minimize an objective function. A brief description is provided in Chapter 16.

1.3   INSTRUMENTATION

An understanding of the physical elements of the control system is necessary for designing the controller. For example, the time delay, caused by a sensor, of a fluid moving through a pipe or by transmission lines may adversely influence the performance of the controller.

The control system is composed of the *process*, *primary elements*, a *controller*, and the *final control elements*. Figure 1.12 shows the hardware elements in a liquid-level control system. A liquid with flow rate F_i and temperature T_i enters the tank. The liquid level (h) in the tank (i.e., the process) is measured by a level transmitter (LT) such as a differential pressure cell (primary element). Information from the sensor is sent to the level controller (LC) along a transmission line. This reading is compared to the set point (h_{SP}) and the controller calculates a signal based on the error ($h - h_{SP}$) and a control law. The output from the controller adjusts the valve (final control element) in order to get the liquid level as close as possible to the set point.

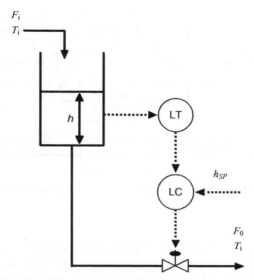

Figure 1.12. Diagram of a liquid-level control system including the hardware elements.

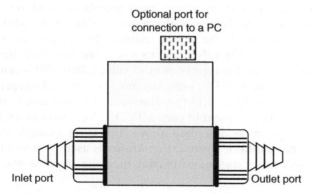

Figure 1.13. Diagram of a typical mass flow controller.

1.3.1 MFCs

MFCs are often used in bioprocess technology to measure and control the gas flow rate into bioreactors. Desired levels of DO and carbon dioxide are achieved during the cultivation of mammalian cells via computer-coupled MFCs (Fig. 1.13). The gas enters the device through the inlet port where it is measured by a *sensor*. Note that only a fraction of the gas goes through the

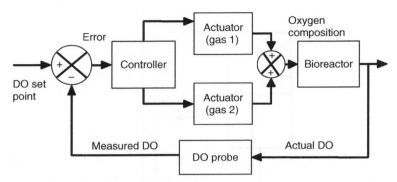

Figure 1.14. Block diagram of the feedback control of the dissolved oxygen using two MFCs.

sensor because of the presence of a *bypass* in the line. The voltage signal from the sensor is transmitted to a controller (built into the device) that compares the sensor output to a set point programmed by an operator or sent to the device by a computer. A proportional valve, placed in the line, receives information from the controller to adjust the outlet flow, following a control law, until it equals the set point. Depending on the model, MFCs are equipped with solenoid, piezoelectric or thermal actuator valves. Data acquisition and interface cards are needed in cases where input signals are sent via a computer [11]. In terms of process dynamics, the *response time* of the MFC is usually determined by how quickly the sensor responds to a change in the flow rate (usually a few seconds). The valve exhibits a very fast response time (less than 2 seconds, depending on the type of control valve). Two MFCs can be used to control the concentration of DO using the molar fraction of oxygen. Note that the *controller block* in Figure 1.14 (programmed in a computer) calculates the set point flow rates that are sent to each MFC. Each *actuator block* is equipped with its control system as described above. This is an example of a *cascade control* where the output of a *primary controller* is used to manipulate the set points of two *secondary controllers* playing the roles of actuators.

1.3.2   DO Probes

The concentration of DO in a bioreactor is measured using a DO probe. For animal cells, the specific rate of oxygen demand is in the range of 0.5–5.0 10^{-10} mmol O_2/cell/h. Culture-DO is generally controlled between 30% and 50% (relative to air saturation). Two types of sensors are the *polarographic* and *galvanic DO electrodes*. In both cases, a membrane (highly permeable to oxygen transport), an anode, cathode, and an electrolyte solution are used. However, an external voltage source is required for polarographic probes contrary to the galvanic DO sensors.

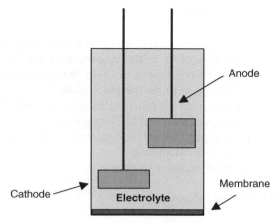

Figure 1.15. Principle of dissolved oxygen measurement.

When the galvanic DO probe is placed in a medium, oxygen diffuses through the membrane and is reduced at the inert cathode (Fig. 1.15):

$$O_2 + 2H_2O + 4e^- \rightarrow 4OH^-. \tag{1.5}$$

At the anode (assuming lead), the following reaction takes place:

$$2Pb + 4OH^- \rightarrow 2Pb(OH)_2 + 4e^-. \tag{1.6}$$

The cathode, usually made of gold or silver, is separated from the external fluid by a semipermeable membrane (e.g., fluorinated ethylene propylene [FEP] Teflon, polyethylene) that allows oxygen to pass through but blocks water and large solute molecules [12]. At steady state, the current signal from the probe is proportional to the flux of oxygen through the membrane and the electrolyte layer located between the membrane and the cathode [12]:

$$I = F \times n \times f, \tag{1.7}$$

where I is the current intensity (amp), F is the Faraday constant (96,500 C/mol), f is the flux of oxygen (mol/s) and n is the number of electrons transferred/mole of O_2 reduced. By considering the oxygen flux from the external phase to cathode Q (g/s), we have

$$Q = \frac{D_m S_m D_e S_e A P_{O2}}{L_m D_e S_e + L_e D_m S_m} \tag{1.8}$$

The parameters of Equation (1.8) are [12]

D_m (cm^2/s):   diffusion coefficient of O_2 in the membrane material
S_m (g/cm^3/atm):   solubility of O_2 in the membrane material
D_e (cm^2/s):   diffusion coefficient of O_2 in the electrolyte
S_e (g/cm^3/atm):   solubility of O_2 in the electrolyte
A (cm^2):   surface area for both the membrane and electrolyte layer perpendicular to the oxygen flux
P_{O2} (atm):   partial pressure of oxygen in the external phase
L_m (cm):   membrane thickness
L_e (cm):   electrolyte layer thickness.

Since $f = Q/32$, the current intensity is

$$I = \frac{96,500 \times 4}{32} \frac{D_m S_m D_e S_e A P_{O2}}{L_m D_e S_e + L_e D_m S_m} = 12,062 \frac{D_m S_m D_e S_e A P_{O2}}{L_m D_e S_e + L_e D_m S_m}. \qquad (1.9)$$

The temperature of the system should be considered when calibrating the device because of the dependence of oxygen partial pressure on temperature. DO probes can be calibrated in nitrogen (zero oxygen) and air-saturated media. These sensors usually exhibit a response time that can be incorporated into the design of the controller.

1.3.3 pH Probes

The pH is another important cell culture environmental parameter that needs to be carefully controlled in industrial fermentation processes. The optimal pH range when growing host mammalian cells, such as hybridoma, CHO, and myeloma, is 7.1–7.4. Production of carbon dioxide (CO_2) and lactic acid ($C_3H_6O_3$) is responsible for the decrease in pH detected during batch cell culture:

$$CO_2 + H_2O \rightleftarrows H_2CO_3 \rightleftarrows H^+ + HCO_3^-$$
$$C_6H_{12}O_6 \rightleftarrows 2CH_3 - CH_2O - COO^- + 2H^+. \qquad (1.10)$$

In the first reaction, carbon dioxide reacts with water to produce carbonic acid (H_2CO_3), which dissociates into a proton (H^+) and the bicarbonate ion (HCO_3^-). Lactic acid (weak acid), which partially dissociates in water to give lactate ion ($CH_3 - CH_2O - COO^-$) and H^+, is generated from glucose ($C_6H_{12}O_6$) in the second reaction. For control purposes, a calibrated pH glass electrode is routinely used to inform the operator (or a computer) when to introduce CO_2 or HCl into the bioreactor, lowering the pH, or to direct the addition of NaOH and NaHCO$_3$, raising the pH.

pH sensors measure the negative logarithm of the dissolved hydronium ion concentration. In fact, the pH is related to the hydrogen ion activity a_H:

$$a_H = 10^{-pH}; \quad pH = -\log(a_H). \tag{1.11}$$

Since the hydrogen ion activity is equal to the hydrogen ion concentration (c_H) for dilute solutions, we have

$$pH = -\log(c_H). \tag{1.12}$$

A standard pH sensor used in fermentation processes consists of a glass electrode and a reference electrode. The glass electrode is made up of a glass bulb membrane, an internal solution, and an Ag/AgCl electrode. Protons (H^+) are exchanged between the glass membrane (silicon dioxide and metal oxides) and the investigated solution:

$$Si-O^- + H_3O^+ \rightarrow Si-O-H^+ + H_2O. \tag{1.13}$$

A reference electrode helps complete the circuit (Fig. 1.16). The voltage measured between the electrodes is proportional to the solution's pH:

$$V = V_0 + \frac{RT}{F}\ln(a_H) = V_0 + \frac{2.303RT}{F}\log(a_H) = V_0 - \frac{2.303RT}{F}pH, \tag{1.14}$$

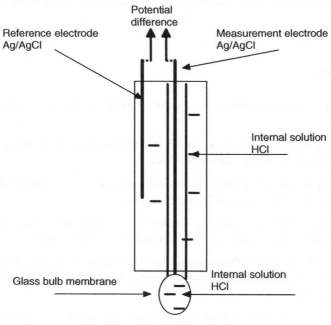

Figure 1.16. pH probe.

where V is the potential of the glass electrode and V_0 is an offset potential. Equation (1.14) reflects equilibrium conditions and does not show how the voltage reading is affected by the pH-response characteristics. Data on the pH response of glass electrodes can be found in Perley [13].

The above online instruments, along with others, such as temperature probes, turbidity meters (for cell mass measurement), and level sensors, affect the performance of the controller. In most cases, their dynamics can be approximated by simple functions that incorporate time-delay phenomena and a time that measures how long it takes to reach an ultimate value (i.e., *time constant*).

1.4 SUMMARY

Applications of process dynamics and control theory extend from biotechnology and physiological systems to drug-delivery processes. In the cases analyzed, an understanding of how the process works and the choices of input variables that have an impact on the controlled variables are important considerations. In addition, a clear statement of the goals of the controller should be included. A good practice is to draw block diagrams to show interconnections among variables and processes involved in the control system. A typical block diagram also includes measuring devices, such as MFCs and DO probes. The operation and dynamics of each piece of equipment must be understood.

PROBLEMS

1.1. Provide two examples of process control applications in biological systems.

1.2. Draw a block diagram of a furnace control system. Identify the main elements (e.g., room, sensor, furnace).

1.3. Provide a block diagram of a feedback control system where the cardiovascular center regulates blood pressure.

1.4. Provide a block diagram of a feedback control system where the vasometer and cardiovascular centers regulate blood pressure.

1.5. Write the ordinary differential equation describing the plasma blood serum concentration for a one-compartment model. Use a single IV bolus dose.

1.6. Solve the following equations governing steady-state oxygen transport across a membrane:

$$D_m \frac{d^2 C_{O_2}}{dx^2} = 0$$

$$C_{O_2}(x = 0) = S_m P_{O_2}$$

$$C_{O_2}(x = L_m) = 0.$$

1.7. Derive the steady-state flux for diffusive oxygen transport across a membrane. Hint: Use the results from Problem 1.6.

1.8. The dynamics of a turbidity meter (nephelometric turbidity unit) is given by

$$y(t) = a\left(1 - e^{-6.67(t-0.23)}\right)\psi(t - 0.23),$$

where $\psi(t - t_d)$ is the unit step function and t_d is the time delay (minute). Calculate $y(1.0)$.

1.9. The dynamics of an analyzer is described by

$$y(t) = a\left(1 - e^{-6.67t}\right),$$

What is the steady-state value of y?

1.10. The dynamics of an enzyme analyzer (g/L) is described by

$$y(t) = \left(1 - e^{-1.3(t-3.0)}\right)\psi(t - 3.0),$$

where $\psi(t - t_d)$ is the unit step function and t_d is the time delay (minute). At what time is $y(t) = 0.98 y(t \to \infty)$? Note: $y(t \to \infty)$ represents the steady-state value of y.

REFERENCES

1. Boudreau MA, McMillan GK. *New Directions in Bioprocess Modeling and Control.* Research Triangle Park, NC: ISA, 2007.

2. Gomes J, Menawat AS. Precise control of dissolved oxygen in bioreactors—a model-based geometric algorithm. *Chemical Engineering Science* 2000; 55:67–78.

3. Takuma S, Hirashima C, Piret JM. Dependence on glucose limitation of the pCO₂ influences on CHO cell growth, metabolism and IgG production. *Biotechnology and Bioengineering* 2007; 97:1479–1488.

4. Glacken MW, Adema E, Sinskey AJ. Mathematical descriptions of hybridoma culture kinetics: II. The relationship between thiol chemistry and the degradation of serum activity. *Biotechnology and Bioengineering* 1989; 33:440–450.

5. Northrop RB. *Endogenous and Exogenous Regulation and Control of Physiological Systems*. Boca Raton, FL: Chapman & Hall/CRC, 2000.

6. Goodman L. Regulation and control in physiological systems: 1960–1980. *Annals of Biomedical Engineering* 1980; 8:281–290.

7. Sherman PM, Stark L. A servoanalytic study of consensual pupil reflex to light. *Journal of Neurophysiology* 1957; 20:17–26.

8. DiStefano JJ, Stubberud AR, Williams IJ. *Schaum's Outline of Theory and Problems of Feedback and Control Systems*. New York: McGraw-Hill, 1995.

9. Simon L. Timely drug delivery from controlled-release devices: dynamic analysis and novel design concepts. *Mathematical Biosciences* 2009; 217:151–158.

10. Simon L. Repeated applications of a transdermal patch: analytical solution and optimal control of the delivery rate. *Mathematical Biosciences* 2007; 209:593–607.

11. Simon L, Karim MN. Identification and control of dissolved oxygen in hybridoma cell culture in a shear sensitive environment. *Biotechnology Progress* 2001; 17: 634–642.

12. Kok R, Zajik J. Transient measurement of low dissolved oxygen concentrations. *The Canadian Journal of Chemical Engineering* 1973; 51:782–787.

13. Perley GA. pH response of glass electrodes. *Analytical Chemistry* 1949; 21: 559–562.

CHAPTER 2

MATHEMATICAL MODELS

A mathematical model uses equations to describe a phenomenon. It is a representation, or abstraction of the reality, constructed after making certain assumptions. For example, the transport of a drug through the skin is often explained by the mathematical theory of diffusion, which postulates that the flux, or the rate of transfer, of a substance through a unit area is proportional to the concentration gradient in the direction normal to the section (*Fick's first law*). Most studies concentrate on a constant *diffusion coefficient* even though its value may be a function of the solute concentration in the skin layer.

One-compartment models are used to show how drugs are absorbed in the body and then eliminated. In this framework, the body is represented by a well-stirred vessel. Such an approach has been shown to successfully predict plasma drug concentrations after a rapid intravenous injection. In reality, the process is much more complex as it involves convection and diffusion mechanisms that are better represented by partial differential equations (PDEs) in space and time.

There are many benefits associated with introducing a mathematical framework to examine process behavior and to support experimental data. In studying biological systems, for example, models are useful for "organizing disparate information into a coherent whole" [1]. They provide a way to investigate individual components of a system and to help understand how the various parts work together. Application of the conservation laws helps discover new

Control of Biological and Drug-Delivery Systems for Chemical, Biomedical, and Pharmaceutical Engineering, First Edition. Laurent Simon.
© 2013 John Wiley & Sons, Inc. Published 2013 by John Wiley & Sons, Inc.

strategies and identify important factors that drive the process and influence product quality. This approach may lead to deeper insights into the advantages, disadvantages, and limitations of advanced technologies.

From a process control viewpoint, modeling guides the choices of manipulated variables and controller tuning settings and helps to operate the plant.

2.1   BACKGROUND

Equations, based on principles of momentum transfer, material and energy balances (i.e., *conservation principles*), are generally written to solve problems encountered in chemical and biological fields.

2.1.1   Macroscopic Momentum Balance

The linear momentum principle for a fixed control volume states that the rate of momentum accumulation equals *the rate of momentum into the control volume minus the rate of momentum out of the control volume plus the sum of all the forces acting on the system*:

$$\frac{d(m\mathbf{v})_{cv}}{dt} = \sum_{i=\text{inlet}} \dot{m}_i \mathbf{v}_i - \sum_{j=\text{outlet}} \dot{m}_j \mathbf{v}_j + \sum_k \mathbf{F}_k, \qquad (2.1)$$

where m is the mass of fluid in the control volume and $\mathbf{v}$ is its average velocity; $\dot{m}_i$ and $\mathbf{v}_i$ are the mass flow rate and average velocity of stream i, respectively; $\mathbf{F}_k$ is an external force that acts on the system (Fig. 2.1). The fixed control volume is a very useful concept applied in fluid mechanics and refers to a region in space bounded by an imaginary surface. Momentum balances help explain fluids flowing through pipes and blood vessels. Since the velocity is a vector, with magnitude and direction, Equation (2.1) can be written for the x-, y-, and z-components of momentum in the Cartesian coordinate system. Momentum transport occurs by *convection* (*bulk fluid motion*) and *molecular transport* (*diffusion*). $\mathbf{F}_k$ is the sum of the force exerted on the surface (i.e., *surface force*), specifically the pressure, and the *body force*, which acts on the control volume.

2.1.2   Macroscopic Mass Balance

Total mass conservation in a fixed control volume can be described as *the rate of mass accumulation equals the rate of mass into the control volume minus the rate of mass out of the control volume*:

$$\frac{d(m)_{cv}}{dt} = \sum_{i=\text{inlet}} \dot{m}_i - \sum_{j=\text{outlet}} \dot{m}_j \qquad (2.2)$$

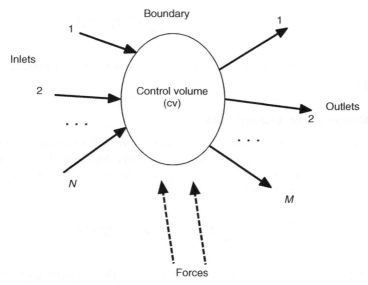

Figure 2.1. A control volume (cv) and the forces acting on the fluid system inside cv.

or

$$\frac{d(\rho V)_{cv}}{dt} = \sum_{i=\text{inlet}} \rho_i F_i - \sum_{j=\text{outlet}} \rho_j F_j, \qquad (2.3)$$

where ρ is the density of the fluid in the control volume; V is the total volume of the system; ρ_i is the density of the fluid in the ith inlet stream; ρ_j is the density of the fluid in the jth outlet stream; F_i and F_j are the volumetric flow rates of the ith inlet stream and the jth outlet stream, respectively.

The *macroscopic species mass balance* includes a net generation term:

$$\frac{d(\rho_A V)_{cv}}{dt} = \sum_{i=\text{inlet}} \rho_{Ai} F_i - \sum_{j=\text{outlet}} \rho_{Aj} F_j + r_A V, \qquad (2.4)$$

where r_A is the mass rate of production (or consumption) of A by reaction. Note that $A = 1, 2, 3, \ldots, N$. In addition, by considering the following definitions $\sum_{A=1}^{N} r_A = 0$, $\rho_i = \sum_{A=1}^{N} \rho_{Ai}$, and $\rho_j = \sum_{A=1}^{N} \rho_{Aj}$, Equation (2.3) can be obtained by summing up Equation (2.4) over all species. Dividing Equation (2.4) by the molecular weight of species A, we have

$$\frac{d(C_A V)_{\text{cv}}}{dt} = \sum_{i=\text{inlet}} C_{Ai} F_i - \sum_{j=\text{outlet}} C_{Aj} F_j + R_A V, \qquad (2.5)$$

where C_A is the molar concentration of A and R_A is the molar production (or consumption) rate of A.

2.1.3 Macroscopic Energy Balance

The macroscopic energy balance states that *the rate of accumulation of energy in a fixed control volume equals the net rate at which total energy is transferred across the surface of the fixed control volume by mass flow and the transfer of energy by heat flow and work*:

$$\frac{d(mE_{\text{cv}})}{dt} = \sum_{i=\text{inlet}} F_i E_i - \sum_{j=\text{outlet}} F_j E_j + \dot{Q} - \dot{W}. \qquad (2.6)$$

Equation (2.6) is an expression of the first law of thermodynamics for open systems (Fig. 2.2). The energy (E) consists of the internal (E_{in}), kinetic (E_{k}), and potential (E_{p}) energies:

$$E = E_{\text{in}} + E_{\text{k}} + E_{\text{p}} = \hat{U} + \frac{u^2}{2} + gz, \qquad (2.7)$$

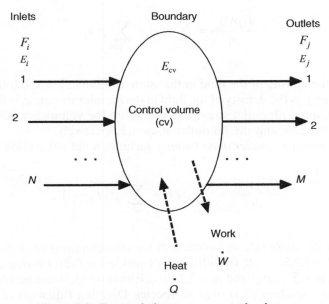

Figure 2.2. Energy balance on a control volume.

where F is a mass flow rate, u is the velocity, z is the height relative to a frame of reference, m is the mass of fluid in the control volume, and $\hat{U}$ is the specific internal energy. The terms $\dot{Q}$ and $\dot{W}$ are the rate at which heat flows to the system from its surroundings and the rate of work done by the system on the surroundings, respectively. The work term accounts for *flow work* due to materials crossing the boundary of the control volume and a *shaft work* $\dot{W}_s$ (e.g., moving parts):

$$\dot{W} = \sum_{j=\text{outlet}} F_j P_j \tilde{V}_j - \sum_{i=\text{inlet}} F_i P_i \tilde{V}_i + \dot{W}_s, \qquad (2.8)$$

where $\tilde{V}_i$ and $\tilde{V}_j$ are specific volumes associated with inlet stream i and outlet stream j, respectively. The pressures are P_i and P_j. Equation (2.6) becomes

$$\frac{d\left[m\left(\hat{U}+\dfrac{u^2}{2}+gz\right)\right]_{\text{cv}}}{dt} = \begin{cases} \sum\limits_{i=\text{inlet}} F_i\left(\hat{U}_i+\dfrac{u_i^2}{2}+gz_i\right) - \sum\limits_{j=\text{outlet}} F_j\left(\hat{U}_j+\dfrac{u_j^2}{2}+gz_j\right) + \dot{Q} \\ - \sum\limits_{j=\text{outlet}} F_j P_j \tilde{V}_j + \sum\limits_{i=\text{inlet}} F_i P_i \tilde{V}_i - \dot{W}_s \end{cases} \qquad (2.9)$$

or

$$\frac{d\left[m\left(\hat{U}+\dfrac{u^2}{2}+gz\right)\right]_{\text{cv}}}{dt} = \begin{cases} \sum\limits_{i=\text{inlet}} F_i\left(\hat{U}_i+P_i\tilde{V}_i+\dfrac{u_i^2}{2}+gz_i\right) \\ - \sum\limits_{j=\text{outlet}} F_j\left(\hat{U}_j+P_j\tilde{V}_j+\dfrac{u_j^2}{2}+gz_j\right) + \dot{Q} - \dot{W}_s. \end{cases} \qquad (2.10)$$

By using the definition of the specific enthalpy

$$\hat{H} = \hat{U} + P\tilde{V}, \qquad (2.11)$$

Equation (2.10) becomes

$$\frac{d\left[m\left(\hat{U}+\dfrac{u^2}{2}+gz\right)\right]_{\text{cv}}}{dt} = \sum_{i=\text{inlet}} F_i\left(\hat{H}_i+\dfrac{u_i^2}{2}+gz_i\right) - \sum_{j=\text{outlet}} F_j\left(\hat{H}_j+\dfrac{u_j^2}{2}+gz_j\right) + \dot{Q} - \dot{W}_s. \qquad (2.12)$$

Equations (2.3), (2.5), and (2.12) can be used to describe the behavior of a processing system. However, changes in variables that can be easily observed and measured, such as the temperature, height, pressure, and concentration, are usually monitored instead of the enthalpy and internal energy. These

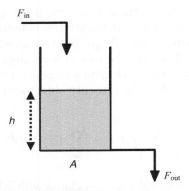

Figure 2.3. Flow in a liquid vessel.

variables constitute the *state variables*. Differential equations that govern how the state variables change in the process as a function of time are called *state equations*. For example, mass accumulation in a liquid vessel with one inlet and one outlet stream (Fig. 2.3) is derived from Equation (2.3):

$$\frac{d(\rho V)}{dt} = \rho_{in} F_{in} - \rho_{out} F_{out}. \tag{2.13}$$

Assuming that the density is constant, we have $\rho = \rho_{in} = \rho_{out}$. As a result, Equation (2.13) becomes

$$\frac{d(V)}{dt} = F_{in} - F_{out}. \tag{2.14}$$

Volume V is related to height ($V = Ah$), where A is the cross-sectional area of the vessel. The state equation describing how the liquid level changes in the tank is

$$\frac{dh}{dt} = \frac{F_{in} - F_{out}}{A}. \tag{2.15}$$

More information about the processing system is required to help choose appropriate relationships necessary to write a complete mathematical model. For instance, the growth of a microorganism may obey Monod kinetics:

$$\mu = \frac{\mu_{max} C_s}{K_s + C_s}, \tag{2.16}$$

where μ is the specific growth rate (h^{-1}), C_s is the limiting substrate concentration (g/L), μ_{max} is the maximum specific growth rate (h^{-1}), and K_s is the Monod constant (g/L). The cell growth is expressed as

$$r_x = \mu C_x, \tag{2.17}$$

with C_x being the cell concentration.

2.2 DYNAMICS OF BIOREACTORS

A bioreactor is a process unit in which biological material is created through an enzymatic reaction. As with reactors used in other industries, bioreactors can operate in batch, continuous or semicontinuous mode. Cells are often described as an open system where substrates are added to the unit. The output consists of the heat generated, chemical products, and additional cells. Information about cell kinetics can be used to design bioreactors.

2.2.1 Batch Bioreactors

In a batch-type culture, microorganisms are placed in a sterilized medium. There is no material crossing the system boundary. The main stages of growth are the *lag*, *exponential*, *stationary*, and *death* phases. The specific growth is affected by the substrate concentration (i.e., Monod kinetics). In the Monod kinetic, given by Equation (2.16), K_s is the concentration of nutrient when the specific growth rate is half its maximum value (Fig. 2.4). The equation is empirical and experiments are conducted to estimate the model parameters.

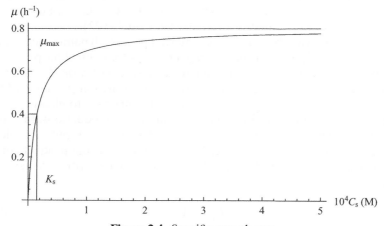

Figure 2.4. Specific growth rate.

The mass balance equation for the cells can be written from Equation (2.5):

$$\frac{d(VC_x)}{dt} = V(\mu C_x),$$

(2.18)

where V is the liquid volume in the vessel. Since the volume is constant, the state equation is

$$\frac{dC_x}{dt} = \mu C_x.$$

(2.19)

Similarly, the substrate concentration is

$$\frac{dC_s}{dt} = -\frac{\mu C_x}{Y_{x/s}}.$$

(2.20)

The parameter $Y_{x/s}$ is a *growth yield coefficient* defined as

$$Y_{x/s} = \frac{C_x - C_{x0}}{C_{s0} - C_s},$$

(2.21)

where C_{x0} and C_{s0} are the initial cell and substrate concentrations, respectively.

If the product (P) is *growth* and *non-growth-associated*, we have

$$\frac{d(C_P)}{dt} = \alpha \mu C_x + \beta C_x,$$

(2.22)

in which α and β are growth and non-growth-associated product formation constants, respectively. The state variables are C_x, C_s, and C_p and the state equations are given by Equations (2.19), (2.20), and (2.22).

In general, state variables define the internal process dynamics and may not be measured directly. When representing a process for control purposes, there is usually an *output equation*, containing measured *output variables*. An output variable is expressed, in its generalized form, as a function of *state* and *input variables*. An input variable can cause the state variables to change. For the batch process, no input variable is included in the state equations. In general, some input variables can be manipulated to control an output variable. An input variable that is not manipulated is a *disturbance*. If the product concentration is being measured, the output equation may be written, using vector notation, as

$$C_p = \mathbf{M}\mathbf{x},$$

(2.23)

where $\mathbf{M} = [0\ 0\ 1]$ and $\mathbf{x} = [C_x\ C_s\ C_p]$.

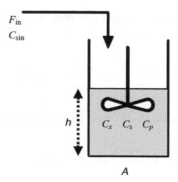

F_{in}
C_{sin}

h

C_x C_s C_p

A

Figure 2.5. Fed-batch bioreactor.

2.2.2 Fed-Batch Bioreactors

With fed-batch operations, it is possible to maintain a desired level of substrate concentration in the bioreactor (Fig. 2.5). This is advantageous in cases where the excess of a particular substrate results in a decrease in cell yield [2]. The total mass balance equation can be derived from Equation (2.13):

$$\frac{d(\rho V)}{dt} = \rho_{in} F_{in} - \rho_{out} F_{out}.$$

Assuming a constant density (i.e., close to the density of water), we have

$$\frac{d(V)}{dt} = F_{in} - F_{out}. \tag{2.24}$$

In addition, for the fed-batch reactor $F_{out} = 0$,

$$\frac{d(V)}{dt} = F_{in}. \tag{2.25}$$

To write the state equations in terms of the liquid level, we use the formula $V = Ah$, where A is the cross-sectional area and h the height:

$$\frac{d(h)}{dt} = \frac{F_{in}}{A}. \tag{2.26}$$

The material balance for the cell concentration is

$$\frac{d(VC_x)}{dt} = V(\mu C_x) \tag{2.27}$$

because the biomass is neither added nor removed from the bioreactor. Since the volume changes in the bioreactor, Equation (2.27) becomes

$$V\frac{d(C_x)}{dt} + C_x\frac{d(V)}{dt} = V\mu C_x. \tag{2.28}$$

When substituting Equation (2.25) into Equation (2.28), we have

$$V\frac{d(C_x)}{dt} + C_x F_{in} = V\mu C_x \tag{2.29}$$

or

$$\frac{d(C_x)}{dt} = \frac{(V\mu - F_{in})C_x}{V}. \tag{2.30}$$

The state equation is written as

$$\frac{d(C_x)}{dt} = \left(\mu - \frac{F_{in}}{Ah}\right)C_x. \tag{2.31}$$

Similarly, the material balance for the substrate concentration is

$$\frac{d(VC_s)}{dt} = F_{in}C_{sin} - \frac{\mu}{Y_{x/s}}VC_x, \tag{2.32}$$

which is expanded to

$$C_s\frac{dV}{dt} + V\frac{dC_s}{dt} = F_{in}C_{sin} - \frac{\mu}{Y_{x/s}}VC_x, \tag{2.33}$$

or after using Equation (2.25),

$$V\frac{dC_s}{dt} = F_{in}C_{sin} - F_{in}C_s - \frac{\mu}{Y_{x/s}}VC_x. \tag{2.34}$$

The state equation for the substrate concentration is

$$\frac{dC_s}{dt} = \frac{F_{in}}{Ah}(C_{sin} - C_s) - \frac{\mu}{Y_{x/s}}C_x. \tag{2.35}$$

The mass balance around the bioreactor gives

$$\frac{d(VC_P)}{dt} = \alpha\mu VC_x \tag{2.36}$$

for a growth-associated product. Following a procedure similar to what is used for the cell concentration, the following equation is obtained:

$$V\frac{dC_P}{dt} + C_P F_{in} = \alpha\mu VC_x, \tag{2.37}$$

which yields

$$\frac{dC_P}{dt} = \alpha\mu C_x - \frac{F_{in}C_P}{Ah}. \tag{2.38}$$

The state equations are given by Equations (2.26), (2.31), (2.35), and (2.38); the input variables are F_{in} and C_{sin}. Model parameters are also included in the equations: μ_{max}, K_s, $Y_{x/s}$, A, and α. The state variables are h, C_x, C_s, and C_P.

2.2.3 Continuous Bioreactors

Continuous stirred-tank bioreactors (CSTRs) are used to maintain microbial populations in the exponential growth phase for a long period of time (Fig. 2.6). The total mass balance is

$$\frac{d(\rho V)}{dt} = \rho_{in}F_{in} - \rho_{out}F_{out}. \tag{2.39}$$

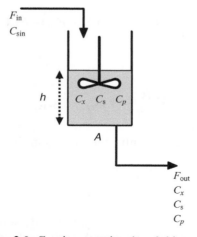

Figure 2.6. Continuous stirred-tank bioreactor.

Assuming a constant density, we have

$$\frac{dV}{dt} = F_{in} - F_{out} \tag{2.40}$$

or

$$\frac{dh}{dt} = \frac{F_{in} - F_{out}}{A}. \tag{2.41}$$

The material balance for the cell concentration is

$$\frac{d(VC_x)}{dt} = V(\mu C_x) - F_{out}C_x. \tag{2.42}$$

The inlet feed is assumed sterile in this case. There will be an input term $F_{in}C_{xin}$ in the equation if the inlet feed contains microorganisms with a concentration of C_{xin}. Equation (2.42) can be written as

$$V\frac{dC_x}{dt} + C_x\frac{dV}{dt} = V\mu C_x - F_{out}C_x, \tag{2.43}$$

which becomes

$$V\frac{dC_x}{dt} = V\mu C_x - C_x\frac{dV}{dt} - F_{out}C_x. \tag{2.44}$$

Substitution of Equation (2.40) yields

$$V\frac{dC_x}{dt} = V\mu C_x - C_x(F_{in} - F_{out}) - F_{out}C_x \tag{2.45}$$

or

$$V\frac{dC_x}{dt} = V\mu C_x - F_{in}C_x. \tag{2.46}$$

The state equation for the concentration of cells in the bioreactor is

$$\frac{dC_x}{dt} = \left(\mu - \frac{F_{in}}{Ah}\right)C_x. \tag{2.47}$$

Equation (2.47) helps to understand why, at steady state, the inlet flow rate can be used to control the specific growth rate: $\mu = F_{in}/Ah$ (i.e., a chemostat).

The ratio F_{in}/Ah is the *dilution rate* and the inverse of the dilution rate is the *residence time*.

The material balance for the substrate concentration is

$$\frac{d(VC_s)}{dt} = F_{in}C_{sin} - \frac{\mu}{Y_{x/s}}VC_x - F_{out}C_s. \tag{2.48}$$

Further expansion gives

$$V\frac{dC_s}{dt} + C_s\frac{dV}{dt} = F_{in}C_{sin} - \frac{\mu}{Y_{x/s}}VC_x - F_{out}C_s \tag{2.49}$$

or

$$V\frac{dC_s}{dt} = F_{in}C_{sin} - \frac{\mu}{Y_{x/s}}VC_x - F_{out}C_s - C_s\frac{dV}{dt}, \tag{2.50}$$

which leads to

$$V\frac{dC_s}{dt} = F_{in}C_{sin} - \frac{\mu}{Y_{x/s}}VC_x - F_{out}C_s - C_s(F_{in} - F_{out}), \tag{2.51}$$

and finally,

$$V\frac{dC_s}{dt} = F_{in}(C_{sin} - C_s) - \frac{\mu}{Y_{x/s}}VC_x. \tag{2.52}$$

The state equation for the substrate concentration, similar to the one obtained in the case of the fed-batch operation, is

$$\frac{dC_s}{dt} = \frac{F_{in}}{Ah}(C_{sin} - C_s) - \frac{\mu}{Y_{x/s}}C_x. \tag{2.53}$$

The mass balance around the bioreactor gives

$$\frac{d(VC_P)}{dt} = \alpha\mu VC_x - F_{out}C_P \tag{2.54}$$

for a growth-associated product or

$$V\frac{dC_P}{dt} + C_P\frac{dV}{dt} = \alpha\mu VC_x - F_{out}C_P. \tag{2.55}$$

Further simplification leads to

$$V\frac{dC_p}{dt} = \alpha\mu VC_x - F_{out}C_P - C_p\frac{dV}{dt} \tag{2.56}$$

or

$$V\frac{dC_p}{dt} = \alpha\mu VC_x - F_{out}C_P - C_p(F_{in} - F_{out}), \tag{2.57}$$

and finally,

$$V\frac{dC_p}{dt} = \alpha\mu VC_x - F_{in}C_p, \tag{2.58}$$

which yields

$$\frac{dC_p}{dt} = \alpha\mu C_x - \frac{F_{in}C_p}{Ah}. \tag{2.59}$$

The state equations are represented by Equations (2.41), (2.47), (2.53), and (2.59). The input variables are F_{in} and C_{sin}.

2.3   ONE- AND TWO-COMPARTMENT MODELS

The pharmacokinetics of drug absorption, distribution, metabolism, and excretion (ADME) is represented by compartmental models.

2.3.1   One-Compartment Model

The one-compartment model describes the kinetics of drug absorption and elimination in the body. This representation presumes that the body behaves like a well-stirred vessel (Fig. 2.7). The pharmaceutic is assumed to distribute only to rapidly perfused tissues. Control can be used to maintain the plasma drug concentration within a target therapeutic range and to develop optimum drug-dosage regimens.

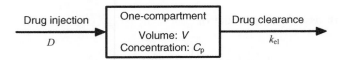

Figure 2.7. Schematic of a one-compartment model.

A mass balance around the process in Figure 2.7 results in the following differential equation:

$$\frac{dVC_p}{dt} = D\delta(t) - k_{el}VC_p,\tag{2.60}$$

where D is the loading dose, V is the distribution volume, k_{el} is the first-order elimination rate constant, C_p is the plasma drug concentration at time t, and $\delta(t)$ is the *Dirac* or *unit impulse function* ($\delta(t = 0) = 1$ and $\delta(t > 0) = 0$). Equation (2.60) is appropriate for a rapid intravenous injection (i.e., *IV bolus*). Because V is constant, Equation (2.60) becomes

$$\frac{dC_p}{dt} = \frac{D}{V}\delta(t) - k_{el}C_p.\tag{2.61}$$

The initial value C_p^0 is defined as D/V; Equation (2.61) is the state equation and C_p is the state variable. In the case of multiple IV boluses, the dosage is the input variable and it can be manipulated to maintain a desired average C_p.

Drugs are also infused relatively slowly through a vein at a constant or zero-order rate. *IV infusion* is used to achieve an effective plasma drug concentration, thereby eliminating the oscillations observed in bolus IV dosing. In this case, Equation (2.61) is modified to account for the constant rate of infusion (k_0 in unit of mass/time):

$$\frac{dVC_p}{dt} = k_0 - k_{el}VC_p\tag{2.62}$$

or

$$\frac{dC_p}{dt} = \frac{k_0}{V} - k_{el}C_p.\tag{2.63}$$

Equation (2.63) is the state equation and k_0 is the input variable. The model parameters are V and k_{el}.

2.3.2 Two-Compartment Model

A two-compartment model is shown in Figure 2.8. The human body is composed of a central compartment consisting of the blood/plasma and well-perfused tissues (e.g., liver, heart) and a peripheral compartment consisting of poorly perfused tissues (e.g., skeletal muscles).

A mass balance around the first compartment, in Figure 2.5, results in the following:

$$\frac{d(C_1V_1)}{dt} = D\delta(t) - k_{el}C_1V_1 - k_{12}C_1V_1 + k_{21}C_2V_2,\tag{2.64}$$

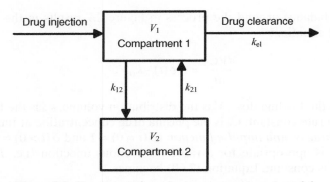

Figure 2.8. Representation of a two-compartment model.

where C is the drug concentration, V is the volume, and k is a mass transfer rate constant. The subscripts 1 and 2 are used to denote the central and peripheral compartments, respectively. Drug elimination in the first compartment is shown by the subscript "el." In addition, the subscript 12 represents a transfer of material from compartment 1 to compartment 2, while drug being transported in the opposite direction is shown by 21. The parameter k_{el} is a first-order elimination rate constant, which is often used to represent drug clearance.

Equation (2.64) can be rewritten as

$$V_1 \frac{dC_1}{dt} = D\delta(t) - k_{el}C_{el}V_1 - k_{12}C_1V_1 + k_{21}C_2V_2. \qquad (2.65)$$

The state equation for C_1 is

$$\frac{dC_1}{dt} = \frac{D}{V_1}\delta(t) - (k_{el} + k_{12})C_1 + k_{21}C_2\frac{V_2}{V_1}. \qquad (2.66)$$

The mass balance around the peripheral compartment gives

$$\frac{d(C_2V_2)}{dt} = k_{12}C_1V_1 - k_{21}C_2V_2 \qquad (2.67)$$

or

$$V_2 \frac{dC_2}{dt} = k_{12}C_1V_1 - k_{21}C_2V_2. \qquad (2.68)$$

The state equation for C_2 is

$$\frac{dC_2}{dt} = k_{12}C_1 \frac{V_1}{V_2} - k_{21}C_2. \tag{2.69}$$

For an IV infusion, the state equations become

$$\frac{dC_1}{dt} = \frac{k_0}{V_1} - (k_{el} + k_{12})C_1 + k_{21}C_2 \frac{V_2}{V_1} \tag{2.70}$$

and Equation (2.69).

2.4 ENZYME KINETICS

Enzymes are biological catalysts that are used to lower the activation energy of biochemical reactions and are produced by microorganisms grown by the methods outlined in Section 2.2. Their catalytic properties are related to the structure or enzyme conformation. Enzymes are preferred over chemical catalysts because they are biodegradable and highly reactive under mild conditions of temperature, pressure, and pH. The study of enzyme kinetics focuses on the rate of enzyme reaction and how this rate is influenced by chemical and physical conditions [3]. An enzyme catalyst (E) is very specific and only targets one or a small number of reactions that involve a substrate (S), to yield a product (P):

$$E + S \underset{k_2}{\overset{k_1}{\rightleftharpoons}} E \cdot S \xrightarrow{k_3} E + P, \tag{2.71}$$

where $E \cdot S$ is an enzyme–substrate complex.

An *enzyme reactor* may operate in batch, fed-batch, and continuous modes just as reactors that use microorganisms. In a batch reactor, the state equations are

$$\frac{dC_S}{dt} = -k_1 C_S C_E + k_2 C_{ES}, \tag{2.72}$$

$$\frac{dC_p}{dt} = k_3 C_{ES}, \tag{2.73}$$

and

$$\frac{dC_{ES}}{dt} = k_1 C_S C_E - k_2 C_{ES} - k_3 C_{ES}. \tag{2.74}$$

Because it is usually assumed that the total enzyme concentration remains constant, $C_{E0} = C_E + C_{ES}$, it is not necessary to write the equation of the enzyme concentration.

Several approaches have been taken to yield rate expressions in terms of measurable quantities (i.e., S and P). For example, the Briggs–Haldane approach assumes that $dC_{ES} / dt \cong 0$. Adding Equation (2.74) to Equation (2.72), with $dC_{ES} / dt \cong 0$, leads to

$$\frac{dC_S}{dt} + \frac{dC_{ES}}{dt} = \frac{dC_S}{dt} = -k_3 C_{ES}, \tag{2.75}$$

which means that $dC_S/dt = -dC_P/dt$.

With $dC_{ES}/dt \cong 0$, we have

$$0 = k_1 C_S C_E - k_2 C_{ES} - k_3 C_{ES}. \tag{2.76}$$

Now, inserting $C_{E0} = C_E + C_{ES}$ into Equation (2.76) yields

$$0 = k_1 C_S (C_{E0} - C_{ES}) - k_2 C_{ES} - k_3 C_{ES} \tag{2.77}$$

or

$$C_{ES} = \frac{k_1 C_{E0} C_S}{k_2 + k_3 + k_1 C_S}, \tag{2.78}$$

and finally,

$$C_{ES} = \frac{C_{E0} C_S}{\dfrac{k_2 + k_3}{k_1} + C_S} \tag{2.79}$$

$$-\frac{dC_S}{dt} = \frac{dC_P}{dt} = \frac{k_3 C_{E0} C_S}{\dfrac{k_2 + k_3}{k_1} + C_S}. \tag{2.80}$$

In this case, two state equations are written:

$$\frac{dC_S}{dt} = -\frac{r_{max} C_S}{K_M + C_S} \tag{2.81}$$

and

$$\frac{dC_P}{dt} = \frac{r_{max}C_S}{K_M + C_S},$$ (2.82)

where

$$r_{max} = k_3 C_{E0}$$ (2.83)

and

$$K_M = \frac{k_2 + k_3}{k_1}.$$ (2.84)

State equations can be derived for batch and continuous stirred-tank reactors.

2.5 SUMMARY

Mathematical models are obtained from the application of mass, momentum, and energy balances. These equations describe macroscopic system behavior and are derived considering the flow of materials into and out of an arbitrary fixed control volume. Total and species balances are applied to cell cultivations, one- and two-compartmental pharmacokinetic models, and enzyme reactors. State variables and ordinary differential equations that describe how these state variables change in the process as a function of time (i.e., the state equations) were also discussed.

PROBLEMS

2.1. A well-stirred batch reactor is heated by an electrical heating tape. The heat capacity of the system (i.e., liquid reactant and reactor) at constant volume is C_v. If the heating rate is $\dot{Q}$, write the state equation describing the temperature of the mixture.

Note: $U = m\hat{U} = m\left[\hat{U}(T_{ref}) + C_v(T - T_{ref})\right]$, where m is the mass of the system and T_{ref} is a reference temperature.

2.2. A continuous stirred-tank is heated by an electrical coil. A feed at T_i, which has a heat capacity C_p, is added to the tank at a mass flow rate $\dot{m}$ and is withdrawn at the same rate. The rate of heat loss through the coil is $\dot{Q}$. Write the state equation describing the temperature of the liquid in the vessel.

Note:

$$\frac{dU}{dt} \simeq \frac{dH}{dt} \quad \text{and} \quad H = M\Big[\hat{H}(T_{\text{ref}}) + C_{\text{p}}(T - T_{\text{ref}})\Big],$$

where M is the mass of liquid in the tank and T_{ref} is a reference temperature.

2.3. Write the linear momentum principle for a closed system (i.e., no mass crosses the system boundary).

2.4. Derive the state equation describing how the drug concentration changes in a one-compartment constant-infusion model. Assume a Michaelis–Menten elimination kinetics.

2.5. The following reaction takes place in a batch enzyme reactor:

$$S + E \underset{k_2}{\overset{k_1}{\rightleftharpoons}} ES$$
$$E + I \underset{k_4}{\overset{k_3}{\rightleftharpoons}} EI$$
$$ES \xrightarrow{k_5} P + E,$$

where S, E, ES, P, I, and EI represent the substrate, enzyme catalyst, enzyme–substrate complex, product, inhibitor, and enzyme–inhibitor complex, respectively. Write the rate equations for the concentrations of P, ES, EI, I, and S.

2.6. A chemostat is equipped with a cell-recycle system where α is the recycle ratio and F is the sterile feed. A centrifuge is in the recycling stream. Write the material balances for the substrate (S) and cell (X) around the reactor. Assumption: The reactor volume remains unchanged.

2.7. For Problem 2.6, write the steady-state material balances for the cell around the centrifuge.

2.8. Derive the state equations for a three-compartment model (i.e., 1, 2, and 3). First-order drug elimination and constant-rate drug infusion occur in the central compartment (1). Compartments 2 and 3 are also connected.

2.9. Derive the substrate concentrations in two continuous stirred-tank enzyme reactors connected in series. The reaction mechanism in each tank can be represented by the Michaelis–Menten equation. Assume that the volume in each vessel is kept constant.

2.10. Propose a method of controlling nutrient concentration in a fed-batch bioreactor.

REFERENCES

1. Bailey JE. Mathematical modeling and analysis in biochemical engineering: past accomplishments and future opportunities. *Biotechnology Progress* 1998; 14:8–20.
2. Blanch HW, Clark DS. *Biochemical Engineering*. Boca Raton, FL: CRC/Taylor and Francis, 1997.
3. Lee JM. *Biochemical Engineering*. Englewood Cliffs, NJ: Prentice Hall, 1992.

CHAPTER 3

LINEARIZATION AND DEVIATION VARIABLES

Most of the state equations derived in Chapter 2 include nonlinear terms. In general, analytical solutions (i.e., closed-form expressions) are only available for some classes of nonlinear equations. Computer simulations offer the ability to obtain numerical results and to visualize actual process behaviors. In addition, simulations make it possible to test a number of designs and to study the influence of an array of conditions on the process.

Linearization is another tool that can be used to describe a system in the vicinity of certain operating conditions. However, there is a drawback to this approach. The performance of the approximation degrades further away from the linearization point. Some of the advantages of the linear approximation are (1) the local stability of a dynamic system can be analyzed and (2) controllers can be tuned using the approximated solution. This technique is appropriate for situations where the plant does not deviate significantly from a desired operating condition.

3.1 COMPUTER SIMULATIONS

Several simulation software packages have been developed for representing real-world processes and for assessing the effects of a number of factors on system performance. MATLAB® (MathWorks, Inc.) and *Mathematica*® (Wolfram Research, Inc.) are two packages, among others, that can be used to

Control of Biological and Drug-Delivery Systems for Chemical, Biomedical, and Pharmaceutical Engineering, First Edition. Laurent Simon.
© 2013 John Wiley & Sons, Inc. Published 2013 by John Wiley & Sons, Inc.

simulate biological processes. A working knowledge of governing engineering principles and the ability to visualize dynamic behaviors are equally important to the development of an efficient control strategy.

The first step in process simulation is to derive a mathematical model of the physical system, as discussed in Chapter 2. Most computational tools are only able to numerically solve the differential equations provided by the user. Model parameter values, rate laws, and certain assumptions have to be entered into the program. Other computational software packages, such as Aspen Plus® (Aspen Technology, Inc.), contain a database of pure components and phase equilibrium information for a range of chemicals. Libraries of equipment models are available and allow the user to build very complex systems and run simulations to analyze the nonlinear behavior of these systems. The performance of a controller, based on a linear representation of the plant, can be evaluated using the original process model.

Simulation platforms can be exploited to solve the nonlinear differential equations included in this book. Most graphics, presented in the subsequent chapters, are generated using MATLAB (http://www.mathworks.com/) or *Mathematica* (http://www.wolfram.com/). Since both software resources provide extensive documentation and support information on their websites, instructions on the use of these tools are not provided in the text.

3.2 LINEARIZATION OF SYSTEMS

3.2.1 Function of One Variable

The Taylor series expansion of functions is applied to linearize nonlinear systems. A function $f(x)$ can be approximated by the following expression:

$$f(x) = f(x_0) + \frac{(x-x_0)}{1!}\left(\frac{df}{dx}\right)_{x_0} + \frac{(x-x_0)^2}{2!}\left(\frac{d^2f}{dx^2}\right)_{x_0} + \ldots + \frac{(x-x_0)^n}{n!}\left(\frac{d^nf}{dx^n}\right)_{x_0},$$

(3.1)

where f is continuous on an interval containing the reference point x_0. Equation (3.1) is known as the Taylor series for the function $f(x)$ around $x = x_0$. The linear approximation of $f(x)$ is

$$f(x) \approx f(x_0) + \frac{(x-x_0)}{1!}\left(\frac{df}{dx}\right)_{x_0}.$$

(3.2)

The error from the linear approximation has the same order of magnitude as

$$\frac{(x-x_0)^2}{2!}\left(\frac{d^2f}{dx^2}\right)_{x_0}.$$

The expression $(df/dx)_{x_0}$ means the derivative of f evaluated at $x = x_0$.

Example 3.1   Linearize the function $f(x) = x^3 + 1$ around $x_0 = 1$.

Solution   The first derivative of f with respect to x is calculated:

$$\frac{df(x)}{dx} = f'(x) = 3x^2. \tag{3.3}$$

The function and its derivative evaluated at $x_0 = 1$ are given by

$$f(1) = 1^3 + 1 = 2; \quad \left(\frac{df}{dx}\right)_{x_0=1} = 3(1)^2 = 3. \tag{3.4}$$

As a result, the linear approximation is

$$\begin{aligned} f(x) &\approx 2 + \frac{(x-1)}{1!} \times 3 \\ f(x) &\approx 2 + 3x - 3 \\ f(x) &\approx 3x - 1. \end{aligned} \tag{3.5}$$

The function and its estimation are plotted in Figure 3.1. It is clear that the linearization produces good results around the reference point.

Example 3.2   Linearize the function $f(x) = 5x^2 + 2x + 2$ around $x_0 = 0$.

Solution   The first derivative of f with respect to x is

$$\frac{df(x)}{dx} = f'(x) = 10x + 2. \tag{3.6}$$

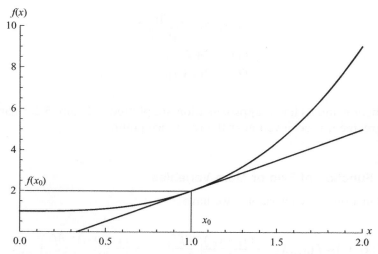

Figure 3.1. Approximation of the function $f(x) = x^3 + 1$ around $x_0 = 1$.

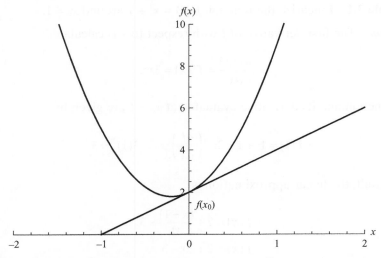

Figure 3.2. Approximation of the function $f(x) = 5x^2 + 2x + 2$ around $x_0 = 0$.

The function and its derivative evaluated at $x_0 = 0$ are given by

$$f(0) = 5(0)^2 + 2(0) + 2 = 2; \quad \left(\frac{df}{dx}\right)_{x_0=0} = 10(0) + 2 = 2. \tag{3.7}$$

As a result, the linear approximation is

$$\begin{aligned} f(x) &\approx 2 + \frac{(x-0)}{1!} \times 2 \\ f(x) &\approx 2 + 2x \\ f(x) &\approx 2(x+1). \end{aligned} \tag{3.8}$$

The function and its linear approximation are plotted in Figure 3.2. Again, the best agreement is achieved near the reference point.

3.2.2 Function of Two or More Variables

For a function of two variables, we have

$$f(x_1, x_2) \approx f(x_{10}, x_{20}) + \frac{(x_1 - x_{10})}{1!}\left(\frac{\partial f}{\partial x_1}\right)_{x_{10},x_{20}} + \frac{(x_2 - x_{20})}{1!}\left(\frac{\partial f}{\partial x_2}\right)_{x_{10},x_{20}}. \tag{3.9}$$

Equation (3.9) can be generalized for n variables:

$$f(x_1, x_2, \ldots, x_n) \approx \begin{cases} f(x_{10}, x_{20}, \ldots, x_{n0}) + \dfrac{(x_1 - x_{10})}{1!} \left(\dfrac{\partial f}{\partial x_1}\right)_{x_{10}, x_{20}, \ldots, x_{n0}} \\ + \dfrac{(x_2 - x_{20})}{1!} \left(\dfrac{\partial f}{\partial x_2}\right)_{x_{10}, x_{20}, \ldots, x_{n0}} + \ldots + \dfrac{(x_n - x_{n0})}{1!} \left(\dfrac{\partial f}{\partial x_n}\right)_{x_{10}, x_{20}, \ldots, x_{n0}} \end{cases}$$

$$(3.10)$$

Example 3.3   Linearize the function $f(x, y) = 5x^2y + xy^2 + 1$ around $(x_0, y_0) = (1,1)$.

Solution   The first derivative of f with respect to x is

$$\frac{\partial f(x, y)}{\partial x} = \frac{\partial\left(5x^2y + xy^2 + 1\right)}{\partial x} = 10xy + y^2. \tag{3.11}$$

The first derivative of f with respect to y is

$$\frac{\partial f(x, y)}{\partial y} = \frac{\partial\left(5x^2y + xy^2 + 1\right)}{\partial y} = 5x^2 + 2xy. \tag{3.12}$$

The function and its derivatives evaluated at $(x_0, y_0) = (1,1)$ are given by

$$\begin{aligned} f(1,1) &= f(1,1) = 5(1)^2(1) + (1)(1)^2 + 1 = 7; \\ \left(\frac{\partial f}{\partial x}\right)_{(x_0, y_0)=(1,1)} &= 10(1)(1) + (1)^2 = 11; \\ \left(\frac{\partial f}{\partial y}\right)_{(x_0, y_0)=(1,1)} &= 5(1)^2 + 2(1)(1) = 7. \end{aligned} \tag{3.13}$$

Application of Equation (3.9) with $(x_1, x_2) = (x, y)$ gives

$$\begin{aligned} f(x, y) &\approx f(x_0, y_0) + \frac{(x - x_0)}{1!}\left(\frac{\partial f}{\partial x}\right)_{x_0, y_0} + \frac{(y - y_0)}{1!}\left(\frac{\partial f}{\partial y}\right)_{x_0, y_0} \\ f(x, y) &\approx f(1,1) + \frac{(x-1)}{1!}\left(\frac{\partial f}{\partial x}\right)_{(x_0, y_0)=(1,1)} + \frac{(y-1)}{1!}\left(\frac{\partial f}{\partial y}\right)_{(x_0, y_0)=(1,1)} \\ f(x, y) &\approx 7 + (x-1)\times(11) + (y-1)\times(7) \\ f(x, y) &\approx 7 + 11x - 11 + 7y - 7 \\ f(x, y) &\approx 11x + 7y - 11. \end{aligned} \tag{3.14}$$

The function and its linear approximation are plotted in Figure 3.3.

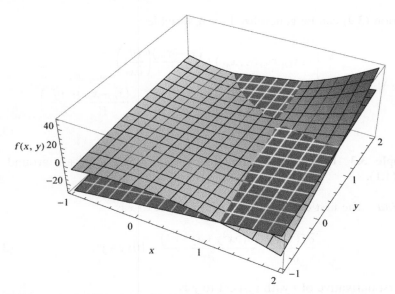

Figure 3.3. Approximation of the function $f(x, y) = 5x^2y + xy^2 + 1$ around $(x_0, y_0) = (1,1)$.

3.2.3 Nonlinear Ordinary Differential Equations (ODEs)

Linearization methods also work for nonlinear ODEs. For example, considering the first-order nonlinear differential equation $dy/dt = f(y, t)$, the linearization technique can be used to obtain an approximate solution: $y(t)$.

Example 3.4   Linearize the differential equation $dy/dt = y^2 - 2y$ around the steady-state point (s).

Solution   The steady-state point is obtained by setting $dy/dt = 0$:

$$y^2 - 2y = 0. \tag{3.15}$$

The steady-state points are $y_0 = 2$ and $y_0 = 0$.

(a) When $y_0 = 2$, the first derivative of $f(y) = y^2 - 2y$ with respect to y is

$$\frac{df(y)}{dy} = 2y - 2, \tag{3.16}$$

where f is only a function of y. The function and its derivative evaluated at $y_0 = 2$ are given by

$$f(2) = (2)^2 - 2(2) = 0; \quad \left(\frac{df}{dy}\right)_{y_0=2} = 2(2) - 2 = 2. \tag{3.17}$$

The result for $f(2)$ is not surprising because $y_0 = 2$ is a steady-state value. The linear approximation around $y_0 = 2$ is

$$f(y) \approx 0 + \frac{(y-2)}{1!} \times 2$$
$$f(y) \approx 2y - 4 \tag{3.18}$$
$$\frac{dy}{dt} \approx 2y - 4.$$

As a check, verify that $2y - 4$ has the same value as $y^2 - 2y$ when $y = 2$.

(b)   When $y_0 = 0$, the derivative evaluated at $y_0 = 2$ is given by

$$\left(\frac{df}{dy} \right)_{y_0=0} = 2(0) - 2 = -2. \tag{3.19}$$

As a result, the linear approximation around $y_0 = 0$ is

$$f(y) \approx 0 - \frac{(y-0)}{1!} \times 2$$
$$f(y) \approx -2y \tag{3.20}$$
$$\frac{dy}{dt} \approx -2y.$$

since $f(0) = 0$ (steady state). Verify that $-2y$ has the same value as $y^2 - 2y$ when $y = 0$.

Figure 3.4 shows that the linear approximations are very good around the points $y_0 = 2$ and $y_0 = 0$.
    Note:

1. The lines $f_1(y) = 2y - 4$ and $f_2(y) = -2y$ are *equations of the tangent lines* to the curve $f(y) = y^2 - 2y$ at the points $y_0 = 2$ and $y_0 = 0$, respectively.
2. Some local properties of the original nonlinear system $dy/dt = y^2 - 2y$ can be analyzed using the linear ODEs $dy/dt \approx 2y - 4$ and $dy/dt \approx -2y$.

Example 3.5   Linearize the differential equation $dy/dt = y^2 - 2y + u$ (where both y and u are functions of t) around the steady-state point: $(u_0, y_0) = (0, 2)$.

Solution   The first derivative of $f(y, u) = y^2 - 2y + u$ with respect to y is

$$\frac{\partial f(y)}{\partial y} = 2y - 2. \tag{3.21}$$

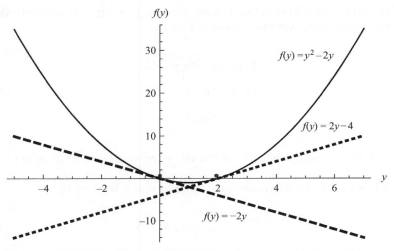

Figure 3.4. Approximation of the function $f(y) = y^2 - 2y$ around the *equilibrium points* $y_0 = 2$ and $y_0 = 0$.

The first derivative of $f(y) = y^2 - 2y + u$ with respect to u is

$$\frac{\partial f(y)}{\partial u} = 1. \tag{3.22}$$

The first derivatives of f evaluated at $u_0 = 0$ and $y_0 = 2$ are given by

$$\left(\frac{\partial f}{\partial y}\right)_{(y_0,u_0)=(2,0)} = 2(2) - 2 = 2$$

$$\left(\frac{\partial f}{\partial u}\right)_{(y_0,u_0)=(2,0)} = 1. \tag{3.23}$$

Finally, the approximation is

$$f(y,u) \approx f(y_0,u_0) + \frac{(y-y_0)}{1!}\left(\frac{\partial f}{\partial y}\right)_{y_0,u_0} + \frac{(u-u_0)}{1!}\left(\frac{\partial f}{\partial u}\right)_{y_0,u_0}$$

$$f(y,u) \approx f(2,0) + \frac{(y-2)}{1!}\left(\frac{\partial f}{\partial y}\right)_{(y_0,u_0)=(2,0)} + \frac{(u-0)}{1!}\left(\frac{\partial f}{\partial u}\right)_{(y_0,u_0)=(2,0)}$$

$$f(x,y) \approx 0 + (y-2) \times (2) + (u-0) \times (1)$$

$$f(x,y) \approx 2(y-2) + (u-0)$$

$$\frac{dy}{dt} \approx 2(y-2) + (u-0). \tag{3.24}$$

Verify that $f(2, 0) = 0$. If *deviation variables* are defined by $\tilde{y} = y - 2$ and $\tilde{u} = u - 0$ (i.e., *deviation from the equilibrium point*), we have

$$\frac{d\tilde{y}}{dt} \approx 2\tilde{y} + \tilde{u} \tag{3.25}$$

because

$$\frac{d\tilde{y}}{dt} = \frac{d(y-2)}{dt} = \frac{dy}{dt} - \frac{d(2)}{dt} = \frac{dy}{dt}$$

$$\frac{d\tilde{y}}{dt} = \frac{dy}{dt}. \tag{3.26}$$

Deviation variables (i.e., $d\tilde{y}/dt \approx 2\tilde{y} + \tilde{u}$ instead of $dy/dt \approx 2(y-2) + (u-0)$) are frequently preferred in control applications where the focus is placed on how the system deviates from an operating point under the effects of disturbances.

The system $dy/dt = y^2 - 2y + u$ may describe a process with state variable y and input variable u. If we consider the system to be originally at *steady state* at time $t = 0$, a relevant question is: How does y vary with time if the input variable u changes suddenly from $u_0 = 0$ or $\tilde{u}_0 = 0$ to a point (u_{new0} or $\tilde{u}_{\text{new0}}$) in a stepwise fashion (also called *step change*)? Based on the linearization result, the system represented by

$$\frac{dy}{dt} \approx 2(y-2) + (u-0) \quad \text{or} \quad \frac{d\tilde{y}}{dt} \approx 2\tilde{y} + \tilde{u}$$

can be used to approximate the response.

3.2.4   Nonlinear Systems of ODEs

If we consider the following dynamic system:

$$\frac{dx_1}{dt} = f_1(x_1, x_2, \ldots, x_n, u_1, u_2, \ldots, u_m)$$

$$\frac{dx_2}{dt} = f_2(x_1, x_2, \ldots, x_n, u_1, u_2, \ldots, u_m)$$

$$\vdots \tag{3.27}$$

$$\frac{dx_n}{dt} = f_n(x_1, x_2, \ldots, x_n, u_1, u_2, \ldots, u_m)$$

with n state variables (x_n) and m input variables (u_m), a linear approximation gives

$$\frac{dx_1}{dt} \approx \begin{cases} f_1(x_{10}, x_{20}, \ldots, x_{n0}, u_{10}, u_{20}, \ldots, u_{m0}) + (x_1 - x_{10})\left(\frac{\partial f_1}{\partial x_1}\right)_{(x_{10}, x_{20}, \ldots, x_{n0}, u_{10}, u_{20}, \ldots, u_{m0})} \\[2ex] + (x_2 - x_{20})\left(\frac{\partial f_1}{\partial x_2}\right)_{(x_{10}, x_{20}, \ldots, x_{n0}, u_{10}, u_{20}, \ldots, u_{m0})} + \ldots \\[2ex] + (x_n - x_{n0})\left(\frac{\partial f_1}{\partial x_n}\right)_{(x_{10}, x_{20}, \ldots, x_{n0}, u_{10}, u_{20}, \ldots, u_{m0})} \\[2ex] + (u_1 - u_{10})\left(\frac{\partial f_1}{\partial u_1}\right)_{(x_{10}, x_{20}, \ldots, x_{n0}, u_{10}, u_{20}, \ldots, u_{m0})} \\[2ex] + (u_2 - u_{20})\left(\frac{\partial f_1}{\partial u_2}\right)_{(x_{10}, x_{20}, \ldots, x_{n0}, u_{10}, u_{20}, \ldots, u_{m0})} + \ldots \\[2ex] + (u_m - u_{m0})\left(\frac{\partial f_1}{\partial u_m}\right)_{(x_{10}, x_{20}, \ldots, x_{n0}, u_{10}, u_{20}, \ldots, u_{m0})} \end{cases}$$

$$\frac{dx_2}{dt} \approx \begin{cases} f_2(x_{10}, x_{20}, \ldots, x_{n0}, u_{10}, u_{20}, \ldots, u_{m0}) + (x_1 - x_{10})\left(\frac{\partial f_2}{\partial x_1}\right)_{(x_{10}, x_{20}, \ldots, x_{n0}, u_{10}, u_{20}, \ldots, u_{m0})} \\[2ex] + (x_2 - x_{20})\left(\frac{\partial f_2}{\partial x_2}\right)_{(x_{10}, x_{20}, \ldots, x_{n0}, u_{10}, u_{20}, \ldots, u_{m0})} + \ldots \\[2ex] + (x_n - x_{n0})\left(\frac{\partial f_2}{\partial x_n}\right)_{(x_{10}, x_{20}, \ldots, x_{n0}, u_{10}, u_{20}, \ldots, u_{m0})} \\[2ex] + (u_1 - u_{10})\left(\frac{\partial f_2}{\partial u_1}\right)_{(x_{10}, x_{20}, \ldots, x_{n0}, u_{10}, u_{20}, \ldots, u_{m0})} \\[2ex] + (u_2 - u_{20})\left(\frac{\partial f_2}{\partial u_2}\right)_{(x_{10}, x_{20}, \ldots, x_{n0}, u_{10}, u_{20}, \ldots, u_{m0})} + \ldots \\[2ex] + (u_m - u_{m0})\left(\frac{\partial f_2}{\partial u_m}\right)_{(x_{10}, x_{20}, \ldots, x_{n0}, u_{10}, u_{20}, \ldots, u_{m0})} \end{cases} \tag{3.28}$$

$$\vdots$$

$$\frac{dx_n}{dt} \approx \begin{cases} f_n(x_{10}, x_{20}, \ldots, x_{n0}, u_{10}, u_{20}, \ldots, u_{m0}) + (x_1 - x_{10})\left(\frac{\partial f_n}{\partial x_1}\right)_{(x_{10}, x_{20}, \ldots, x_{n0}, u_{10}, u_{20}, \ldots, u_{m0})} \\[2ex] + (x_2 - x_{20})\left(\frac{\partial f_n}{\partial x_2}\right)_{(x_{10}, x_{20}, \ldots, x_{n0}, u_{10}, u_{20}, \ldots, u_{m0})} + \ldots \\[2ex] + (x_n - x_{n0})\left(\frac{\partial f_n}{\partial x_n}\right)_{(x_{10}, x_{20}, \ldots, x_{n0}, u_{10}, u_{20}, \ldots, u_{m0})} \\[2ex] + (u_1 - u_{10})\left(\frac{\partial f_n}{\partial u_1}\right)_{(x_{10}, x_{20}, \ldots, x_{n0}, u_{10}, u_{20}, \ldots, u_{m0})} \\[2ex] + (u_2 - u_{20})\left(\frac{\partial f_n}{\partial u_2}\right)_{(x_{10}, x_{20}, \ldots, x_{n0}, u_{10}, u_{20}, \ldots, u_{m0})} + \ldots \\[2ex] + (u_m - u_{m0})\left(\frac{\partial f_n}{\partial u_m}\right)_{(x_{10}, x_{20}, \ldots, x_{n0}, u_{10}, u_{20}, \ldots, u_{m0})} \end{cases}$$

Since the linearization is done around the steady state, we have

$$f_i(x_{10}, x_{20}, \ldots, x_{n0}, u_{10}, u_{20}, \ldots, u_{m0}) = f_i(x_{1s}, x_{2s}, \ldots, x_{ns}, u_{1s}, u_{2s}, \ldots, u_{ms}) = 0,$$
(3.29)

where the subscript s is used to represent a steady-state point. Using deviation variables leads to

$$\frac{d\tilde{x}_1}{dt} \approx \begin{aligned} & \tilde{x}_1\left(\frac{\partial f_1}{\partial x_1}\right)_P + \tilde{x}_2\left(\frac{\partial f_1}{\partial x_2}\right)_P + \ldots + \tilde{x}_n\left(\frac{\partial f_1}{\partial x_n}\right)_P \\ & + \tilde{u}_1\left(\frac{\partial f_1}{\partial u_1}\right)_P + \tilde{u}_2\left(\frac{\partial f_1}{\partial u_2}\right)_P + \ldots + \tilde{u}_m\left(\frac{\partial f_1}{\partial u_m}\right)_P \end{aligned}$$

$$\frac{d\tilde{x}_2}{dt} \approx \begin{aligned} & \tilde{x}_1\left(\frac{\partial f_2}{\partial x_1}\right)_P + \tilde{x}_2\left(\frac{\partial f_2}{\partial x_2}\right)_P + \ldots + \tilde{x}_n\left(\frac{\partial f_2}{\partial x_n}\right)_P \\ & + \tilde{u}_1\left(\frac{\partial f_2}{\partial u_1}\right)_P + \tilde{u}_2\left(\frac{\partial f_2}{\partial u_2}\right)_P + \ldots + \tilde{u}_m\left(\frac{\partial f_2}{\partial u_m}\right)_P \end{aligned}$$
(3.30)

$$\frac{d\tilde{x}_n}{dt} \approx \begin{aligned} & \tilde{x}_1\left(\frac{\partial f_n}{\partial x_1}\right)_P + \tilde{x}_2\left(\frac{\partial f_n}{\partial x_2}\right)_P + \ldots + \tilde{x}_n\left(\frac{\partial f_n}{\partial x_n}\right)_P \\ & + \tilde{u}_1\left(\frac{\partial f_n}{\partial u_1}\right)_P + \tilde{u}_2\left(\frac{\partial f_n}{\partial u_2}\right)_P + \ldots + \tilde{u}_m\left(\frac{\partial f_n}{\partial u_m}\right)_P, \end{aligned}$$

where P represents the steady-state point $(x_{1s}, x_{2s}, \ldots, x_{ns}, u_{1s}, u_{2s}, \ldots, u_{ms})$. The continuous-time state-space representation of Equation (3.30) is

$$\frac{d\tilde{\mathbf{x}}}{dt} \approx \mathbf{A}\tilde{\mathbf{x}} + \mathbf{B}\tilde{\mathbf{u}}$$
(3.31)

with

$$\mathbf{A} = \begin{bmatrix} \left(\frac{\partial f_1}{\partial x_1}\right)_P & \left(\frac{\partial f_1}{\partial x_1}\right)_P & \cdots & \left(\frac{\partial f_1}{\partial x_n}\right)_P \\ \left(\frac{\partial f_2}{\partial x_1}\right)_P & \left(\frac{\partial f_2}{\partial x_1}\right)_P & \cdots & \left(\frac{\partial f_2}{\partial x_n}\right)_P \\ \vdots & \vdots & \vdots & \vdots \\ \left(\frac{\partial f_n}{\partial x_1}\right)_P & \left(\frac{\partial f_n}{\partial x_1}\right)_P & \cdots & \left(\frac{\partial f_n}{\partial x_n}\right)_P \end{bmatrix},$$
(3.32)

$$\mathbf{B} = \begin{bmatrix} \left(\frac{\partial f_1}{\partial u_1}\right)_P & \left(\frac{\partial f_1}{\partial u_2}\right)_P & \cdots & \left(\frac{\partial f_1}{\partial u_m}\right)_P \\ \left(\frac{\partial f_2}{\partial u_1}\right)_P & \left(\frac{\partial f_2}{\partial u_2}\right)_P & \cdots & \left(\frac{\partial f_2}{\partial u_m}\right)_P \\ \vdots & \vdots & \vdots & \vdots \\ \left(\frac{\partial f_n}{\partial u_1}\right)_P & \left(\frac{\partial f_n}{\partial u_2}\right)_P & \cdots & \left(\frac{\partial f_n}{\partial u_m}\right)_P \end{bmatrix},$$
(3.33)

$$\tilde{\mathbf{x}} = \begin{bmatrix} \tilde{x}_1 \\ \tilde{x}_2 \\ \vdots \\ \tilde{x}_n \end{bmatrix} \tag{3.34}$$

and

$$\tilde{\mathbf{u}} = \begin{bmatrix} \tilde{u}_1 \\ \tilde{u}_2 \\ \vdots \\ \tilde{u}_m \end{bmatrix}. \tag{3.35}$$

In addition, if the output equation is defined as

$$
\begin{aligned}
y_1 &= g_1(x_1, x_2, \ldots, x_n, u_1, u_2, \ldots, u_m) \\
y_2 &= g_2(x_1, x_2, \ldots, x_n, u_1, u_2, \ldots, u_m) \\
&\vdots \\
y_q &= g_q(x_1, x_2, \ldots, x_n, u_1, u_2, \ldots, u_m),
\end{aligned}
\tag{3.36}
$$

it can be shown that

$$\tilde{\mathbf{y}} \approx \mathbf{C}\tilde{\mathbf{x}} + \mathbf{D}\tilde{\mathbf{u}}, \tag{3.37}$$

where

$$\mathbf{C} = \begin{bmatrix}
\left(\dfrac{\partial g_1}{\partial x_1}\right)_P & \left(\dfrac{\partial g_1}{\partial x_2}\right)_P & \cdots & \left(\dfrac{\partial g_1}{\partial x_n}\right)_P \\
\left(\dfrac{\partial g_2}{\partial x_1}\right)_P & \left(\dfrac{\partial g_2}{\partial x_2}\right)_P & \cdots & \left(\dfrac{\partial g_2}{\partial x_n}\right)_P \\
\vdots & \vdots & \vdots & \vdots \\
\left(\dfrac{\partial g_q}{\partial x_1}\right)_P & \left(\dfrac{\partial g_q}{\partial x_2}\right)_P & \cdots & \left(\dfrac{\partial g_q}{\partial x_n}\right)_P
\end{bmatrix} \tag{3.38}$$

and

$$\mathbf{D} = \begin{bmatrix}
\left(\dfrac{\partial g_1}{\partial u_1}\right)_P & \left(\dfrac{\partial g_1}{\partial u_2}\right)_P & \cdots & \left(\dfrac{\partial g_1}{\partial u_m}\right)_P \\
\left(\dfrac{\partial g_2}{\partial u_1}\right)_P & \left(\dfrac{\partial g_2}{\partial u_2}\right)_P & \cdots & \left(\dfrac{\partial g_2}{\partial u_m}\right)_P \\
\vdots & \vdots & \vdots & \vdots \\
\left(\dfrac{\partial g_q}{\partial u_1}\right)_P & \left(\dfrac{\partial g_q}{\partial u_2}\right)_P & \cdots & \left(\dfrac{\partial g_q}{\partial u_m}\right)_P
\end{bmatrix} \tag{3.39}$$

The q output variables are

$$\tilde{\mathbf{y}} = \begin{bmatrix} \tilde{y}_1 \\ \tilde{y}_2 \\ \vdots \\ \tilde{y}_q \end{bmatrix}. \tag{3.40}$$

The size of the matrices (called *Jacobian* matrices) are $\mathbf{A}(n \times n)$, $\mathbf{B}(n \times m)$, $\mathbf{C}(q \times n)$, and $\mathbf{D}(q \times m)$. In this notation, the number of rows in a matrix of size $i \times j$ is i and the number of columns is j.

3.3 GLYCOLYTIC OSCILLATION

The glycolytic pathway is used for the degradation of glucose to pyruvate and is employed by all major groups of microorganisms. *Adenosine triphosphate* (ATP, energy currency of the cell) is also produced. The process can be represented by the following equation [1]:

$$C_6H_{12}O_6 + 2NAD^+ + 2ADP + 2P_i \rightarrow 2C_3H_4O_3 + 2ATP + 2NADH + 2H^+ \tag{3.41}$$

Glucose: $C_6H_{12}O_6$
Nicotinamide adenine dinucleotide: NAD^+
Adenosine diphosphate: ADP
Phosphate group: P_i
Pyruvate: $CH_3COCOOH$
Adenosine triphosphate: ATP
Nicotinamide adenine dinucleotide, reduced: NADH
Hydrogen ion: H^+.

Because the reactions occur in the presence or absence of oxygen, the pathway is utilized by yeast in the production of alcohol. The mechanism is usually invoked to illustrate sustained oscillation in a metabolic pathway. A simplified system of two differential equations, involving adenosine diphosphate (normalized concentration x), a glycolytic intermediate, and fructose-6-phosphate (normalized concentration y), is used here to describe the oscillation [2, 3]:

$$\frac{dx}{dt} = -x + \alpha y + x^2 y \tag{3.42}$$

and

$$\frac{dy}{dt} = \beta - \alpha y + x^2 y, \tag{3.43}$$

where α and β are positive numbers. A first step in studying process stability (to be discussed in later chapters) is to linearize the model around an equilibrium point. In this case, the matrix **A** from Equation (3.32) is

$$\mathbf{A} = \begin{bmatrix} \left(\dfrac{\partial f_1}{\partial x}\right)_{x_s, y_s} & \left(\dfrac{\partial f_1}{\partial y}\right)_{x_s, y_s} \\ \left(\dfrac{\partial f_2}{\partial x}\right)_{x_s, y_s} & \left(\dfrac{\partial f_2}{\partial y}\right)_{x_s, y_s} \end{bmatrix}, \tag{3.44}$$

where

$$f_1(x, y) = -x + \alpha y + x^2 y \tag{3.45}$$

and

$$f_2(x, y) = \beta - \alpha y + x^2 y. \tag{3.46}$$

To calculate the equilibrium points, we set $f_1(x, y) = 0$ and $f_2(x, y) = 0$ to give

$$-x + \alpha y + x^2 y = 0 \tag{3.47}$$

and

$$\beta - \alpha y - x^2 y = 0. \tag{3.48}$$

After adding Equations (3.47) and (3.48), we obtain $x_s = \beta$; y is calculated by replacing x in Equation (3.47):

$$y_s = \frac{\beta}{\alpha + \beta^2},$$

where "s" denotes a steady-state value. The matrix **A** is

$$\mathbf{A} = \begin{bmatrix} -1 + 2x_s y_s & x_s^2 + \alpha \\ -2x_s y_s & -x_s^2 - \alpha \end{bmatrix}. \tag{3.49}$$

Replacing the steady-state point in **A** gives

$$\mathbf{A} = \begin{bmatrix} -1 + \dfrac{2\beta^2}{\alpha + \beta^2} & \alpha + \beta^2 \\ -\dfrac{2\beta^2}{\alpha + \beta^2} & -\alpha - \beta^2 \end{bmatrix} = \begin{bmatrix} \dfrac{-\alpha + \beta^2}{\alpha + \beta^2} & \alpha + \beta^2 \\ -\dfrac{2\beta^2}{\alpha + \beta^2} & -\alpha - \beta^2 \end{bmatrix}. \tag{3.50}$$

Therefore,

$$\frac{d\tilde{\mathbf{x}}}{dt} \approx \begin{bmatrix} \dfrac{-\alpha+\beta^2}{\alpha+\beta^2} & \alpha+\beta^2 \\ -\dfrac{2\beta^2}{\alpha+\beta^2} & -\alpha-\beta^2 \end{bmatrix} \tilde{\mathbf{x}}$$

$$\begin{bmatrix} \dfrac{d\tilde{x}}{dt} \\ \dfrac{d\tilde{y}}{dt} \end{bmatrix} \approx \begin{bmatrix} \dfrac{-\alpha+\beta^2}{\alpha+\beta^2} & \alpha+\beta^2 \\ -\dfrac{2\beta^2}{\alpha+\beta^2} & -\alpha-\beta^2 \end{bmatrix} \begin{bmatrix} \tilde{x} \\ \tilde{y} \end{bmatrix}$$

(3.51)

or

$$\frac{d\tilde{x}}{dt} = \left(\frac{-\alpha+\beta^2}{\alpha+\beta^2}\right)\tilde{x} + (\alpha+\beta^2)\tilde{y}$$

$$\frac{d\tilde{y}}{dt} = \left(-\frac{2\beta^2}{\alpha+\beta^2}\right)\tilde{x} - (\alpha+\beta^2)\tilde{y}$$

(3.52)

with $\tilde{x} = x - x_s$ and $\tilde{y} = y - y_s$.

3.4   HODGKIN–HUXLEY MODEL

The flow of current through the membrane of a nerve axon can be investigated using the Hodgkin and Huxley (HH) model. This mathematical model was proposed in 1952 to describe the electrical excitation of the squid giant axon [4]. Current is transported through the membrane either by charging the membrane capacity or by transporting ions through resistances connected in parallel with the capacity [4]. Four differential equations are written to describe the model:

$$C_M \frac{dV}{dt} = I_{ext} - \left[\bar{g}_{Na}m^3h(V+V_E-V_{Na}) + \bar{g}_K n^4(V+V_E-V_K) + g_l(V+V_E-V_l)\right]$$

$$\frac{dm}{dt} = [\alpha_m(V)(1-m) - \beta_m(V)m]$$

$$\frac{dh}{dt} = [\alpha_h(V)(1-h) - \beta_h(V)h]$$

$$\frac{dn}{dt} = [\alpha_n(V)(1-n) - \beta_n(V)n],$$

(3.53)

where the parameters and variables are

V: membrane potential (mV)

t: time (ms)

m: dimensionless sodium activation

h: dimensionless sodium inactivation

n: dimensionless potassium activation

V_{Na}: equilibrium potential of sodium (mV)

V_K: equilibrium potential of potassium (mV)

V_I: equilibrium potential of leak current (mV)

V_E: induced transmembrane potential (mV)

$\bar{g}_{Na}$: maximum sodium conductance (mmho/cm^2)

$\bar{g}_K$: maximum potassium conductance (mmho/cm^2)

g_I: maximum conductance for the leak current I (mmho/cm^2)

I_{ext}: externally applied current (μA/cm^2)

C_M: membrane capacity per unit area (μF/cm^2).

The parameters $\alpha_m(V)$, $\beta_m(V)$, $\alpha_h(V)$, $\beta_h(V)$, $\alpha_n(V)$, and $\beta_n(V)$ are nonlinear functions of the membrane potential [4]:

$$
\begin{aligned}
\alpha_m(V) &= 0.1(25.0-V)/[\exp((25.0-V)/10.0)-1.0]\\
\beta_m(V) &= 4.0\exp(-V/18.0)\\
\alpha_h(V) &= 0.07\exp(-V/20.0)\\
\beta_h(V) &= 1.0/[\exp((-V+30.0)/10.0)+1.0]\\
\alpha_n(V) &= 0.01(10.0-V)/[\exp((10.0-V)/10.0)-1.0]\\
\beta_n(V) &= 0.125\exp(-V/80.0).
\end{aligned}
\tag{3.54}
$$

This model has been used to study the dynamic behavior of neurons and to reveal oscillatory or nonoscillatory patterns depending on the model parameter values. The state variables are $V, m, h,$ and n; I_{ext} is the input variable. A simpler set of equations that captures the dynamic characteristics of the HH model is the FitzHugh–Nagumo model [3]:

$$
\begin{aligned}
\frac{dV}{dt} &= c\left(V+W-\frac{V^3}{3}+I_{ext}\right)\\
\frac{dW}{dt} &= -\frac{1}{c}(V-a+bW),
\end{aligned}
\tag{3.55}
$$

where W is a recovery parameter corresponding to combined forces that are apt to return the axonal membrane to a resting state [3]. The parameters are subject to the following constraints [5]:

$$1 - \frac{2b}{3} < a < 1, \quad 0 < b < 1, \quad b < c^2. \tag{3.56}$$

Consider the following parameters: $a = 0.7$, $b = 0.8$, and $c = 3$; then Equation (3.55) becomes

$$\frac{dV}{dt} = 3\left(V + W - \frac{V^3}{3} + I_{\text{ext}}\right)$$
$$\frac{dW}{dt} = -\frac{1}{3}(V - 0.7 + 0.8W). \tag{3.57}$$

To calculate the steady-state values for $I_{\text{ext}} = 0$, Equation (3.57) is written as

$$f_1 = 3\left(V + W - \frac{V^3}{3} + I_{\text{ext}}\right) = 0$$
$$f_2 = -\frac{1}{3}(V - 0.7 + 0.8W) = 0 \tag{3.58}$$

or

$$V + W - \frac{V^3}{3} = 0$$
$$V + 0.8W = 0.7. \tag{3.59}$$

The steady-state points are

$$(V_s, W_s, I_{\text{exts}}) = (1.19941, -0.62426, 0)$$
$$(V_s, W_s, I_{\text{exts}}) = (-0.599704 + 1.35238i, 1.62463 - 1.69048i, 0) \tag{3.60}$$
$$(V_s, W_s, I_{\text{exts}}) = (-0.599704 - 1.35238i, 1.62463 + 1.69048i, 0).$$

Linearization of Equation (3.57) around $(V_s, W_s, I_{\text{exts}}) = (1.19941, -0.62426, 0)$ leads to the following **A** matrix:

$$\mathbf{A} = \begin{bmatrix} \left(\dfrac{\partial f_1}{\partial V}\right)_{V_s, W_s, I_{\text{exts}}} & \left(\dfrac{\partial f_1}{\partial W}\right)_{V_s, W_s, I_{\text{exts}}} \\ \left(\dfrac{\partial f_2}{\partial V}\right)_{V_s, W_s, I_{\text{exts}}} & \left(\dfrac{\partial f_2}{\partial W}\right)_{V_s, W_s, I_{\text{exts}}} \end{bmatrix} \tag{3.61}$$

or

$$\mathbf{A} = \begin{bmatrix} 3(1 - V_s^2) & 3 \\ -\dfrac{1}{3} & -\dfrac{4}{15} \end{bmatrix}. \tag{3.62}$$

Similarly, the matrix **B** is

$$\mathbf{B} = \begin{bmatrix} \left(\dfrac{\partial f_1}{\partial V} \right)_{V_s, W_s, I_{exts}} \\[2ex] \left(\dfrac{\partial f_2}{\partial W} \right)_{V_s, W_s, I_{exts}} \end{bmatrix} \tag{3.63}$$

or

$$\mathbf{B} = \begin{bmatrix} 3 \\ 0 \end{bmatrix}. \tag{3.64}$$

After substituting $(V_s, W_s, I_{exts}) = (1.19941, -0.62426, 0)$ into **A** and **B**, we obtain

$$\frac{d\tilde{\mathbf{x}}}{dt} \approx \begin{bmatrix} -1.31574 & 3 \\ -\dfrac{1}{3} & -\dfrac{4}{15} \end{bmatrix} \tilde{\mathbf{x}} + \begin{bmatrix} 3 \\ 0 \end{bmatrix} \tilde{u}$$

$$\begin{bmatrix} \dfrac{d\tilde{V}}{dt} \\[2ex] \dfrac{d\tilde{W}}{dt} \end{bmatrix} \approx \begin{bmatrix} -1.31574 & 3 \\ -\dfrac{1}{3} & -\dfrac{4}{15} \end{bmatrix} \begin{bmatrix} \tilde{V} \\ \tilde{W} \end{bmatrix} + \begin{bmatrix} 3 \\ 0 \end{bmatrix} \tilde{u}, \tag{3.65}$$

and finally,

$$\frac{d\tilde{V}}{dt} \approx -1.31574\tilde{V} + 3\tilde{W} + 3\tilde{u}$$

$$\frac{d\tilde{W}}{dt} \approx -\frac{1}{3}\tilde{V} - \frac{4}{15}\tilde{W} \tag{3.66}$$

with $\tilde{V} = V - V_s$, $\tilde{W} = W - W_s$ and $\tilde{u} = I_{ext} - I_{exts}$.

3.5 SUMMARY

The concepts of linearization and deviations were introduced. These methods help convert a set of nonlinear equations into a linear system. For a dynamic system, the original process is linearized around a steady-state (or equilibrium) point and can be written in a state-space form that involves Jacobian matrices evaluated at this point. Two examples, the glycolytic pathway and a Hodgkin–Huxley model, were presented to illustrate how to apply the method.

PROBLEMS

3.1. Linearize the function $f(x) = 2x^3 + 3$ around $x_0 = 2$.

3.2. Linearize the function $f(x, y) = xy^3 + 5xy + 3$ around $(x_0, y_0) = (0,1)$.

3.3. Linearize the differential equation

$$A\frac{dh}{dt} = F_{in} - \beta\sqrt{h}$$

around the steady-state point $F_{in,ss}$ and h_{ss}.

3.4. Linearize the differential equation

$$\frac{dx}{dt} = y - \alpha\frac{x^2}{x^2 + 1}$$

$$\frac{dy}{dt} = x - y$$

around the steady-state point (x_0, y_0).

3.5. The dynamic equation for the concentration in a continuous stirred-tank reactor is given by

$$V\frac{dC}{dt} = F_{in}(C_{in} - C) - k_0 e^{-E/RT}CV,$$

where F_{in}, V, and C_{in} are constant. The temperature (T) and concentration (C) are state variables. Linearize the differential equation around the steady-state point (C_0, T_0).

3.6. The component balance equations in a continuous enzyme reactor are given by

$$V\frac{dS}{dt} = F_S S_{in} - F_1 S - r_s V$$

$$V\frac{dP}{dt} = -F_1 P + r_P V$$

$$V\frac{dE}{dt} = F_E E_{in} - F_1 E$$

with

$$r_S = v_m \frac{S}{K_M + S}$$

$$v_m = kE$$

$$r_P = 2r_S,$$

where the concentrations of substrate (S), enzyme (E), and product (P) represent the state variables. Linearize the differential equations around the steady-state point (S_0, P_0, E_0). The flows are held constant; the volume is V_0.

3.7. A kinetic model of enzyme catalysis is given by

$$\frac{d\sigma_1}{d\theta} = v_1 - \frac{\sigma_1 \sigma_2^{\gamma}}{1 + \sigma_2^{\gamma}(1 + \sigma_1)}$$

$$\frac{d\sigma_2}{d\theta} = \alpha_2 \left(\frac{\sigma_1 \sigma_2^{\gamma}}{1 + \sigma_2^{\gamma}(1 + \sigma_1)} - \chi_2 \sigma_2 \right),$$

where σ_1 and σ_2 are the relative substrate and product concentrations, respectively. Linearize the differential equations around the steady-state point $(\sigma_{10}, \sigma_{20})$.

3.8. A system describing the concentrations of an activator (x) and an inhibitor (y) is given by (Gierer–Meinhardt model)

$$\frac{dx}{dt} = \rho + \frac{cx^2}{y} - \mu x$$

$$\frac{dy}{dt} = dx^2 - \gamma y.$$

(a) Calculate the steady-state concentrations.
(b) Linearize the differential equations around the steady-state concentrations.

3.9. The cell (C_x) and substrate (C_s) concentrations in a chemostat are described by

$$\frac{dC_x}{dt} = (\mu - D)C_x$$

and

$$\frac{dC_s}{dt} = D(C_{s0} - C_s) - \frac{\mu C_x}{Y_{x/s}},$$

where $Y_{x/s}$ is the yield coefficient. The specific growth rate is

$$\mu = \frac{\mu_{\max} C_s}{K_s + C_s + \frac{C_s^2}{K_I}}.$$

Linearize the differential equations around the steady-state point $(C_{x,ss}, C_{s,ss})$.

3.10. Redo Problem 3.9 when the specific growth rate is

$$\mu = \frac{\mu_{max}C_s}{K_s + C_s}\left(1 - \frac{C_s}{C_{sm}}\right)^n.$$

REFERENCES

1. Prescott LM, Harley JP, Klein DA. *Microbiology*, 3rd ed. Dubuque, IA: Wm. C. Brown Publishers, 1996.
2. Sel'kov EE. Self-oscillations in glycolysis. 1. A simple kinetic model. *European Journal of Biochemistry* 1968; 4:79–86.
3. Edelstein-Keshet L. *Mathematical Models in Biology*. New York: Random House, 1988.
4. Hodgkin A, Huxley A. A quantitative description of membrane current and its application to conduction and excitation in nerve. *Journal of Physiology* 1952; 117:500–544.
5. Fitzhugh R. Impulses and physiological states in theoretical models of nerve membrane. *Biophysical Journal* 1961; 1:445–466.

CHAPTER 4

STABILITY CONSIDERATIONS

Stability is an important concept in process dynamics and control. The notion of stability, elaborated in this chapter for linear and nonlinear dynamic systems, can provide valuable insight into the behaviors of several processes ranging from bioreactors operating in a continuous mode to predator–prey interactions. Methods for computing eigenvalues of a matrix are discussed. These numbers help determine whether an equilibrium point is asymptotically stable, marginally stable, or unstable.

4.1 DEFINITION OF STABILITY

Consider, for example, a plant operating at steady state. Process variables, such as the temperature (T), remain unchanged. If T deviates from its initial value (T_0) at time $t = t_0$ as a result of a disturbance, two scenarios are possible: (1) The system returns to T_0 as time passes, and (2) the system deviates further and farther from T_0 and does not return to its original value. The first scenario describes a *stable process*. The second refers to an *unstable* plant.

Control of Biological and Drug-Delivery Systems for Chemical, Biomedical, and Pharmaceutical Engineering, First Edition. Laurent Simon.
© 2013 John Wiley & Sons, Inc. Published 2013 by John Wiley & Sons, Inc.

4.1.1 Stability Analysis for Linear Systems

Consider a *linear time invariant* (LTI) system:

$$\frac{d\mathbf{x}}{dt} = \mathbf{A}\mathbf{x} + \mathbf{B}\mathbf{u} \tag{4.1}$$
$$\mathbf{y} = \mathbf{C}\mathbf{x} + \mathbf{D}\mathbf{u}.$$

The name LTI designates a system, represented in a state-space form, that is linear and its matrices, $\mathbf{A}, \mathbf{B}, \mathbf{C}$, and $\mathbf{D}$, are not functions of time. The complete solution $\mathbf{x}(t)$ is given by

$$\mathbf{x}(t) = e^{\mathbf{A}(t-t_0)}\mathbf{x}_0 + \int_{t_0}^{t} e^{\mathbf{A}(t-t_0)}\mathbf{B}\mathbf{u}(\tau)d\tau. \tag{4.2}$$

Consequently,

$$\mathbf{y} = \mathbf{C}e^{\mathbf{A}(t-t_0)}\mathbf{x}_0 + \mathbf{C}\int_{t_0}^{t} e^{\mathbf{A}(t-t_0)}\mathbf{B}\mathbf{u}(\tau)d\tau + \mathbf{D}\mathbf{u}. \tag{4.3}$$

The first term, $\mathbf{C}e^{\mathbf{A}(t-t_0)}\mathbf{x}_0$, is the *zero-input system response* (obtained when $\mathbf{u} = \mathbf{0}$). The second term, $\mathbf{C}\int_{t_0}^{t} e^{\mathbf{A}(t-t_0)}\mathbf{B}\mathbf{u}(\tau)d\tau + \mathbf{D}\mathbf{u}$, is the *zero-state system response* component (obtained when $\mathbf{x}_0 = \mathbf{0}$). The use of the matrix exponential e^{Ψ}, where Ψ is a matrix,

$$e^{\Psi} = \sum_{k=0}^{\infty} \frac{1}{k!}\Psi^k, \tag{4.4}$$

will not be pursued in this book to generate a solution $\mathbf{x}(t)$. Other techniques, such as Laplace transforms, will be used. However, the notation is introduced here to show the *zero-input* and *zero-state components* of the response $\mathbf{y}(t)$ and to lay the foundation for defining the stability of a system. In *Mathematica*®, the matrix exponential is implemented as "MatrixExp[Ψ]"; in MATLAB®, it is "expm[Ψ]."

There are several types of stability. In this chapter, the focus is placed on *internal stability* and *bounded-input bounded-output* (BIBO) stability. Internal stability is based on the zero-input response:

$$\frac{d\mathbf{x}}{dt} = \mathbf{A}\mathbf{x}. \tag{4.5}$$

The system is *asymptotically stable* at the equilibrium point $\mathbf{x}_s = \mathbf{0}$ if the real part of any eigenvalue of $\mathbf{A}$ is less than 0: Re $(\lambda_i) < 0$ for all i. The steady-state point $\mathbf{x}_s = \mathbf{0}$ is *marginally stable* if Re $(\lambda_i) \le 0$ for all i, Re $(\lambda_i) = 0$ for at least one eigenvalue and the eigenvalues with Re $(\lambda_i) = 0$ are simple roots. Multiple roots located on the imaginary axis may or may not lead to unbounded solutions [1]. The system is unstable if it is neither stable nor marginally stable.

Eigenvalues of $\mathbf{A}$ are obtained by solving the equation

$$\det(\mathbf{A} - \lambda\mathbf{I}) = 0, \tag{4.6}$$

where $\mathbf{I}$ is the identity matrix and "det" represents the determinant. Equation (4.6) is known as the *characteristic equation* of $\mathbf{A}$. These eigenvalues are solutions of the *characteristic polynomial* and can be easily obtained in *Mathematica* by using the command "Eigenvalues[$\mathbf{A}$]." In MATLAB, the command is "eig $(\mathbf{A})$." For a 2×2 matrix,

$$\mathbf{A} = \begin{pmatrix} a_{11} & a_{12} \\ a_{21} & a_{22} \end{pmatrix},$$

the two eigenvalues are

$$\lambda_{1,2} = \frac{\alpha \pm \sqrt{\delta}}{2} \tag{4.7}$$

with

$$\begin{aligned} \alpha &= \text{Trace}(\mathbf{A}) = a_{11} + a_{22} \\ \gamma &= \det(\mathbf{A}) = a_{11}a_{22} - a_{12}a_{21} \\ \delta &= \text{disc}(\mathbf{A}) = \alpha^2 - 4\gamma. \end{aligned} \tag{4.8}$$

It can be shown that for a 2×2 system, the equilibrium state is asymptotically stable (i.e., Re $(\lambda_i) < 0$) if $\alpha < 0$ and $\gamma > 0$.

For a clearer understanding of the notion of stability, it is useful to consider the following example:

$$\frac{dx}{dt} = ax; \quad x(t = 0) = x_0, \tag{4.9}$$

where a is a real or complex number. The solution of $x(t)$ is

$$x(t) = x_0 e^{at}. \tag{4.10}$$

At $x_0 = x_s = 0$, the system is at equilibrium. However, what happens to x *(t)* if there is a small random disturbance? With a deviation in the steady state, Equation (4.10) becomes

$$x(t) = (x_s + \Delta x_0)e^{at}, \tag{4.11}$$

where Δx_0 is the size of the perturbation. There are two cases:

1. If a is real and positive, x *(t)*$\rightarrow\infty$ as $t\rightarrow\infty$ (i.e., unbounded), the equilibrium point is unstable.
2. If a is real and negative, x *(t)*$\rightarrow x_s = 0$ as $t \rightarrow \infty$, the equilibrium point is stable.

In Equation (4.10), the eigenvalue is a. Higher-dimension systems will result in more than one eigenvalue. However, the solution will still contain exponential parts (i.e., $e^{\lambda_i t}$). The condition Re $(\lambda_i) < 0$ guarantees that $\mathbf{x}$ *(t)*$\rightarrow \mathbf{x}_s = \mathbf{0}$ as $t \rightarrow \infty$. In other words, *all trajectories (different finite initial states) tend to converge to the origin as $t \rightarrow \infty$.* For a *marginally stable system, all trajectories lead to a bounded response.*

Another definition of stability is the following: a dynamic system is stable if *every bounded-input signal* $(|u(t)| \leq u_i < \infty \; \forall t \geq 0)$ *produces a bounded output* $(|y(t)| \leq y_i < \infty \; \forall t \geq 0)$. This definition is referred to as *BIBO stability*. Otherwise, the system is unstable. A bounded input has a lower and upper value and does not increase in magnitude to $\pm\infty$ as time passes. Values that are very large (i.e., large deviation from their nominal point) or, mathematically, approaches $\pm\infty$ are considered *unbounded*. According to this definition, a stable system may reach a new operating point after a bounded change in the input variable. The response is only due to the input $\mathbf{u}$ because the initial state $\mathbf{x}_0$ is set equal to zero (i.e., $\mathbf{x}_0 = \mathbf{0}$, *zero-state response* component). *Asymptotic stability implies BIBO stability. However, the converse is not always true.*

Example 4.1   Consider the system

$$\frac{dx_1}{dt} = -3x_1 + \frac{1}{3}x_2$$
$$\frac{dx_2}{dt} = 3x_1 - \frac{1}{3}x_2. \tag{4.12}$$

The matrix $\mathbf{A}$ is

$$\mathbf{A} = \begin{bmatrix} -3 & \dfrac{1}{3} \\ 3 & -\dfrac{1}{3} \end{bmatrix}. \tag{4.13}$$

The equilibrium point $(x_{1s}, x_{2s}) = (0, 0)$ is asymptotically stable if Trace$(\mathbf{A}) < 0$ and det $(\mathbf{A}) > 0$. Calculation of Trace$(\mathbf{A})$ gives

$$\text{Trace}(\mathbf{A}) = a_{11} + a_{22} = -3 - \frac{1}{3} = -\frac{10}{3} < 0. \tag{4.14}$$

The determinant of $\mathbf{A}$ is

$$\det(\mathbf{A}) = a_{11}a_{22} - a_{12}a_{21} = (-3)\left(-\frac{1}{3}\right) - (3)\left(\frac{1}{3}\right) = 0. \tag{4.15}$$

As a result, the equilibrium point is not asymptotically stable. The eigenvalues of $\mathbf{A}$ are

$$\lambda_1 = -\frac{10}{3}, \quad \lambda_2 = 0, \tag{4.16}$$

which also confirms that the equilibrium point $(x_{1s}, x_{2s}) = (0, 0)$ is indeed a marginally stable point. The system behavior for two initial states, $(x_{10}, x_{20}) = (0.1, 0.2)$ and $(x_{10}, x_{20}) = (0.3, 0.4)$, are shown in Figure 4.1. As t increases, the system response remains bounded without returning to the equilibrium point $(x_{1s}, x_{2s}) = (0, 0)$.

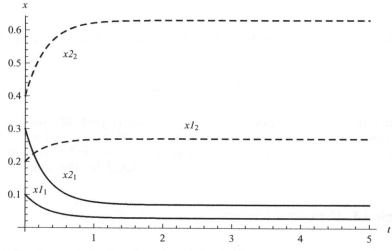

Figure 4.1. System behavior when the origin of a 2×2 linear system is an asymptotically stable point. The state variables, $x_1(t)$ and $x_2(t)$, are denoted by $x1_1(t)$ and $x1_2(t)$, respectively, when the initial condition is $(x_{10}, x_{20}) = (0.1, 0.2)$. The state variables are labeled $x2_1(t)$ and $x2_2(t)$ when $(x_{10}, x_{20}) = (0.3, 0.4)$.

Example 4.2   Consider the system

$$
\begin{aligned}
\frac{dx_1}{dt} &= -3x_1 + \frac{1}{3}x_2 \\
\frac{dx_2}{dt} &= 3x_1 - \frac{2}{3}x_2.
\end{aligned}
\tag{4.17}
$$

The matrix is

$$
\mathbf{A} = \begin{bmatrix} -3 & \dfrac{1}{3} \\ 3 & -\dfrac{2}{3} \end{bmatrix}.
\tag{4.18}
$$

The equilibrium point $(x_{1s}, x_{2s}) = (0, 0)$ is asymptotically stable if $\text{Trace}(\mathbf{A}) < 0$ and $\det(\mathbf{A}) > 0$. Calculation of $\text{Trace}(\mathbf{A})$ and $\det(\mathbf{A})$ yields

$$
\text{Trace}(\mathbf{A}) = a_{11} + a_{22} = -3 - \frac{2}{3} = -\frac{11}{3} < 0
\tag{4.19}
$$

and

$$
\det(\mathbf{A}) = a_{11}a_{22} - a_{12}a_{21} = (-3)\left(-\frac{2}{3}\right) - (3)\left(\frac{1}{3}\right) = 1 > 0.
\tag{4.20}
$$

As a result, the equilibrium point is asymptotically stable. The eigenvalues of $\mathbf{A}$,

$$
\lambda_1 = -3.3699, \quad \lambda_2 = -0.29674,
\tag{4.21}
$$

show that the equilibrium point $(x_{1s}, x_{2s}) = (0, 0)$ is asymptotically stable. Plots of the state variables are shown for two initial conditions: $(x_{10}, x_{20}) = (0.1, 0.2)$ and $(x_{10}, x_{20}) = (0.3, 0.4)$ (Fig. 4.2). As t increases, $x_1(t)$ and $x_2(t)$ approach the equilibrium point $(x_{1s}, x_{2s}) = (0, 0)$, as predicted by the signs of the eigenvalues.

Example 4.3   Consider the system

$$
\begin{aligned}
\frac{dx_1}{dt} &= 3x_1 - 9x_2 \\
\frac{dx_2}{dt} &= 4x_1 - 3x_2.
\end{aligned}
\tag{4.22}
$$

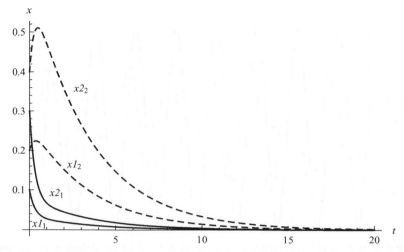

Figure 4.2. System behavior when the origin of a 2×2 linear system is an asymptotically stable point. The state variables, $x_1(t)$ and $x_2(t)$, are denoted by $x1_1(t)$ and $x1_2(t)$, respectively, when $(x_{10}, x_{20}) = (0.1, 0.2)$. The state variables are labeled $x2_1(t)$ and $x2_2(t)$ when $(x_{10}, x_{20}) = (0.3, 0.4)$.

The matrix **A** is

$$\mathbf{A} = \begin{bmatrix} 3 & -9 \\ 4 & -3 \end{bmatrix}. \tag{4.23}$$

The eigenvalues of **A** are

$$\lambda_1 = 3i\sqrt{3}, \quad \lambda_2 = -3i\sqrt{3}. \tag{4.24}$$

As a result, the equilibrium point $(x_{1s}, x_{2s}) = (0, 0)$ is marginally stable. Sustained oscillations (i.e., amplitude neither decays nor grows) are shown for two trajectories: $(x_{10}, x_{20}) = (0.01, 0.02)$ and $(x_{10}, x_{20}) = (0.03, 0.04)$ (Figs. 4.3 and 4.4, respectively).

Example 4.4   Consider the system

$$\frac{dx_1}{dt} = -3x_1 + \frac{1}{3}x_2 + u$$

$$\frac{dx_2}{dt} = 3x_1 - \frac{2}{3}x_2 \tag{4.25}$$

$$y = x_1.$$

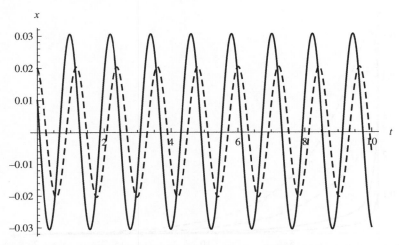

Figure 4.3. System behavior when the origin of a 2×2 linear system is a marginally stable point. The initial condition is $(x_{10}, x_{20}) = (0.01, 0.02)$.

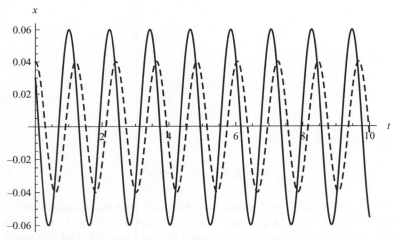

Figure 4.4. System behavior when the origin of a 2×2 linear system is a marginally stable point. The initial condition is $(x_{10}, x_{20}) = (0.03, 0.04)$.

This model is similar to the one presented in Example 4.2 except for the input u. The equilibrium point was found to be asymptotically stable. As a result, the system is BIBO stable. Let $u(t) = 0.1$ and $(x_{10}, x_{20}) = (0, 0)$. As t increases, the response $y(t)$ approaches a steady-state value (Fig. 4.5). Figure 4.6 shows the system behavior when $u(t) = \sin(t)$ and $(x_{10}, x_{20}) = (0, 0)$. A sustained oscillation is obtained.

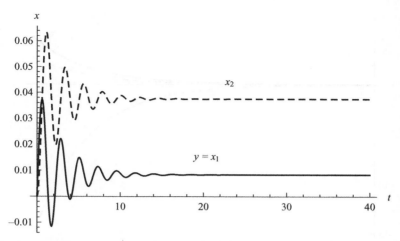

Figure 4.5. Behavior of a 2×2 BIBO stable system. The process responds to a step change of size 0.1.

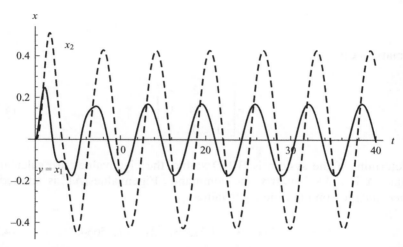

Figure 4.6. Behavior of a 2×2 BIBO stable system. The process responds to a change $u(t) = \sin(t)$.

Example 4.5   Consider the system

$$\frac{dx_1}{dt} = -3x_1 + 2x_2 - x_3 + u$$

$$\frac{dx_2}{dt} = 3x_1 - 4x_2 \qquad\qquad (4.26)$$

$$\frac{dx_3}{dt} = -x_1 + x_3$$

$$y = x_1.$$

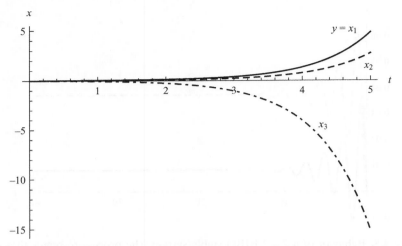

Figure 4.7. Behavior of a 3×3 BIBO unstable system. The process responds to a step change $u(t) = 0.1$.

The matrix $\mathbf{A}$ is

$$\mathbf{A} = \begin{bmatrix} -3 & 2 & -1 \\ 3 & -4 & 0 \\ -1 & 0 & 1 \end{bmatrix}. \tag{4.27}$$

To determine if the system is BIBO stable, the eigenvalues are calculated. Because $\mathbf{A}$ is a 3×3 matrix, the command "Eigenvalues[$\mathbf{A}$]" is applied in *Mathematica* to compute the eigenvalues:

$$\lambda_1 = -6.05765, \quad \lambda_2 = 1.31398, \quad \lambda_3 = -1.25634. \tag{4.28}$$

The equilibrium point $(x_{1s}, x_{2s}, x_{3s}) = (0, 0, 0)$ is not asymptotically stable. Therefore, the system is BIBO unstable. Let $u(t) = 0.1$ and $(x_{10}, x_{20}, x_{30}) = (0, 0, 0)$. Figure 4.7 shows that as t increases, the magnitude of the response $y(t)$ becomes very large.

4.1.2 Stability Analysis for Nonlinear Systems

Stability analysis for nonlinear dynamic systems is similar to the study of linear systems after performing linearization of the problem around a steady-state point. For a 2×2 system, the equilibrium point (x_{1s}, x_{2s}) is asymptotically stable if $\mathrm{Trace}(\mathbf{A}) < 0$ and $\det(\mathbf{A}) > 0$, where

$$\mathbf{A} = \begin{bmatrix} \left(\dfrac{\partial f_1}{\partial \tilde{x}_1}\right)_{x_{1s},x_{2s}} & \left(\dfrac{\partial f_1}{\partial \tilde{x}_2}\right)_{x_{1s},x_{2s}} \\[3ex] \left(\dfrac{\partial f_2}{\partial \tilde{x}_1}\right)_{x_{1s},x_{2s}} & \left(\dfrac{\partial f_2}{\partial \tilde{x}_2}\right)_{x_{1s},x_{2s}} \end{bmatrix}, \tag{4.29}$$

which is equivalent to the statement that the real part of any eigenvalue of $\mathbf{A}$ is less than 0: Re $(\lambda_i) < 0$ for all i. Similar to linear systems, the equilibrium point may be marginally stable. BIBO stability also applies to nonlinear systems and can be verified using the approach discussed for linear processes: *asymptotic stability implies BIBO stability*. If the state variables for the original nonlinear system are plotted after slight perturbations in the steady-state points, the state variables should return to the equilibrium point, which is not necessarily the point $(0, 0)$. The steady-state point is marginally stable if the state variables remain bounded after a small increase or decrease in the equilibrium point.

Example 4.6   Consider the system

$$\frac{dx}{dt} = -3x - xy$$
$$\frac{dy}{dt} = xy - y. \tag{4.30}$$

The functions f_1 and f_2 in the matrix $\mathbf{A}$ are

$$f_1(x, y) = -3x - xy \tag{4.31}$$

and

$$f_2(x, y) = xy - y. \tag{4.32}$$

To calculate the equilibrium points, we set $f_1(x, y) = 0$ and $f_2(x, y) = 0$ to give

$$-3x - xy = 0 \tag{4.33}$$

and

$$xy - y = 0. \tag{4.34}$$

The solutions are $(x_s, y_s) = (0, 0)$ and $(x_s, y_s) = (1, -3)$, where s stands for a steady-state value. The matrix $\mathbf{A}$ is

$$\mathbf{A} = \begin{bmatrix} -3 - y_s & -x_s \\ y_s & -1 + x_s \end{bmatrix}. \tag{4.35}$$

When the steady states are substituted in $\mathbf{A}$, we have

$$\mathbf{A} = \begin{bmatrix} -3 & 0 \\ 0 & -1 \end{bmatrix} \tag{4.36}$$

and

$$\mathbf{A} = \begin{bmatrix} 0 & -1 \\ -3 & 0 \end{bmatrix} \tag{4.37}$$

for $(x_s, y_s) = (0, 0)$ and $(x_s, y_s) = (1, -3)$, respectively. The eigenvalues are $\lambda_1 = -3$, $\lambda_2 = -1$ when $(x_s, y_s) = (0, 0)$ and $\lambda_1 = -\sqrt{3}$, $\lambda_2 = \sqrt{3}$ when $(x_s, y_s) = (1, -3)$. Consequently, $(x_s, y_s) = (0, 0)$ is an asymptotically stable steady-state point, while $(x_s, y_s) = (1, -3)$ is an asymptotically unstable steady-state point. According to this analysis, the original nonlinear dynamic system should return to $(x_s, y_s) = (0, 0)$ after a small perturbation. Also, a minor deviation from $(x_s, y_s) = (1, -3)$ should lead to an unbounded response. These behaviors are shown in Figures 4.8 and 4.9.

Example 4.7 Consider the system

$$\frac{dx}{dt} = -3x - xy - 2$$
$$\frac{dy}{dt} = xy - y. \tag{4.38}$$

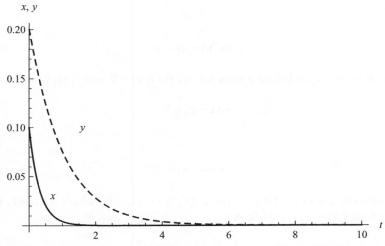

Figure 4.8. Behavior of a 2×2 nonlinear system in the vicinity of an asymptotically stable steady-state point $(0, 0)$. The initial condition was set at $(x_{10}, x_{20}) = (0.1, 0.2)$.

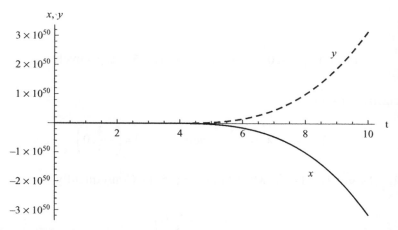

Figure 4.9. Behavior of a 2×2 nonlinear system in the vicinity of an unstable steady-state point $(1, -3)$. The initial condition was set at $(x_0, y_0) = (1.01, -2.99)$.

The functions f_1 and f_2 in the matrix **A** are

$$f_1(x, y) = -3x - xy - 2 \tag{4.39}$$

and

$$f_2(x, y) = xy - y. \tag{4.40}$$

After setting $f_1(x, y) = 0$ and $f_2(x, y) = 0$, we obtain

$$(x_s, y_s) = \left(-\frac{2}{3}, 0\right) \quad \text{and} \quad (x_s, y_s) = (1, -5).$$

The matrix **A** is the same as the one calculated in Example 4.6:

$$\mathbf{A} = \begin{bmatrix} -3 - y_s & -x_s \\ y_s & -1 + x_s \end{bmatrix}. \tag{4.41}$$

Substituting the steady states in **A** gives

$$\mathbf{A} = \begin{bmatrix} -3 & \dfrac{2}{3} \\ 0 & -\dfrac{5}{3} \end{bmatrix} \tag{4.42}$$

and

$$\mathbf{A} = \begin{bmatrix} 2 & -1 \\ -5 & 0 \end{bmatrix} \tag{4.43}$$

for

$$(x_s, y_s) = \left(-\frac{2}{3}, 0\right) \quad \text{and} \quad (x_s, y_s) = (1, -5), \text{respectively.}$$

The eigenvalues are

$$\lambda_1 = -3, \quad \lambda_2 = -\frac{5}{3} \quad \text{when} \quad (x_s, y_s) = \left(-\frac{2}{3}, 0\right)$$

and $\lambda_1 = 1 + \sqrt{6}, \lambda_2 = 1 - \sqrt{6}$ when $(x_s, y_s) = (-5, 1)$. Consequently,

$$(x_s, y_s) = \left(-\frac{2}{3}, 0\right)$$

is an asymptotically stable steady-state point, while $(x_s, y_s) = (1, -5)$ is unstable. The system should return to

$$(x_s, y_s) = \left(-\frac{2}{3}, 0\right)$$

after a small deviation. Similarly, a fluctuation near $(x_s, y_s) = (1, -5)$ should lead to an unbounded response. These behaviors are shown in Figures 4.10 and 4.11.

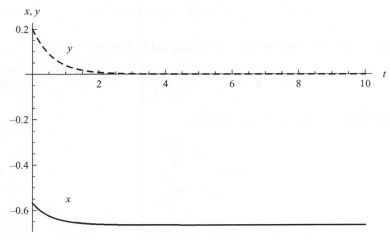

Figure 4.10. Behavior of a 2×2 nonlinear system in the neighborhood of an asymptotically stable steady-state point $(x_s, y_s) = \left(-\frac{2}{3}, 0\right)$. The initial condition is $(x_0, y_0) = (-0.567, 0.200)$.

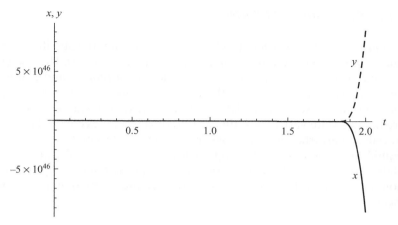

Figure 4.11. Behavior of a 2×2 nonlinear system in the neighborhood of an asymptotically stable steady-state point $(x_s, y_s) = (1, -5)$. The initial condition is $(x_0, y_0) = (1.01, -4.99)$.

4.2 STEADY-STATE CONDITIONS AND EQUILIBRIUM POINTS

The equilibrium or steady-state points of a system are obtained mathematically by setting accumulation terms in the equations for mass, energy, and momentum balance equal to zero. In reality, these points constitute operating conditions that satisfy requirements of safety, economics, environmental regulations, product specifications, and operational constraints [2]. Asymptotic stability ensures that the process is self-regulating. No control action is required to ensure that the process will return to this condition. External influences are not likely to affect product specifications set by the designers. BIBO stability verifies whether the response will settle to a new steady state (or a bounded output) with respect to a change in the input variable. The question of how long it will take the process to reach the new steady state is not addressed. Control strategies may be necessary to minimize the time it takes to reach a new equilibrium.

Why is it necessary to operate a plant at an unstable steady state? For some processes, such as a continuous stirred-tank reactor, the decision may be based on the desire to improve the yield and to create safe operational conditions. Some types of fluid catalytic cracking (FCC), a unit used in the petroleum refining industry to convert heavy gas oil to gasoline and light hydrocarbons, need to operate at an unstable steady state in order to achieve the maximum gasoline yield. Polymerization, conducted in a continuous stirred-tank reactor, exhibits steady-state multiplicity. Operating in an unstable region may be necessary to achieve an intermediate level of monomer conversion [3]. In all of these cases, a controller is required to ensure that the process can operate at the unstable steady state.

4.3 PHASE-PLANE DIAGRAMS

Phase-plane diagrams provide another method to study the stability of equilibrium points for a system of nonlinear differential equations. The analysis is performed in unforced (zero-input) *two-dimensional models* and consists of plotting the two state variables in the Cartesian plane as they evolve in time. A point on the phase plane is defined by $(x_1 (t_i), x_2 (t_i))$, where $x_1 (t_i)$ and $x_2 (t_i)$ are the two state variables at time t_i. Based on examples covered in Section 4.1, variables perturbed about an asymptotically stable steady state return to the equilibrium point as the time approaches ∞. For a marginally stable system, the state variables have to remain bounded. The phase-plane analysis is useful for nonlinear systems because it does not require the computation of Jacobian matrices.

Example 4.8   Consider the system presented in Example 4.6:

$$\frac{dx}{dt} = -3x - xy$$
$$\frac{dy}{dt} = xy - y. \tag{4.44}$$

The phase-plane diagram is shown in Figure 4.12. Observe how the trajectories converge toward the equilibrium point $(0, 0)$ and diverge away from the steady-state point $(1, -3)$. The command "*ParametricPlot*" from *Mathematica* can be used to produce the graph. The user would need to define several initial conditions.

Example 4.9   The marginally stable system presented in Example 4.3 is

$$\frac{dx_1}{dt} = 3x_1 - 9x_2$$
$$\frac{dx_2}{dt} = 4x_1 - 3x_2. \tag{4.45}$$

Even though the system is linear, a phase-plane analysis can be conducted. The sustained oscillation predicted by the complex number is observed (i.e., *neutral center*) (Fig. 4.13).

4.4 POPULATION KINETICS

The interactions between the spruce budworm and balsam fir forests of the Northeastern United States and Canada are analyzed using the following dynamic system [4]:

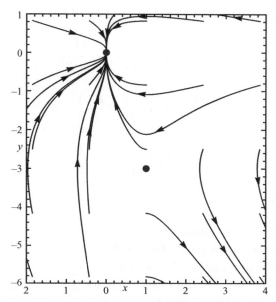

Figure 4.12. Phase-plane diagram representing the behavior of a 2×2 nonlinear system in the neighborhood of an unstable steady-state point $(1, -3)$ and a stable steady-state point $(0, 0)$. The plot was generated using a *Mathematica*® notebook by Gianluca Gorni: *CurvesGraphics6:* http://users.dimi.uniud.it/~gianluca.gorni/Mma/Mma.html, accessed August 20, 2010.

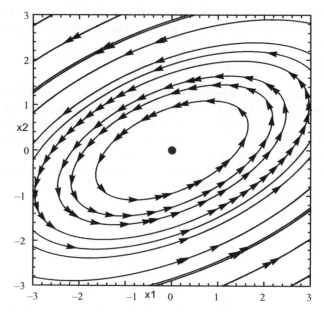

Figure 4.13. Phase-plane diagram representing the behavior of a 2×2 linear system in the neighborhood of a marginally stable steady-state point $(0, 0)$. The plot was generated using a *Mathematica*® notebook by Gianluca Gorni: *CurvesGraphics6:* http://users.dimi.uniud.it/~gianluca.gorni/Mma/Mma.html, accessed August 20, 2010.

$$\frac{dB}{dt} = r_B B \left(1 - \frac{B}{K_B} \right) - \beta \frac{B^2}{\alpha^2 + B^2} \tag{4.46}$$

$$\frac{dS}{dt} = r_s S \left(1 - \frac{S}{K_S} \times \frac{K_E}{E} \right) \tag{4.47}$$

$$\frac{dE}{dt} = r_E E \left(1 - \frac{E}{K_E} \right) - P \frac{B}{S}. \tag{4.48}$$

The variables and parameters are

S: branch surface area (branches/acre)
E: tree energy reserves (i.e., foliage and health)
B: spruce budworm density (larvae/acre)
r_B: intrinsic budworm growth rate (per year)
r_s: intrinsic branch growth rate (per year)
r_E: intrinsic E growth rate (per year)
K_E: maximum E level
K_S: maximum branch density (branch/acre)
K_B: carrying capacity (larvae/acre)
P: rate of energy consumption by budworm (branch/larva/year)
α: scale of budworm densities at which saturation begins to occur (larvae/acre)
β: maximum budworm predated (larvae/acre/year).

Studies of stability and ecological system behaviors about equilibrium points are often pursued to help identify conditions under which a system returns to this initial state after a temporary disturbance. Such investigations may help limit the threat posed by insects, prevent the destruction of a forest, and identify cyclic stabilities following a pest outbreak.

If B is defined as a constant (i.e., near steady state), the steady-state equations, Equations (4.47) and (4.48), are

$$r_s S \left(1 - \frac{S}{K_S} \times \frac{K_E}{E} \right) = 0 \tag{4.49}$$

and

$$r_E E \left(1 - \frac{E}{K_E} \right) - P \frac{B}{S} = 0. \tag{4.50}$$

The equilibrium points are found by solving Equations (4.49) and (4.50).

Using deviation variables: $\tilde{S} = S - S_e$ and $\tilde{E} = E - E_e$, the linear approximation of the system is

$$
\begin{pmatrix} \dfrac{\partial \tilde{S}}{\partial t} \\ \dfrac{\partial \tilde{E}}{\partial t} \end{pmatrix} = \begin{pmatrix} r_S - 2\dfrac{r_S K_E S_e}{E_e K_S} & \dfrac{BP}{S_e^2} \\ \dfrac{r_S K_E S_e^2}{E_e^2 K_S} & r_E - 2\dfrac{r_E E_e}{K_E} \end{pmatrix} \begin{pmatrix} \tilde{S} \\ \tilde{E} \end{pmatrix},
\tag{4.51}
$$

where (S_e, E_e) is an equilibrium point. The determinant of the Jacobian matrix is

$$
\det(\mathbf{A}) = -\frac{r_S\left(BPK_E^2 + E_E(2E_E - K_E)r_E(E_E K_S - 2K_E S_E)\right)}{E_E^2 K_E K_S},
\tag{4.52}
$$

where

$$
\mathbf{A} = \begin{pmatrix} r_S - 2\dfrac{r_S K_E S_e}{E_e K_S} & \dfrac{BP}{S_e^2} \\ \dfrac{r_S K_E S_e^2}{E_e^2 K_S} & r_E - 2\dfrac{r_E E_e}{K_E} \end{pmatrix}.
\tag{4.53}
$$

The trace of $\mathbf{A}$ is

$$
\text{Trace}(\mathbf{A}) = r_E - \frac{2E_E r_E}{K_E} + r_S - \frac{2K_E r_S S_E}{E_E K_S}.
\tag{4.54}
$$

The equilibrium point is asymptotically stable if $\text{Trace}(\mathbf{A}) < 0$ and $\det(\mathbf{A}) > 0$.

4.5 DYNAMICS OF BIOREACTORS

A continuous stirred-tank bioreactor (CSTR) was described in Section 2.2.3. The state equations for the cell and substrate concentrations are

$$
\frac{dC_x}{dt} = (\mu - D)C_x
\tag{4.55}
$$

and

$$\frac{dC_s}{dt} = D(C_{\text{sin}} - C_s) - \frac{\mu}{Y_{x/s}} C_x, \tag{4.56}$$

respectively, where $D = F_{\text{in}}/Ah$. The microorganism obeys Monod kinetics:

$$\mu = \frac{\mu_{\text{max}} C_s}{K_s + C_s}. \tag{4.57}$$

In this example, the volume is held constant and the product concentration is not included in the model. The parameters for a fermentation are [5]: $F_{\text{in}} = 500\,\text{L/h}$, $V = Ah = 1429\,\text{L}$, $\mu_{\text{max}} = 0.7/\text{hour}$, $K_s = 5.0\,\text{g/L}$, $C_{\text{sin}} = 85\,\text{g/L}$, and $Y_{x/s} = 0.65$.

The steady-state concentrations are obtained by setting the time derivatives in Equations (4.55) and (4.56) equal to zero:

$$f_1 = (\mu - D)C_x = 0 \tag{4.58}$$

and

$$f_2 = (C_{\text{sin}} - C_s) - \frac{\mu C_x}{Y_{C_x/C_s}} = 0. \tag{4.59}$$

Equation (4.58) indicates that $C_x = 0$ or $\mu = D$. When $C_x = 0$, Equation (4.59) leads to $C_s = C_{\text{sin}}$ (trivial solution). For $\mu = D$, we have

$$C_s = \frac{DK_s}{\mu_{\text{max}} - D} \tag{4.60}$$

and

$$C_x = Y_{C_x/C_s}(C_{\text{sin}} - C_s) = Y_{C_x/C_s}\left(C_{\text{sin}} - \frac{DK_s}{\mu_{\text{max}} - D}\right). \tag{4.61}$$

Consequently, the equilibrium points are

$$(C_{xe}, C_{se}) = (0, C_{\text{sin}}) \quad \text{and} \quad (C_{xe}, C_{se}) = \left(Y_{C_x/C_s}\left(C_{\text{sin}} - \frac{DK_s}{\mu_{\text{max}} - D}\right), \frac{DK_s}{\mu_{\text{max}} - D}\right).$$

Substituting the parameter values yields $(C_{xe}, C_{se}) = (0, 85)$ and $(C_{xe}, C_{se}) = (52.0, 5.0)$.

The functions f_1 and f_2 become

$$f_1 = \left(-\frac{500}{1429} + \frac{0.7C_s}{5.0 + C_s}\right)C_x \tag{4.62}$$

and

$$f_2 = \frac{500(85 - C_s)}{1429} - \frac{1.077C_sC_x}{5.0 + C_s}. \tag{4.63}$$

The linearized system is

$$\begin{pmatrix} \dfrac{d\tilde{C}_x}{\partial t} \\ \dfrac{d\tilde{C}_s}{\partial t} \end{pmatrix} = \begin{pmatrix} \dfrac{-1.75 + 0.35C_{se}}{5.0 + C_{se}} & -\dfrac{1.077C_{se}}{5.0 + C_{se}} \\ \dfrac{3.5C_{xe}}{(5.0 + C_{se})^2} & \dfrac{-0.35(5.0 + C_{se})(5.0 + C_{se}) - 5.38C_{xe}}{(5.0 + C_{se})^2} \end{pmatrix} \begin{pmatrix} \tilde{C}_x \\ \tilde{C}_s \end{pmatrix}. \tag{4.64}$$

When $(C_{xe}, C_{se}) = (0, 85)$,

$$\mathbf{A} = \begin{pmatrix} \dfrac{-1.75 + 0.35C_{se}}{5.0 + C_{se}} & -\dfrac{1.077C_{se}}{5.0 + C_{se}} \\ \dfrac{3.5C_{xe}}{(5.0 + C_{se})^2} & \dfrac{-0.35(5.0 + C_{se})(5.0 + C_{se}) - 5.38C_{xe}}{(5.0 + C_{se})^2} \end{pmatrix} = \begin{pmatrix} 0.31 & -1.02 \\ 0 & -0.35 \end{pmatrix}. \tag{4.65}$$

The eigenvalues are $\lambda_1 = -0.35$ and $\lambda_2 = 0.31$. The steady-state point is unstable because one eigenvalue is positive.

When $(C_{xe}, C_{se}) = (52.0, 5.0)$,

$$\mathbf{A} = \begin{pmatrix} 0 & -0.54 \\ 1.82 & -3.15 \end{pmatrix}. \tag{4.66}$$

The eigenvalues are $\lambda_1 = -2.80$ and $\lambda_2 = -0.35$. The steady-state point is stable because both eigenvalues are negative.

4.6 GLYCOLYTIC OSCILLATION

The glycolytic pathway, used in Section 3.3, involves the following differential equations [6], which can be written in a dimensionless form as

$$\frac{dx}{dt} = -x + \alpha y + x^2 y \tag{4.67}$$

and

$$\frac{dy}{dt} = \beta - \alpha y - x^2 y, \tag{4.68}$$

where α and β are positive numbers. The normalized variables x and y are the concentrations of adenosine diphosphate and fructose-6-phosphate, respectively. The following steady-state point was obtained:

$$(x_s, y_s) = \left(\beta, \frac{\beta}{\alpha + \beta^2} \right).$$

The linearized equation was

$$\frac{d\tilde{x}}{dt} = \left(\frac{-\alpha + \beta^2}{\alpha + \beta^2} \right) \tilde{x} + \left(-\frac{2\beta^2}{\alpha + \beta^2} \right) \tilde{y}$$
$$\frac{d\tilde{y}}{dt} = (\alpha + \beta^2) \tilde{x} - (\alpha + \beta^2) \tilde{y} \tag{4.69}$$

with $\tilde{x} = x - x_s$ and $\tilde{y} = y - y_s$. The Jacobian matrix $\mathbf{A}$ is

$$\mathbf{A} = \begin{bmatrix} \dfrac{-\alpha + \beta^2}{\alpha + \beta^2} & -\dfrac{2\beta^2}{\alpha + \beta^2} \\ \alpha + \beta^2 & -\alpha - \beta^2 \end{bmatrix}. \tag{4.70}$$

For $\alpha = 0.08$ and $\beta = 0.6$ [7], matrix$\mathbf{A}$ can be written as

$$\mathbf{A} = \begin{bmatrix} 0.64 & -1.64 \\ 0.44 & -0.44 \end{bmatrix}. \tag{4.71}$$

The trace of $\mathbf{A}$ is

$$\mathrm{Trace}(\mathbf{A}) = \frac{-\alpha - \alpha^2 + \beta^2 - 2\alpha\beta^2 - \beta^4}{\alpha + \beta^2} = 0.196. \tag{4.72}$$

The determinant of $\mathbf{A}$ is

$$\det(\mathbf{A}) = \alpha + \beta^2 = 0.44. \tag{4.73}$$

Consequently, the equilibrium point is unstable because $\text{Trace}(\mathbf{A}) < 0$. A calculation of the eigenvalues shows that $\lambda_1 = 0.098 - 0.656i$ and $\lambda_2 = 0.098 + 0.656i$. The positive real parts of the complex numbers confirm that the equilibrium point is unstable.

4.7 HODGKIN–HUXLEY MODEL

The Hodgkin–Huxley model, discussed in Section 3.4, yields the following differential equations:

$$\frac{dV}{dt} = 3\left(V + W - \frac{V^3}{3} + I_{\text{ext}}\right)$$
$$\frac{dW}{dt} = -\frac{1}{3}(V - 0.7 + 0.8W) \tag{4.74}$$

when the FitzHugh–Nagumo model was used. The membrane potential denoted by V and W is a recovery parameter; when I_{ext} was set to zero, one of the steady-state points was $(V_s, W_s, I_{\text{exts}}) = (1.19941, -0.62426, 0)$. The linearized equation at this equilibrium point was

$$\frac{d\tilde{V}}{dt} \approx -1.31574\tilde{V} + 3\tilde{W} + 3\tilde{u}$$
$$\frac{d\tilde{W}}{dt} \approx -\frac{1}{3}\tilde{V} - \frac{4}{15}\tilde{W} \tag{4.75}$$

with $\tilde{V} = V - V_s$, $\tilde{W} = W - W_s$ and $\tilde{u} = I_{\text{ext}} - I_{\text{exts}}$.
    The Jacobian matrix $\mathbf{A}$ is

$$\mathbf{A} = \begin{bmatrix} -1.31574 & 3 \\ -\dfrac{1}{3} & -\dfrac{4}{15} \end{bmatrix}. \tag{4.76}$$

The trace and determinant of $\mathbf{A}$ are -1.58 and 1.35, respectively. The equilibrium point is asymptotically stable if $\text{Trace}(\mathbf{A}) < 0$ and $\det(\mathbf{A}) > 0$. As a result, the system is BIBO stable at the equilibrium point. Let $I_{\text{ext}} = -0.128$ and the initial state defined by $(V_0, W_0) = (1.19941, -0.62426)$. As t increases, the response approaches a steady-state value (Fig. 4.14).

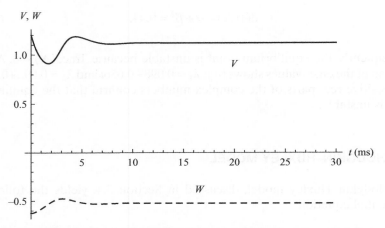

Figure 4.14. Behavior of a 2×2 BIBO stable system. The process responds to a step change of size -0.128.

4.8 SUMMARY

The local stability of linear and nonlinear systems was addressed in this chapter. Conditions for internal and BIBO stability are given. In the context of internal stability, a linear system is asymptotically stable at the equilibrium point $\mathbf{x}_s = \mathbf{0}$ if the real part of any eigenvalue of a Jacobian matrix is less than 0: Re $(\lambda_i) < 0$ for all i. The steady-state point $\mathbf{x}_s = \mathbf{0}$ is marginally stable if Re $(\lambda_i) \leq 0$ for all i, Re $(\lambda_i) = 0$ for at least one eigenvalue, and the eigenvalues with Re $(\lambda_i) = 0$ are simple roots. A dynamic system is BIBO stable if every bounded-input signal produces a bounded output. Otherwise, the system is unstable. Stability analysis for nonlinear dynamic systems is similar to the study of linear systems after linearizing the problem around a steady-state point and using a Jacobian matrix. Examples dealing with population kinetics, bioreactors, glycolytic oscillation, and a Hodgkin–Huxley model were studied to illustrate concepts of stability. Phase-plane diagrams are useful for studying the stability of 2×2 linear and nonlinear systems.

PROBLEMS

4.1. Consider the following dynamic system:

$$\frac{dx_1}{dt} = -x_1 + 2x_2$$
$$\frac{dx_2}{dt} = 3x_1 - x_2.$$

Study the stability of the equilibrium point.

4.2.   A predator–prey model is given by

$$\frac{dx}{dt} = x(\alpha - \gamma_1 y)$$

$$\frac{dx_2}{dt} = -y(\beta - \gamma_2 x),$$

where α, β, γ_1, and γ_2 are parameters that represent species interactions. The variables x and y denote prey and predator population densities, respectively.

(a) Determine the steady states.

(b) Study the stability of the steady states.

4.3.   The populations of two competing species (x, y) are represented by

$$\frac{dx}{dt} = x(1 - x - y)$$

$$\frac{dy}{dt} = y(3 - x - y).$$

(a) Determine the nontrivial steady states.

(b) Study the stability of the nontrivial steady states.

4.4.   In a predator–prey model, the changes in prey (x) and predator (y) populations are represented by

$$\frac{dx}{dt} = rx\left(1 - \frac{x}{K}\right) - \frac{\kappa xy}{(x + D)}$$

$$\frac{dy}{dt} = sy\left(1 - \frac{y}{\gamma x}\right).$$

(a) Determine the positive equilibrium points.

(b) Establish the stability condition for the positive equilibrium points.

4.5.   The state equations for the cell (C_x) and substrate (C_s) concentrations are

$$\frac{dC_x}{dt} = (\mu - D)C_x$$

and

$$\frac{dC_s}{dt} = D(C_{sin} - C_s) - \frac{\mu}{Y_{x/s}}C_x,$$

where $D = F_{in}/Ah$. The microorganism obeys Monod kinetics:

$$\mu = \frac{\mu_{max}C_s}{K_s + C_s}.$$

The model parameters are $F_{in} = 550\,L/h$, $V = Ah = 1429\,L$, $\mu_{max} = 0.7/$ hour, $K_s = 5.0\,g/L$, $C_{sin} = 85\,g/L$, and $Y_{x/s} = 0.65$.

(a) Determine the equilibrium points.

(b) Study the stability of the equilibrium points.

4.6.   Redo Problem 4.5 with $\mu_{max} = 0.5/hour$.

4.7.   The following equations have been used to describe glycolytic oscillation:

$$\frac{dx}{dt} = -x + \alpha y + x^2 y$$

$$\frac{dy}{dt} = \beta - \alpha y - x^2 y,$$

where the normalized variables x and y are the concentrations of adenosine diphosphate and fructose-6-phosphate, respectively. Construct the phase-plane diagram for $\alpha = 0.08$ and $\beta = 0.6$.

4.8.   Redo Problem 4.7 for $\alpha = 0.1$ and $\beta = 0.2$.

4.9.   Consider Problem 4.7:

(a) Find the nontrivial steady state of the system for $\alpha = 0.1$ and $\beta = 0.2$.

(b) Find the Jacobian matrix of the system at the nontrivial steady state.

(c) Is the nontrivial steady state stable?

4.10. The Hodgkin–Huxley model is described by the following differential equations:

$$\frac{dV}{dt} = 3\left(V + W - \frac{V^3}{3} + I_{ext}\right)$$

$$\frac{dW}{dt} = -\frac{1}{3}(V - 0.7 + 0.8W).$$

When $I_{ext} = 0$, a steady-state point is $(V_s, W_s, I_{exts}) = (1.19941, -0.62426, 0)$. Construct the phase-plane diagram for the model.

REFERENCES

1. Bellman R. *Stability Theory of Differential Equations*. New York: McGraw-Hill, 1953.
2. Stephanopoulos G. *Chemical Process Control: an Introduction to Theory and Practice*. Englewood Cliffs, NJ: Prentice Hall, 1984.
3. Méndez-Acosta HO, Femat R, González-Âalvarez V, eds. *Selected Topics in Dynamics and Control of Chemical and Biological Processes*. Berlin: Springer, 2007.
4. Ludwig D, Jones D, Holling C. Qualitative analysis of insect outbreak systems: the spruce budworm and forest. *Journal of Animal Ecology* 1978; 47:315–332.
5. Lee JM. *Biochemical Engineering*. Englewood Cliffs, NJ: Prentice Hall, 1992.
6. Sel'kov EE. Self-oscillations in glycolysis. 1. A simple kinetic model. *European Journal of Biochemistry* 1968; 4:79–86.
7. Strogatz SH. *Nonlinear Dynamics and Chaos: with Applications in Physics, Biology, Chemistry, and Engineering*. Reading, MA: Addison-Wesley, 1994.

CHAPTER 5

LAPLACE TRANSFORMS

Laplace transforms will be applied to solve linear differential equations with constant coefficients. In Chapter 4, graphs were generated to study the stability of steady-state points. In the case of a linear system, Laplace transform-based techniques may be used to obtain an analytical solution, which can be implemented to study the dynamic behavior of a process. Note that the equations described in this book could be solved using other methods. However, Laplace transforms provide a systematic way to examine the relationship between input and output variables. In addition, these tools become very useful for frequency response analysis (Chapter 10) and controller design.

5.1 DEFINITION OF LAPLACE TRANSFORMS

Let $f(t)$ be a function defined on the interval $[0, \infty)$. The Laplace transform of $f(t)$ denoted by $\mathcal{L}\{f(t)\}$ or $\bar{f}(s)$ is

$$\mathcal{L}\{f(t)\} = \bar{f}(s) = \int_0^\infty f(t)e^{-st}dt. \qquad (5.1)$$

Control of Biological and Drug-Delivery Systems for Chemical, Biomedical, and Pharmaceutical Engineering, First Edition. Laurent Simon.
© 2013 John Wiley & Sons, Inc. Published 2013 by John Wiley & Sons, Inc.

The integral with the infinite limit $\int_0^\infty f(t)e^{-st}dt$ is an *improper integral*, which can be written as $\lim_{b\to\infty}\int_0^b f(t)e^{-st}dt$. The Laplace transform exists if $\lim_{b\to\infty}\int_0^b f(t)e^{-st}dt$ converges (i.e., is a finite value), where s is a complex number.

Example 5.1   Let $f(t) = t^2$, the Laplace transform of $f(t)$ is

$$\mathcal{L}\{t^2\} = \overline{f}(s) = \int_0^\infty t^2 e^{-st}dt. \tag{5.2}$$

By using integration by parts,

$$\int u\,dv = uv - \int v\,du; \tag{5.3}$$

with $u = t^2$ and $dv = e^{-st}dt$, we have

$$\int t^2 e^{-st}dt = -\frac{t^2 e^{-st}}{s} + \frac{2}{s}\int t e^{-st}dt \tag{5.4}$$

since $du = 2t\,dt$ and $v = -e^{-st}/s$. Integration by parts of $\int t e^{-st}dt$ gives

$$\int t e^{-st}dt = -\frac{t e^{-st}}{s} - \frac{e^{-st}}{s^2}. \tag{5.5}$$

Consequently,

$$
\begin{aligned}
\int t^2 e^{-st}dt &= -\frac{t^2 e^{-st}}{s} + \frac{2}{s}\int t e^{-st}dt \\
\int t^2 e^{-st}dt &= -\frac{t^2 e^{-st}}{s} + \frac{2}{s}\left(-\frac{t e^{-st}}{s} - \frac{e^{-st}}{s^2}\right) \\
\int t^2 e^{-st}dt &= -\frac{t^2 e^{-st}}{s} - \frac{2t e^{-st}}{s^2} - \frac{2e^{-st}}{s^3} \\
\int t^2 e^{-st}dt &= -\frac{e^{-st}\left(2 + 2st + s^2 t^2\right)}{s^3}.
\end{aligned}
\tag{5.6}
$$

The Laplace transform is

$$\mathcal{L}\{t^2\} = \left[-\frac{e^{-st}\left(2 + 2st + s^2 t^2\right)}{s^3}\right]_0^\infty. \tag{5.7}$$

To evaluate $\overline{f}(s)$, we notice that

$$\lim_{t \to \infty}\left(-\frac{e^{-st}(2+2st+s^2t^2)}{s^3}\right) = 0 \tag{5.8}$$

if $s > 0$. For $t = 0$, we have

$$-\frac{e^{-st}(2+2st+s^2t^2)}{s^3} = -\frac{2}{s^3}. \tag{5.9}$$

As a result, the Laplace transform of t^2 is only defined for $s > 0$ and $\mathcal{L}\{t^2\} = 2/s^3$. In this book, we will not be deriving Laplace transforms of functions. Tables will be used instead.

5.2 PROPERTIES OF LAPLACE TRANSFORMS

The Laplace transform operator satisfies several properties:

- *Linearity Property*
  If a and b are constants while $f_1(t)$ and $f_2(t)$ are two functions of t, then

$$\mathcal{L}\{af_1(t) + bf_2(t)\} = a\mathcal{L}\{f_1(t)\} + b\mathcal{L}\{f_2(t)\} = a\overline{f_1}(s) + b\overline{f_2}(s). \tag{5.10}$$

- *Change of Scale Property*
  If $\mathcal{L}\{f(t)\} = \overline{f}(s)$, then

$$\mathcal{L}\{f(at)\} = \frac{1}{a}\overline{f}\left(\frac{s}{a}\right). \tag{5.11}$$

- *First Translation (or Shifting) Property*
  If $\mathcal{L}\{f(t)\} = \overline{f}(s)$, then

$$\mathcal{L}\{e^{at}f(t)\} = \overline{f}(s-a). \tag{5.12}$$

- *Second Translation (or Shifting) Property*
  If $\mathcal{L}\{f(t)\} = \overline{f}(s)$ and $g(t) = \begin{cases} f(t-a) & t > a \\ 0 & t < a \end{cases}$, then

$$\mathcal{L}\{g(t)\} = e^{-as}\overline{f}(s). \tag{5.13}$$

5.3 LAPLACE TRANSFORMS OF FUNCTIONS, DERIVATIVES, AND INTEGRALS

The Laplace transform of some basic functions are given below.

5.3.1 Constant function

If $f(t) = a$, where a is a constant, we have

$$\mathcal{L}\{f(t)\} = \mathcal{L}\{a\} = \frac{a}{s}. \tag{5.14}$$

To show the results of Equation (5.14), the definition of the Laplace transform is used:

$$\mathcal{L}\{f(t)\} = \bar{f}(s) = \int_0^\infty a e^{-st} dt$$

$$\mathcal{L}\{f(t)\} = -\frac{a}{s} e^{-st} \Big|_0^\infty = 0 + \frac{a}{s} = \frac{a}{s}. \tag{5.15}$$

5.3.2 Exponential Function

If $f(t) = e^{-at}$, $t \geq 0$, then

$$\mathcal{L}\{f(t)\} = \mathcal{L}\{e^{-at}\} = \int_0^\infty e^{-at} e^{-st} dt$$

$$\mathcal{L}\{f(t)\} = \int_0^\infty e^{-(a+s)t} dt = -\frac{1}{s+a} \left[e^{-(a+s)t} \right]_0^\infty \tag{5.16}$$

$$\mathcal{L}\{f(t)\} = -\frac{1}{s+a}(0-1) = \frac{1}{s+a}$$

$$\mathcal{L}\{f(t)\} = \frac{1}{s+a}$$

$$\mathcal{L}\{e^{-at}\} = \frac{1}{s+a}. \tag{5.17}$$

5.3.3 Step Function

The step function of size a is defined as

$$f(t) = \begin{cases} 0 & t < 0 \\ a & t \geq 0. \end{cases} \tag{5.18}$$

The Laplace transform of $f(t)$ is

$$\mathcal{L}\{f(t)\} = \mathcal{L}[a] = \frac{a}{s}. \tag{5.19}$$

5.3.4 Ramp Function

The ramp function is defined as

$$f(t) = at, \quad t \geq 0, \tag{5.20}$$

where a is a constant. Then,

$$\mathcal{L}\{f(t)\} = \mathcal{L}\{at\} = \int_0^\infty ate^{-st} dt, \tag{5.21}$$

but

$$\int te^{-st} dt = -\frac{te^{-st}}{s} - \frac{e^{-st}}{s^2} \tag{5.22}$$

after applying integration by parts. As a result,

$$\mathcal{L}\{f(t)\} = a\left[-\frac{te^{-st}}{s} - \frac{e^{-st}}{s^2} \right]_0^\infty. \tag{5.23}$$

To evaluate $\bar{f}(s)$, we notice that

$$\lim_{t \to \infty} \left(\frac{te^{-st}}{s} + \frac{e^{-st}}{s^2} \right) = 0 \tag{5.24}$$

if $s > 0$. Then,

$$\mathcal{L}\{f(t)\} = 0 - a\left(0 - \frac{1}{s^2} \right)$$

$$\mathcal{L}\{f(t)\} = \frac{a}{s^2} \tag{5.25}$$

$$\mathcal{L}\{at\} = \frac{a}{s^2}. \tag{5.26}$$

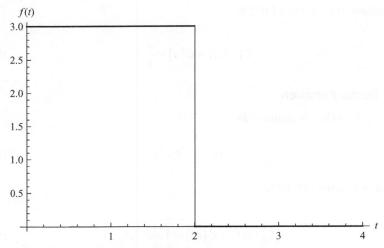

Figure 5.1. A rectangular pulse function with $h = 3$ and $t_1 = 2$.

5.3.5 Rectangular Pulse Function

The rectangular pulse function is defined as (Fig. 5.1)

$$f(t) = \begin{cases} 0 & t < 0 \\ h & 0 \le t < t_1 \\ 0 & t > t_1. \end{cases} \tag{5.27}$$

Then,

$$\mathcal{L}\{f(t)\} = \int_0^{t_1} h e^{-st} dt$$

$$\mathcal{L}\{f(t)\} = -\frac{h}{s} e^{-st} \Big|_0^{t_1} = -\frac{h}{s}\left(e^{-st_1} - 1\right). \tag{5.28}$$

Finally,

$$\mathcal{L}\{f(t)\} = \frac{h}{s}\left(1 - e^{-st_1}\right). \tag{5.29}$$

5.3.6 Translated Function

Consider a function $f(t)$ and let $g(t)$ represent $f(t)$ delayed by t_0 time units (Fig. 5.2a,b). The function $g(t)$ can be written as

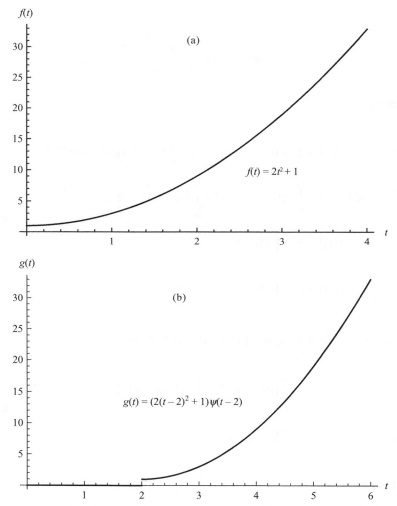

Figure 5.2. Function with and without time delay. (a) Function with no time delay, (b) function with time delay, $t_0 = 2$.

$$g(t) = f(t - t_0)\psi(t - t_0), \tag{5.30}$$

where $\psi(t - t_0)$ is a unit step function. By definition of a unit step function, we have

$$\psi(t - t_0) = \begin{cases} 0 & t - t_0 < 0 \\ 1 & t - t_0 \geq 0 \end{cases} \tag{5.31}$$

or

$$\psi(t - t_0) = \begin{cases} 0 & t < t_0 \\ 1 & t \geq t_0 \end{cases}. \tag{5.32}$$

Therefore,

$$\mathcal{L}\{g(t)\} = \mathcal{L}\{f(t - t_0)\psi(t - t_0)\} = \int_0^\infty f(t - t_0)\psi(t - t_0)e^{-st}dt$$

$$\mathcal{L}\{g(t)\} = \int_0^{t_0} f(t - t_0)(0)e^{-st} + \int_{t_0}^\infty f(t - t_0)e^{-st}dt$$

$$\mathcal{L}\{g(t)\} = e^{-st_0}\int_{t_0}^\infty f(t - t_0)e^{-s(t-t_0)}dt \tag{5.33}$$

$$\mathcal{L}\{g(t)\} = e^{-st_0}\int_{t_0}^\infty f(t - t_0)e^{-s(t-t_0)}d(t - t_0)$$

since $d(t - t_0) = dt$. If $t - t_0$ is replaced by τ, we have

$$\mathcal{L}\{g(t)\} = e^{-st_0}\int_{t_0 - t_0}^\infty f(\tau)e^{-s\tau}d(\tau)$$

$$\mathcal{L}\{g(t)\} = e^{-st_0}\int_0^\infty f(\tau)e^{-s\tau}d(\tau). \tag{5.34}$$

As a result,

$$\mathcal{L}\{g(t)\} = e^{-st_0}\overline{f}(s). \tag{5.35}$$

Laplace transforms of other functions are given in Table 5.1.

5.3.7 Laplace Transforms of Derivatives

- *First-Order Derivative*

$$\mathcal{L}\left\{\frac{df(t)}{dt}\right\} = s\overline{f}(s) - f(0) \tag{5.36}$$

TABLE 5.1 Laplace Transforms of Various Functions

Time Function	Laplace Transform
$f(t)$	$\overline{f}(s)$
1	$\dfrac{1}{s}$
at	$\dfrac{a}{s^2}$
$t^n, n = 1, 2, 3, \cdots$	$\dfrac{n!}{s^{n+1}}$
e^{-at}	$\dfrac{1}{s+a}$
$\delta(t)$(unit impulse)	1
$e^{at}t^n, n = 1, 2, 3, \cdots$	$\dfrac{n!}{(s-a)^{n+1}}$
$\sin(\omega t)$	$\dfrac{\omega}{s^2+\omega^2}$
$\cos(\omega t)$	$\dfrac{s}{s^2+\omega^2}$
$e^{at}\sin(\omega t)$	$\dfrac{\omega}{(s-a)^2+\omega^2}$
$e^{at}\cos(\omega t)$	$\dfrac{s-a}{(s-a)^2+\omega^2}$
$\sinh(\omega t)$	$\dfrac{\omega}{s^2-\omega^2}$
$\cosh(\omega t)$	$\dfrac{s}{s^2-\omega^2}$

- *Second-Order Derivative*

$$\mathcal{L}\left\{\frac{d^2 f(t)}{dt^2}\right\} = s^2\overline{f}(s) - sf(0) - f'(0) \tag{5.37}$$

- *Nth-Order Derivative*

$$\mathcal{L}\left\{\frac{d^n f(t)}{dt^n}\right\} = s^n\overline{f}(s) - s^{n-1}f(0) - s^{n-2}f'(0) - \ldots - sf^{(n-2)}(0) - f^{(n-1)}(0) \tag{5.38}$$

5.3.8 Laplace Transforms of Integrals

$$\mathcal{L}\left\{\int_0^t f(t)\,dt\right\} = \int_0^\infty\left(\int_0^t f(t)\,dt\right)e^{-st}\,dt \tag{5.39}$$

By using integration by parts

$$\int u dv = uv - \int v du$$

and $u = \int_0^t f(\tau)d\tau$, $du = f(t)dt$, $v = e^{-st}$, $dv = -se^{-st}dt$, we have

$$\int_0^\infty \left\{ \int_0^t f(t)dt \right\} e^{-st}dt = -\frac{1}{s}\left\{ \left[e^{-st}\int_0^t f(\tau)d\tau \right]_0^\infty - \int_0^\infty (e^{-st})(f(t)dt) \right\}$$

$$\int_0^\infty \left\{ \int_0^t f(t)dt \right\} e^{-st}dt = -\frac{1}{s}\left\{ 0 - \int_0^\infty (e^{-st})(f(t)dt) \right\}$$

$$\int_0^\infty \left\{ \int_0^t f(t)dt \right\} e^{-st}dt = \frac{1}{s}\int_0^\infty (e^{-st})(f(t)dt) \tag{5.40}$$

$$\int_0^\infty \left\{ \int_0^t f(t)dt \right\} e^{-st}dt = \frac{1}{s}\int_0^\infty e^{-st} f(t)dt.$$

Finally,

$$\mathcal{L}\left\{ \int_0^t f(t)dt \right\} = \frac{1}{s}\overline{f}(s). \tag{5.41}$$

5.3.9 Final Value Theorem

The value of a function $f(t)$ as $t \to \infty$ can be found by using the formula

$$\lim_{t\to\infty} f(t) = \lim_{s\to 0} s\overline{f}(t) \tag{5.42}$$

as long as $\lim_{t\to\infty} f(t)$ exists.

Equation (5.42) can be proven using the relationship

$$\int_0^\infty \frac{df(t)}{dt} e^{-st}dt = s\overline{f}(s) - f(0). \tag{5.43}$$

Taking the limit of both sides of Equation (5.43) as $s \to 0$ yields

$$\lim_{s \to 0} \int_0^\infty \frac{df(t)}{dt} e^{-st} dt = \lim_{s \to 0} \left[s\overline{f}(s) - f(0) \right]$$

$$\int_0^\infty \frac{df(t)}{dt} dt = \lim_{s \to 0} \left[s\overline{f}(s) - f(0) \right] \qquad (5.44)$$

$$\lim_{t \to \infty} f(t) - f(0) = \lim_{s \to 0} \left[s\overline{f}(s) - f(0) \right]$$

$$\lim_{t \to \infty} f(t) = \lim_{s \to 0} \left[s\overline{f}(s) \right].$$

5.3.10   Initial Value Theorem

The value of a function $f(t)$ as $t \to 0$ can be found by using the formula

$$\lim_{t \to 0} f(t) = \lim_{t \to 0} s\overline{f}(t) \qquad (5.45)$$

as long as $\lim_{t \to 0} f(t)$ exists.

Similarly, taking the limit of both sides of Equation (5.43) as $s \to \infty$ gives

$$\lim_{s \to \infty} \int_0^\infty \frac{df(t)}{dt} e^{-st} dt = \lim_{s \to \infty} \left[s\overline{f}(s) - f(0) \right]$$

$$0 = \lim_{s \to \infty} \left[s\overline{f}(s) - f(0) \right] \qquad (5.46)$$

$$0 = \lim_{s \to \infty} s\overline{f}(s) - f(0)$$

$$\lim_{s \to \infty} s\overline{f}(s) = f(0)$$

or

$$\lim_{t \to 0} f(t) = \lim_{s \to \infty} s\overline{f}(s). \qquad (5.47)$$

Example 5.2   Consider the function $f(t)$:

$$f(t) = 5\cos 4t + e^{-t} + 4t. \qquad (5.48)$$

The Laplace transform $\mathcal{L}\{f(t)\}$ can be calculated:

$$
\begin{aligned}
\mathcal{L}\{f(t)\} &= \mathcal{L}\{5\cos 4t + e^{-t} + 4t\} \\
\bar{f}(s) &= 5\mathcal{L}\{\cos 4t\} + \mathcal{L}\{e^{-t}\} + 4\mathcal{L}\{t\} \\
\bar{f}(s) &= \frac{5s}{s^2 + 16} + \frac{1}{s+1} + 4\frac{1}{s^2} \\
\bar{f}(s) &= \frac{64 + 64s + 20s^2 + 9s^3 + 6s^4}{s^2(1+s)(16+s^2)}.
\end{aligned}
\tag{5.49}
$$

Example 5.3   Consider the vector-valued function $\mathbf{f}(t)$:

$$
\mathbf{f}(t) = \begin{bmatrix} 2\sin(t) + 3e^{-4t} \\ 1 + 6t \\ \cos(t) + 8\sin(t - t_0)\psi(t - t_0) \end{bmatrix},
\tag{5.50}
$$

where $\psi(t - t_0)$ is the unit step function. The Laplace transform $\mathcal{L}\{\mathbf{f}(t)\}$ yields the following:

$$
\bar{\mathbf{f}}(s) = \begin{bmatrix} \dfrac{3}{4+s} + \dfrac{2}{1+s^2} \\ \dfrac{1}{s} + \dfrac{6}{s^2} \\ \dfrac{s}{1+s^2} + \dfrac{8e^{-st_0}}{1+s^2} \end{bmatrix}.
\tag{5.51}
$$

In summary,

$$
\bar{\mathbf{f}}(s) = \begin{bmatrix} \dfrac{11 + 2s + 3s^2}{(4+s)(1+s^2)} \\ \dfrac{6 + s}{s^2} \\ \dfrac{s + 8e^{-st_0}}{1+s^2} \end{bmatrix}.
\tag{5.52}
$$

5.4   LAPLACE TRANSFORMS OF LINEAR ORDINARY DIFFERENTIAL EQUATION (ODE) AND PARTIAL DIFFERENTIAL EQUATION (PDE)

Laplace transforms can be applied to solve linear differential equations. Linear ODEs are transformed into algebraic equations containing the complex variable s, while PDEs are transformed into ODEs. In this section, we will focus on how to perform these operations. Methods of inverting the Laplace transforms and obtaining solutions in the time domain will be covered in Chapter 6.

Example 5.4   Consider the following linear ODE:

$$\frac{d^2 y}{dt^2} + y = 2t, \quad y(0) = 1, \quad y'(0) = -2. \tag{5.53}$$

We take the Laplace transform of both sides of Equation (5.53):

$$\mathcal{L}\left\{\frac{d^2 y}{dt^2} + y\right\} = \mathcal{L}\{2t\}. \tag{5.54}$$

After using the formula for Laplace transform of derivatives, we have

$$s^2 \bar{y}(s) - sy(0) - y'(0) + \bar{y}(s) = \frac{2}{s^2}$$

$$\bar{y}(s)(s^2 + 1) = \frac{2}{s^2} + s - 2$$

$$\bar{y}(s) = \frac{2}{s^2(s^2 + 1)} + \frac{s - 2}{s^2 + 1} \tag{5.55}$$

$$\bar{y}(s) = \frac{2 - 2s^2 + s^3}{s^2(1 + s^2)}.$$

Example 5.5   Consider the following linear ODE:

$$\frac{d^2 y}{dt^2} + y = \cos t, \quad y(0) = 0, \quad y'(0) = 1. \tag{5.56}$$

The Laplace transform of both sides of Equation (5.56) yields

$$\mathcal{L}\left\{\frac{d^2 y}{dt^2} + y\right\} = \mathcal{L}\{\cos t\}, \tag{5.57}$$

resulting in

$$s^2 \bar{y}(s) - sy(0) - y'(0) + \bar{y}(s) = \frac{s}{s^2 + 1}$$

$$\bar{y}(s)(s^2 + 1) = \frac{s}{s^2 + 1} + 1$$

$$\bar{y}(s) = \frac{s}{(s^2 + 1)^2} + \frac{1}{s^2 + 1} \tag{5.58}$$

$$\bar{y}(s) = \frac{s^2 + s + 1}{(s^2 + 1)^2}.$$

In the two previous examples, the solution $y(t)$ was not found. Instead, the Laplace transform $\bar{y}(s)$ was obtained using simple algebraic manipulations. Methods on how to invert $\bar{y}(s)$ will be outlined in Chapter 6.

Example 5.6   Consider the PDE

$$\frac{\partial C(x, t)}{\partial x} = \frac{\partial C(x, t)}{\partial t} + C(x, t), \quad C(x, 0) = e^{-2x}, \tag{5.59}$$

where the variable C is a function of position x and time t, hence, $C(x,t)$. In this case, the partial derivative is used to describe the change in C with respect to x or t. The Laplace transform with respect to t is applied to give

$$\mathcal{L}\left\{\frac{\partial C(x, t)}{\partial x}\right\} = \mathcal{L}\left\{\frac{\partial C(x, t)}{\partial t}\right\} + \mathcal{L}\{C(x, t)\}, \tag{5.60}$$

where $x > 0, t > 0$.
  Note that

$$\mathcal{L}\left\{\frac{\partial C(x, t)}{\partial x}\right\} = \int_0^\infty e^{-st} \frac{\partial C(x, t)}{\partial x} dt = \frac{d}{dx} \int_0^\infty e^{-st} C(x, t) dt = \frac{d\bar{C}(x, s)}{dx} \tag{5.61}$$

and

$$\mathcal{L}\left\{\frac{\partial C(x, t)}{\partial t}\right\} = s\bar{C}(x, s) - C(x, 0). \tag{5.62}$$

As a result, Equation (5.60) becomes

$$\frac{d\bar{C}(x, s)}{dx} = s\bar{C}(x, s) - C(x, 0) + \bar{C}(x, s) \tag{5.63}$$

or

$$\frac{d\bar{C}(x, s)}{dx} = s\bar{C}(x, s) - e^{-2x} + \bar{C}(x, s). \tag{5.64}$$

Finally,

$$\frac{d\bar{C}(x, s)}{dx} - (s+1)\bar{C}(x, s) = -e^{-2x}. \tag{5.65}$$

Equation (5.65) is an ODE in s. This equation can be solved for $\bar{C}(x, s)$ using integrating factors. The solution of a first-order ODE of type

$$\frac{dy}{dx} + p(x)y(x) = q(x) \tag{5.66}$$

is obtained after using an integration factor $\int p(x)dx$:

$$y(x) = e^{-\int p(x)dx} \int \left\{ e^{\int p(x)dx} q(x) \right\} dx. \tag{5.67}$$

Consequently, the solution of Equation (5.65) is

$$
\begin{aligned}
\bar{C}(x, s) &= e^{\int (s+1)dx} \int \left\{ e^{-\int (s+1)dx} \left(-e^{-2x} \right) \right\} dx \\
\bar{C}(x, s) &= e^{(s+1)x} \left\{ \int \left(-e^{-(s+1)x} e^{-2x} \right) dx \right\} \\
\bar{C}(x, s) &= e^{(s+1)x} \left\{ \int \left(-e^{-(s+3)x} \right) dx \right\} \\
\bar{C}(x, s) &= e^{(s+1)x} \left\{ \frac{e^{-(s+3)x}}{s+3} + c \right\} \\
\bar{C}(x, s) &= \frac{e^{-2x}}{s+3} + ce^{(s+1)x}.
\end{aligned}
\tag{5.68}
$$

For $C(x,t)$ to have a finite value as $x \to \infty$, c has to be set equal to zero:

$$\bar{C}(x, s) = \frac{e^{-2x}}{s+3}. \tag{5.69}$$

Example 5.7   Consider the PDE

$$
\begin{aligned}
\frac{\partial C(x, t)}{\partial t} &= \frac{\partial^2 C(x, t)}{\partial x^2}, \quad x > 0, \quad t > 0, \\
C(0, t > 0) &= 1, \quad C(x > 0, 0) = 0
\end{aligned}
\tag{5.70}
$$

often used to describe diffusion through a semi-infinite membrane.

The Laplace transform with respect to t can be applied in this case:

$$\mathcal{L}\left\{ \frac{\partial C(x, t)}{\partial t} \right\} = \mathcal{L}\left\{ \frac{\partial^2 C(x, t)}{\partial x^2} \right\} \tag{5.71}$$

to give

$$s\bar{C}(x, s) - C(x, 0) = \frac{d^2\bar{C}(x, s)}{dx^2} \tag{5.72}$$

or

$$\frac{d^2\bar{C}(x, s)}{dx^2} - s\bar{C}(x, s) = 0. \tag{5.73}$$

To solve Equation (5.73), the characteristic equation is

$$m^2 - s = 0, \tag{5.74}$$

which can be factored out:

$$\left(m - \sqrt{s}\right)\left(m + \sqrt{s}\right) = 0. \tag{5.75}$$

Since the roots are real and distinct, we have

$$\bar{C}(x, s) = c_1 e^{-x\sqrt{s}} + c_2 e^{x\sqrt{s}}. \tag{5.76}$$

For $C(x,t)$ to have a finite value as $x \to \infty$, c_2 must set equal to zero:

$$\bar{C}(x, s) = c_1 e^{-x\sqrt{s}}. \tag{5.77}$$

Using the condition $C(0,t) = 1$ or $\bar{C}(0, s) = 1/s$, we obtain $c_1 = 1/s$. Finally,

$$\bar{C}(x, s) = \frac{e^{-x\sqrt{s}}}{s}. \tag{5.78}$$

5.5 CONTINUOUS FERMENTATION

In Section 4.5, the dynamic equation of a bioreactor, after linearization around a stable steady-state point, was characterized as such:

$$\frac{d\tilde{C}_x}{dt} = -0.54\tilde{C}_s$$

$$\frac{d\tilde{C}_s}{dt} = 1.82\tilde{C}_x - 3.15\tilde{C}_s. \tag{5.79}$$

The equilibrium point, written using deviation variables, is $(\tilde{C}_{xe}, \tilde{C}_{se}) = (0.0, 0.0)$. To study the linearized system behavior after a step change of size 1.0 g/L in the initial (or steady-state) substrate concentration, we first use the Laplace transform to obtain expressions in terms of the frequency variable s.

Applying the Laplace transforms to Equation (5.79) yields

$$s\bar{C}_x - \tilde{C}_x(0) = -0.54\bar{C}_s$$
$$s\bar{C}_s - \tilde{C}_s(0) = 1.82\bar{C}_x - 3.15\bar{C}_s \tag{5.80}$$

or

$$s\bar{C}_x = -0.54\bar{C}_s$$
$$s\bar{C}_s - 1.0 = 1.82\bar{C}_x - 3.15\bar{C}_s, \tag{5.81}$$

where $\bar{C}_x$ and $\bar{C}_s$ are the Laplace transforms of $\tilde{C}_x$ and $\tilde{C}_s$, respectively. Equation (5.81) is a system of two equations with two unknowns that can be solved using several methods. In this example, Cramer's rule is implemented. Equation (5.81) is written as

$$s\bar{C}_x + 0.54\bar{C}_s = 0$$
$$-1.82\bar{C}_x + (s + 3.15)\bar{C}_s = 1.0. \tag{5.82}$$

Therefore, $\bar{C}_x$ and $\bar{C}_s$ can be expressed as quotients of two determinants:

$$\bar{C}_x = \frac{\begin{vmatrix} 0 & 0.54 \\ 1.0 & s+3.15 \end{vmatrix}}{\begin{vmatrix} s & 0.54 \\ -1.82 & s+3.15 \end{vmatrix}} \tag{5.83}$$

and

$$\bar{C}_s = \frac{\begin{vmatrix} s & 0 \\ -1.82 & 1.0 \end{vmatrix}}{\begin{vmatrix} s & 0.54 \\ -1.82 & s+3.15 \end{vmatrix}}. \tag{5.84}$$

Consequently,

$$\bar{C}_x = \frac{-0.54}{(s+0.35)(s+2.80)} \tag{5.85}$$

and

$$\bar{C}_s = \frac{s}{(s+0.35)(s+2.80)},\tag{5.86}$$

respectively. Note that roots of the denominator of $\bar{C}_x$ and $\bar{C}_s$ are the same as the eigenvalues ($\lambda_1 = -2.80$ and $\lambda_2 = -0.35$) calculated in Section 4.5.

5.6   TWO-COMPARTMENT MODELS

A two-compartment model was derived in Section 2.3.2 and shown in Fig. 2.8. The drug concentrations in the two compartments are

$$\frac{dC_1}{dt} = \frac{D}{V_1}\delta(t) - (k_{el} + k_{12})C_1 + k_{21}C_2\frac{V_2}{V_1}\tag{5.87}$$

and

$$\frac{dC_2}{dt} = k_{12}C_1\frac{V_1}{V_2} - k_{21}C_2.\tag{5.88}$$

The initial conditions are $C_1(0) = 0$ and $C_2(0) = 0$. Equation (5.87) can also be written as

$$\frac{dC_1}{dt} = -(k_{el} + k_{12})C_1 + k_{21}C_2\frac{V_2}{V_1}\tag{5.89}$$

with the initial value $C_1(0)$ defined as $C_{10} = D/V_1$, where D is the dosage. To find the Laplace transforms $\bar{C}_1(s)$ and $\bar{C}_2(s)$, we start applying the Laplace operator to Equations (5.89) and (5.88):

$$s\bar{C}_1 - C_{10} = -(k_{el} + k_{12})\bar{C}_1 + k_{21}\bar{C}_2\frac{V_2}{V_1}\tag{5.90}$$

and

$$s\bar{C}_2 = k_{12}\bar{C}_1\frac{V_1}{V_2} - k_{21}\bar{C}_2.\tag{5.91}$$

Equations (5.90) and (5.91) can be rewritten as

$$(s + k_{el} + k_{12})\bar{C}_1 - k_{21}\frac{V_2}{V_1}\bar{C}_2 = C_{10}\tag{5.92}$$

and

$$k_{12} \frac{V_1}{V_2} \bar{C}_1 - (s + k_{21}) \bar{C}_2 = 0. \tag{5.93}$$

Application of Cramer's rule gives

$$\bar{C}_1 = -\frac{C_{10}(s + k_{21})}{k_{12}k_{21} - (s + k_{21})(s + k_{12} + k_{el})} \tag{5.94}$$

and

$$\bar{C}_2 = -\frac{C_{10}k_{12}V_1}{(k_{12}k_{21} - (s + k_{21})(s + k_{12} + k_{el}))V_2}. \tag{5.95}$$

If we apply the final value theorem, we notice that

$$\lim_{t \to \infty} C_1(t) = \lim_{s \to 0} s\bar{C}_1 = 0; \quad \lim_{t \to \infty} C_2(t) = \lim_{s \to 0} s\bar{C}_2 = 0, \tag{5.96}$$

which proves that the drug disappears in the body after a bolus IV injection. Also, it can be shown that

$$C_1(0) = \lim_{s \to \infty} s\bar{C}_1 = C_{10}; \quad C_2(0) = \lim_{s \to \infty} s\bar{C}_2 = 0 \tag{5.97}$$

5.7 GENE REGULATION

Living cells, due to their regulatory systems, are effective at adjusting bio-chemical processes, or gene expression programs, to respond to stimuli or changes in the intracellular and/or extracellular environments [1]. Prokaryotic and eukaryotic cells use environmental signals such as temperature changes, nutrient availability, and light intensity. In an example of positive feedback control, an autogenous regulator X is activated by phosphorylation, binds to a promoter, and mediates transcription of the gene that encodes the regulator [1]. The gene is transcribed from two promoters: One is inducible by the phos-phorylated form of X and the other is constitutive (i.e., transcribed continu-ously). In the presence of a signal that promotes the phosphorylation of X, the number of phosphorylated regulator molecules increases, resulting in addi-tional increases in the total regulator concentration and the level of the phos-phorylated form. Without such an activating signal, dephosphorylation of the regulatory protein occurs at a higher rate than phosphorylation, leading to a preponderance of unphosphorylated regulator molecules. The inducible promoter is thus replaced by a constitutive promoter and the total level of the regulator is no longer influenced by the phosphorylation or dephosphorylation rates [1].

A mathematical model of gene regulation is given by [1]

$$\frac{dA}{dt} = k_a P - (k_{-a} + k_a) A \qquad (5.98)$$

and

$$\frac{dP}{dt} = k_1 + k_2 \frac{KA^H}{1 + KA^H} + k_{-a} A - (k_a + k_d) P, \qquad (5.99)$$

where A (μM) and P (μM) are the active (phosphorylated) and inactive (unphosphorylated) forms of a regulatory protein; k_a (5/minute) and k_{-a} (20/minute) are the phosphorylation and dephosphorylation rates for the protein; k_d (0.08/minute) is a degradation (or dilution) rate for both forms of the protein; k_1 (0.01 μM/min) and k_2 (0.3 μM/min) are the rates of protein synthesis due to the constitutive and inducible promoters, respectively. The parameter K (5/μM^2) is the association constant for regulator–promoter interactions and the exponent $H = 2$ is the Hill coefficient.

The steady-state values of A and P are found by setting Equations (5.98) and (5.99) equal to zero: $A_s = 0.028$ μM and $P_s = 0.11$ μM (i.e., the equilibrium point of the model under noninducing conditions). The Jacobian matrix obtained after linearizing the system around the steady-state point is

$$\mathbf{M} = \begin{bmatrix} -20.08 & 5 \\ 20.08 & -5.08 \end{bmatrix} \qquad (5.100)$$

such that

$$\frac{d\tilde{A}}{dt} = -20.08\tilde{A} + 5\tilde{P}$$

$$\frac{d\tilde{P}}{dt} = 20.08\tilde{A} - 5.08\tilde{P} \qquad (5.101)$$

with $\tilde{A} = A - A_s$ and $\tilde{P} = P - P_s$. To determine the system behavior after a small perturbation in the initial conditions, the first step could be the derivation of the Laplace transforms of $\tilde{A}$ and $\tilde{P}$: $\bar{A}$ and $\bar{P}$, respectively.

Applying the Laplace transforms to Equation (5.101) yields

$$s\bar{A} - \tilde{A}_0 = -20.08\bar{A} + 5\bar{p}$$

$$s\bar{P} - \tilde{P}_0 = 20.08\bar{A} - 5.08\bar{P}, \qquad (5.102)$$

where $\tilde{A}_0$ and $\tilde{P}_0$ are the small deviations from the steady state. Using the Cramer's method outlined above, the Laplace transforms are

$$\bar{A} = \frac{(5.08+s)\tilde{A}_0 + 5\tilde{P}_0}{(0.0635+s)(25.1+s)} \tag{5.103}$$

and

$$\bar{P} = \frac{20.1\tilde{A}_0 + (20.1+s)\tilde{P}_0}{(0.0635+s)(25.1+s)}. \tag{5.104}$$

The roots of the denominator of Equation (5.104) are negative, which implies a stable equilibrium point. As a result, we expect $\lim_{t\to\infty} \tilde{A}(t) = 0$ and $\lim_{t\to\infty} \tilde{P}(t) = 0$ (why?).

5.8 SUMMARY

Laplace transforms are applied to solve linear differential equations and to study process stability. This chapter introduces how these tools are implemented to convert a system of dynamic ODEs into algebraic expressions. PDEs can also be transformed into ODEs, which are further solved in the Laplace domain. The variables of interest are usually obtained as a quotient of two functions of s. Examples in continuous fermentation, compartmental models, and gene regulation were presented.

PROBLEMS

5.1. Evaluate $\mathcal{L}\{4t^4 + 5e^{-4t} + \sin(5t)\}$.

5.2. Evaluate $\mathcal{L}\{e^{-2t} \sin(3t)\}$.

5.3. Find $\mathcal{L}\{f(t)\}$ if $f(t) = \begin{cases} 0 & 0 < t < 3 \\ 5 & t > 3 \end{cases}$.

5.4. Evaluate $\mathcal{L}\{e^{-2t} \cos(3t)\}$.

5.5. The biomass ($\tilde{C}_x$) and substrate ($\tilde{C}_s$) concentrations in a bioreactor are given by

$$\frac{d\tilde{C}_x}{dt} = -0.40\tilde{C}_s$$

$$\frac{d\tilde{C}_s}{dt} = 2.0\tilde{C}_x - 3.0\tilde{C}_s.$$

Evaluate $\mathcal{L}\{\tilde{C}_x\}$ and $\mathcal{L}\{\tilde{C}_s\}$ after a step change of size 1.5 g/L in the initial substrate concentration.

5.6. The biomass $(\tilde{C}_x)$ and substrate $(\tilde{C}_s)$ concentrations in a bioreactor are given by

$$\frac{d\tilde{C}_x}{dt} = -0.30\tilde{C}_s$$

$$\frac{d\tilde{C}_s}{dt} = 2.5\tilde{C}_x - 4.0\tilde{C}_s.$$

Evaluate $\tilde{C}_x (t \rightarrow \infty)$ and $\tilde{C}_s (t \rightarrow \infty)$ using the final value theorem and after a step change of size 1.0 g/L in the initial substrate concentration.

5.7. The Laplace transforms of drug concentrations in a two-compartment model are given by

$$\overline{C}_1 = -\frac{C_{10}(s+k_{21})}{k_{12}k_{21} - (s+k_{21})(s+k_{12}+k_{el})}$$

and

$$\overline{C}_2 = -\frac{C_{10}k_{12}V_1}{(k_{12}k_{21} - (s+k_{21})(s+k_{12}+k_{el}))V_2}.$$

Using the initial value theorem, show that $C_1(0) = C_{10}$ and $C_2(0) = 0$.

5.8. Show that the Laplace transforms in Problem 5.7 (see Section 5.6) can be obtained using the following system:

$$\frac{dC_1}{dt} = \frac{D}{V_1}\delta(t) - (k_{el}+k_{12})C_1 + k_{21}C_2\frac{V_2}{V_1}$$

and

$$\frac{dC_2}{dt} = k_{12}C_1\frac{V_1}{V_2} - k_{21}C_2,$$

where $C_1(0) = 0$ and $C_2(0) = 0$.

Hint: Use the Laplace transform of the unit impulse function.

5.9. The Laplace transforms of the concentrations of the active (phosphorylated) and inactive (unphosphorylated) forms (A and P, respectively) of a regulatory protein are given by (see Section 5.7)

$$\tilde{A} = \frac{(5.08 + s)\tilde{A}_0 + 5\tilde{P}_0}{(0.0635 + s)(25.1 + s)}$$

and

$$\tilde{P} = \frac{20.1\tilde{A}_0 + (20.1 + s)\tilde{P}_0}{(0.0635 + s)(25.1 + s)}.$$

Show that $\lim_{t \to \infty} \tilde{A}(t) = 0$; $\quad \lim_{t \to \infty} \tilde{P}(t) = 0$.

5.10. Consider the following system:

$$\frac{\partial C(x, t)}{\partial t} = \frac{\partial^2 C(x, t)}{\partial x^2}$$
$$C(0, t) = 1, \quad C(1, t) = 0$$
$$C(x, 0) = 0.$$

Find $\mathcal{L}\{C(x,t)\}$.

REFERENCE

1. Mitrophanov AY, Groisman EA. Positive feedback in cellular control systems. *Bioessays* 2008; 30:542–555.

CHAPTER 6

INVERSE LAPLACE TRANSFORMS

The previous chapter shows that Laplace transforms can be applied to convert a system of dynamic ordinary differential equations (ODEs) into algebraic expressions. The states are then written as explicit functions of the frequency variable s. However, an inversion step, such as the one described in this chapter, is required to solve for the variables of interest in the time domain.

6.1 HEAVISIDE EXPANSIONS

Partial fraction expansion is used to decompose a rational function $P(s)/Q(s)$ into a series of simpler fractions:

$$F(s) = \frac{P(s)}{Q(s)} = \sum_{n=1}^{N} \frac{c_n}{r_n(s)}, \tag{6.1}$$

where $P(s)$ and $Q(s)$ are polynomials with $P(s)$ being a lower order than $Q(s)$; c_n's are constant coefficients; $r_n(s)$ are the factors; and N is the order of $Q(s)$. The inverse Laplace transform of $F(s)$ is

$$f(t) = \mathcal{L}^{-1}\{F(s)\} = \mathcal{L}^{-1}\left\{\frac{P(s)}{Q(s)}\right\} = \sum_{n=1}^{N} \mathcal{L}^{-1}\left\{\frac{c_n}{r_n(s)}\right\} \tag{6.2}$$

Control of Biological and Drug-Delivery Systems for Chemical, Biomedical, and Pharmaceutical Engineering, First Edition. Laurent Simon.
© 2013 John Wiley & Sons, Inc. Published 2013 by John Wiley & Sons, Inc.

or

$$f(t) = \mathcal{L}^{-1}\left\{\frac{c_1}{r_1(s)}\right\} + \mathcal{L}^{-1}\left\{\frac{c_2}{r_2(s)}\right\} + \cdots + \mathcal{L}^{-1}\left\{\frac{c_N}{r_N(s)}\right\}. \tag{6.3}$$

Once c_n and $r_n(s)$ are known, $f(t)$ is obtained by inspection of Tables 5.1 and 6.1.

TABLE 6.1 Inverse Laplace Transforms of Various Functions

Laplace Transform	Time Function
$\bar{f}(s)$	$f(t)$
$\dfrac{1}{s}$	1
$\dfrac{1}{s^n}, n = 1, 2, 3, \cdots$	$\dfrac{t^{n-1}}{(n-1)!}$
$\dfrac{1}{s+a}$	e^{-at}
$\dfrac{1}{(s-a)^n}, n = 1, 2, 3, \cdots$	$\dfrac{t^{n-1}e^{at}}{(n-1)!}$
$\dfrac{1}{s^2+\omega^2}$	$\dfrac{\sin(\omega t)}{\omega}$
$\dfrac{s}{s^2+\omega^2}$	$\cos(\omega t)$
$\dfrac{1}{(s-b)^2+\omega^2}$	$\dfrac{e^{bt}\sin(\omega t)}{\omega}$
$\dfrac{s-b}{(s-b)^2+\omega^2}$	$e^{bt}\cos(\omega t)$
$\dfrac{1}{(s-a)(s-b)}, a \neq b$	$\dfrac{e^{bt}-e^{at}}{b-a}$
$\dfrac{s}{(s-a)(s-b)}, a \neq b$	$\dfrac{be^{bt}-ae^{at}}{b-a}$
$\dfrac{1}{(s^2+\omega^2)^2}$	$\dfrac{\sin(\omega t)-\omega t\cos(\omega t)}{2\omega^3}$
$\dfrac{s}{(s^2+\omega^2)^2}$	$\dfrac{t\sin(\omega t)}{2\omega}$
$\dfrac{s^2-\omega^2}{(s^2+\omega^2)^2}$	$t\cos(\omega t)$
$\dfrac{1}{s^2+bs+c}$	$\dfrac{e^{\left(-(b/2)-1/2\sqrt{b^2-4c}\right)t}-e^{\left(-(b/2)+1/2\sqrt{b^2-4c}\right)t}}{\sqrt{b^2-4c}}$
$\dfrac{1}{s^2+bs}$	$\dfrac{1}{b}-\dfrac{e^{-bt}}{b}$

The *Heaviside expansion formula* is a technique that helps determine c_n. This method is best demonstrated through examples. Two cases, based on the roots of the denominator $Q(s)$, are described as follows:

1. $Q(s)$ has a number of distinct roots.
2. $Q(s)$ has repeated roots (also called *multiple poles* of $F(s)$).

CASE 1: Distinct Roots

Example 6.1   Consider the Laplace transform of a function $f(t)$:

$$\bar{f}(s) = \frac{1}{s^2 - 5s + 6}. \tag{6.4}$$

The denominator can be factorized to give

$$\bar{f}(s) = \frac{1}{(s-3)(s-2)}. \tag{6.5}$$

The roots of the denominator are +2 and +3. As a result, Equation (6.5) can be written as

$$\bar{f}(s) = \frac{1}{(s-3)(s-2)} = \frac{c_1}{s-3} + \frac{c_2}{s-2}. \tag{6.6}$$

According to the Heaviside expansion, we first multiply both sides of

$$\frac{1}{(s-3)(s-2)} = \frac{c_1}{s-3} + \frac{c_2}{s-2}$$

by $(s-3)$ to obtain

$$\frac{s-3}{(s-3)(s-2)} = \frac{c_1(s-3)}{s-3} + \frac{c_2(s-3)}{s-2} \tag{6.7}$$

or

$$\frac{1}{s-2} = c_1 + \frac{c_2(s-3)}{s-2}. \tag{6.8}$$

Equation (6.8) is satisfied for any value of s. If $s = 3$ is chosen to eliminate c_2, we have $c_1 = 1/(3 - 2) = 1$.

Also, to calculate c_2, both sides of

$$\frac{1}{(s-3)(s-2)} = \frac{c_1}{s-3} + \frac{c_2}{s-2}$$

are multiplied by $(s-2)$ to yield

$$\frac{s-2}{(s-3)(s-2)} = \frac{c_1(s-2)}{s-3} + c_2 \qquad (6.9)$$

or

$$\frac{1}{(s-3)} = \frac{c_1(s-2)}{s-3} + c_2. \qquad (6.10)$$

Similarly, if $s = 2$, then $c_2 = -1$. Hence,

$$\overline{f}(s) = \frac{1}{s-3} - \frac{1}{s-2} \qquad (6.11)$$

and

$$f(t) = \mathcal{L}^{-1}\left\{\frac{1}{s-3}\right\} - \mathcal{L}^{-1}\left\{\frac{1}{s-2}\right\}. \qquad (6.12)$$

Table 5.1 gives

$$f(t) = e^{3t} - e^{2t}. \qquad (6.13)$$

The function "*InverseLaplaceTransform*" in *Mathematica*® can be used to invert $\overline{f}(s)$:

$$\text{Inverse Laplace transform}\left[\frac{1}{s^2 - 5s + 6}, s, t\right].$$

Remarks

1. The denominator of $\overline{f}(s)$ should be factored out. This step is important to implement the procedure. It may be necessary to solve explicitly for the roots of a higher-degree polynomial.
2. The denominator of $\overline{f}(s)$ contains expressions in the form of $r_n = s - a_n$ so that the roots are immediately identified. This form will be helpful in later chapters when analyzing the transient behavior of a system.

Example 6.2   Consider the Laplace transform of a function $f(t)$:

$$\overline{f}(s) = \frac{s}{3s^3 + 16s^2 + 23s + 6}. \tag{6.14}$$

The denominator is factorized to give $3(s+\frac{1}{3})(s+2)(s+3)$. In *Mathematica*, the *Factor* command is entered. The Laplace function (Eq. 6.14) becomes

$$\overline{f}(s) = \frac{s}{3(s+\frac{1}{3})(s+2)(s+3)} = \frac{1}{3}\left\{\frac{c_1}{s+\frac{1}{3}} + \frac{c_2}{s+2} + \frac{c_3}{s+3}\right\}. \tag{6.15}$$

For simplicity, the coefficients c_i are multiplied by $\frac{1}{3}$ to give

$$\overline{f}(s) = \frac{c_{11}}{s+\frac{1}{3}} + \frac{c_{12}}{s+2} + \frac{c_{13}}{s+3}, \tag{6.16}$$

where $c_{1i} = \left(\frac{1}{3}\right)c_i$.

According to the Heaviside expansion formula, we first multiply both sides of

$$\frac{s}{3(s+\frac{1}{3})(s+2)(s+3)} = \frac{c_{11}}{s+\frac{1}{3}} = \frac{c_{12}}{s+2} + \frac{c_{13}}{s+3}$$

by $(s+\frac{1}{3})$ to obtain

$$\frac{s}{3(s+2)(s+3)} = c_{11} + \frac{c_{12}(s+\frac{1}{3})}{s+2} + \frac{c_{13}(s+\frac{1}{3})}{s+3}. \tag{6.17}$$

By setting $s = -\frac{1}{3}$, we have

$$c_{11} = \left.\frac{s}{3(s+2)(s+3)}\right|_{s=-1/3} = -\frac{1}{40}. \tag{6.18}$$

Similarly, we multiply both sides of

$$\frac{s}{3(s+\frac{1}{3})(s+2)(s+3)} = \frac{c_{11}}{s+\frac{1}{3}} + \frac{c_{12}}{s+2} + \frac{c_{13}}{s+3}$$

by $(s+2)$ and assign the value of -2 to s to get

$$c_{12} = \frac{s}{3\left(s+\frac{1}{3}\right)(s+3)}\Bigg|_{s=-2} = \frac{2}{5}. \tag{6.19}$$

The coefficient c_{13} is found using the same procedure:

$$c_{13} = \frac{s}{3\left(s+\frac{1}{3}\right)(s+2)}\Bigg|_{s=-3} = -\frac{3}{8}. \tag{6.20}$$

Hence,

$$\bar{f}(s) = \frac{-\frac{1}{40}}{s+\frac{1}{3}} + \frac{\frac{2}{5}}{s+2} + \frac{-\frac{3}{8}}{s+3} \tag{6.21}$$

and

$$f(t) = \mathcal{L}^{-1}\left\{\frac{-\frac{1}{40}}{s+\frac{1}{3}}\right\} + \mathcal{L}^{-1}\left\{\frac{\frac{2}{5}}{s+2}\right\} + \mathcal{L}^{-1}\left\{\frac{-\frac{3}{8}}{s+3}\right\}. \tag{6.22}$$

Again, inspection of Table 5.1 leads to

$$f(t) = -\frac{1}{40}e^{-1/3t} + \frac{2}{5}e^{-2t} - \frac{3}{8}e^{-3t}. \tag{6.23}$$

Example 6.3   Consider a Laplace transform with *complex roots*:

$$\bar{f}(s) = \frac{s+1}{s^2 - 4s + 8}. \tag{6.24}$$

The roots of the denominator are $s = 2 - 2i$ and $s = 2 + 2i$. The function $\bar{f}(s)$ can be written as

$$\frac{s+1}{s^2 - 4s + 8} = \frac{c_1}{s - (2 - 2i)} + \frac{c_2}{s - (2 + 2i)}. \tag{6.25}$$

A closer look at the Heaviside expansion technique shows that computation of the coefficient c_n reduces to the application of the following formula:

$$c_n = (s+a_n)\frac{P(s)}{Q(s)}\bigg|_{s=-a_n}. \tag{6.26}$$

The use of Equation (6.26) gives

$$c_1 = [s-(2-2i)]\frac{(s+1)}{[s-(2-2i)][s-(2+2i)]}\bigg|_{s=2-2i}$$

$$c_1 = \frac{(s+1)}{[s-(2+2i)]}\bigg|_{s=2-2i} = \frac{1}{2}+\frac{3}{4}i. \tag{6.27}$$

Similarly, c_2 is

$$c_2 = [s-(2+2i)]\frac{(s+1)}{[s-(2-2i)][s-(2+2i)]}\bigg|_{s=2+2i}$$

$$c_2 = \frac{(s+1)}{[s-(2-2i)]}\bigg|_{s=2+2i} = \frac{1}{2}-\frac{3}{4}i. \tag{6.28}$$

Therefore, $\overline{f}(s)$ becomes

$$\overline{f}(s) = \frac{\frac{1}{2}+\frac{3}{4}i}{s-(2-2i)} + \frac{\frac{1}{2}-\frac{3}{4}i}{s-(2+2i)} \tag{6.29}$$

and

$$f(t) = \left(\frac{1}{2}+\frac{3}{4}i\right)e^{(2-2i)t} + \left(\frac{1}{2}-\frac{3}{4}i\right)e^{(2+2i)t}$$

$$f(t) = \frac{1}{2}e^{2t}\left[\left(1+\frac{3}{2}i\right)e^{(-2i)t} + \left(1-\frac{3}{2}i\right)e^{(2i)t}\right]. \tag{6.30}$$

From Euler's identity,

$$e^{\alpha i} = \cos\alpha + i\sin\alpha, \tag{6.31}$$

$$e^{(2t)i} = \cos(2t) + i\sin(2t), \tag{6.32}$$

and

$$e^{(-2t)i} = \cos(2t) - i\sin(2t), \tag{6.33}$$

$f(t)$ can be written as

$$f(t) = \frac{1}{2}e^{2t}\left[\left(1+\frac{3}{2}i\right)(\cos(2t)-i\sin(2t))+\left(1-\frac{3}{2}i\right)(\cos(2t)+i\sin(2t))\right]$$

$$f(t) = \frac{1}{2}e^{2t}[2\cos(2t)+3\sin(2t)]. \tag{6.34}$$

CASE 2: Multiple (or Repeated) Roots

Repeated roots are handled differently from simple poles.

Example 6.4 The Laplace transform of a function $f(t)$ is given by

$$\overline{f}(s) = \frac{1}{(s+3)^3(s+1)}. \tag{6.35}$$

The denominator has the following roots: -3 repeated three times and -1. The function $\overline{f}(s)$ is expanded to

$$\overline{f}(s) = \frac{c_1}{(s+3)} + \frac{c_2}{(s+3)^2} + \frac{c_3}{(s+3)^3} + \frac{b}{(s+1)}. \tag{6.36}$$

Constants c_3 and b are found by the method outlined for distinct roots:

$$b = \frac{(s+1)}{(s+3)^3(s+1)}\bigg|_{s=-1} = \frac{1}{(s+3)^3}\bigg|_{s=-1} = \frac{1}{8} \tag{6.37}$$

and

$$c_3 = \frac{(s+3)^3}{(s+3)^3(s+1)}\bigg|_{s=-3} = \frac{1}{(s+1)}\bigg|_{s=-3} = -\frac{1}{2}. \tag{6.38}$$

To calculate c_2, we first multiply both sides of the following equation:

$$\frac{1}{(s+3)^3(s+1)} = \frac{c_1}{(s+3)} + \frac{c_2}{(s+3)^2} + \frac{c_3}{(s+3)^3} + \frac{b}{(s+1)} \tag{6.39}$$

by $(s+3)^3$:

$$\frac{(s+3)^3}{(s+3)^3(s+1)} = \frac{c_1(s+3)^3}{(s+3)} + \frac{c_2(s+3)^3}{(s+3)^2} + \frac{c_3(s+3)^3}{(s+3)^3} + \frac{b(s+3)^3}{(s+1)}$$

$$\frac{1}{(s+1)} = c_1(s+3)^2 + c_2(s+3) + c_3 + \frac{b(s+3)^3}{(s+1)}. \tag{6.40}$$

If we choose $s = -3$, c_2 cannot be calculated because its coefficient is zero. Both sides of Equation (6.40) are differentiated with respect to s:

$$\frac{d}{ds}\left(\frac{1}{(s+1)}\right) = \frac{d}{ds}\left[c_1(s+3)^2\right] + \frac{d}{ds}\left[c_2(s+3)\right] + \frac{d}{ds}(c_3) + \frac{d}{ds}\left[\frac{b(s+3)^3}{(s+1)}\right]$$

$$-\frac{1}{(s+1)^2} = 2(3+s)c_1 + c_2 + 0 + \frac{3b(3+s)^2}{1+s} - \frac{b(3+s)^3}{(1+s)^2}. \tag{6.41}$$

Now, setting $s = -3$ yields

$$c_2 = -\frac{1}{4}.$$

To find c_1, both sides of Equation (6.40) are differentiated twice with respect to s:

$$\frac{2}{(s+1)^3} = 2c_1 + A, \tag{6.42}$$

where A represents terms in $(s+3)$. When

$$s = -3, \quad c_1 = -\frac{1}{8}.$$

As a result,

$$\bar{f}(s) = -\frac{1}{8(s+3)} - \frac{1}{4(s+3)^2} - \frac{1}{2(s+3)^3} + \frac{1}{8(s+1)}. \tag{6.43}$$

Using the formula from Table 6.1,

$$f(t) = \frac{t^{n-1}e^{at}}{(n-1)!}; \quad \bar{f}(s) = \frac{1}{(s-a)^n}, \quad n = 1, 2, 3, \cdots,$$

we have

$$f(t) = -\frac{1}{8}e^{-3t} - \frac{1}{4}te^{-3t} - \frac{1}{4}t^2e^{-3t} + \frac{1}{8}e^{-t}. \tag{6.44}$$

6.2   RESIDUE THEOREM

The residue theorem can also lead to the results in Section 6.1. We have previously seen that $F(s)$ can be written as

$$F(s) = \frac{P(s)}{Q(s)}. \tag{6.45}$$

Two cases, based on the roots of the denominator $Q(s)$, are described as follows:

1. $Q(s)$ has a number of distinct roots (also called *simple poles* of $F(s)$). The inverse transform of $F(s)$ is

$$f(t) = \mathcal{L}^{-1}\{F(s)\} = \sum_{i=1}^{\infty} \rho_n(t), \tag{6.46}$$

where $\rho_n(t)$ is the *residue* of $e^{st}F(s)$ at the pole a_n:

$$\rho_n(t) = \lim_{s \to a_n}\left[(s - a_n)\frac{P(s)}{Q(s)}\right]e^{a_n t}. \tag{6.47}$$

Equation (6.46) can be applied for an infinite number of poles. However, in practice, only a few poles are responsible for the system dynamics:

$$f(t) = \mathcal{L}^{-1}\{F(s)\} = \sum_{i=1}^{N} \rho_n(t), \tag{6.48}$$

where N is the number of simple poles. An equivalent form of the residue is

$$\rho_n(t) = \frac{P(a_n)}{\left.\dfrac{dQ}{ds}\right|_{s=a_n}}e^{a_n t}, \tag{6.49}$$

which can be derived by applying the definition of a function derivative,

$$\left.\frac{dQ}{ds}\right|_{s=a_n} = \lim_{s \to a_n}\left[\frac{Q(s) - Q(a_n)}{s - a_n}\right], \tag{6.50}$$

and using the fact that $Q(a_n) = 0$.

2. $Q(s)$ has repeated roots.

If the function $F(s)$ *contains a pole of order m* at $s = a_n$ (i.e., $Q(s)$ has m repeated roots), Equation (6.46) still applies. However, the residue becomes [1]

$$\rho_n(t) = e^{a_n t} \left\{ A_1 + t A_2 + \frac{t^2}{2!} A_3 + \cdots + \frac{t^{m-1}}{(m-1)!} A_m \right\} \tag{6.51}$$

with A_i defined as

$$A_i = \lim_{s \to a_n} \frac{1}{(m-i)!} \frac{d^{m-i}}{ds^{m-i}} \left\{ (s - a_n)^m F(s) \right\}. \tag{6.52}$$

It should be noted that the method of residues can always be implemented if constants $M > 0$ and $k > 0$ exist such that the transform $F(s)$ satisfies [2]

$$|F(s)| < \frac{M}{R^k}, \tag{6.53}$$

where $s = Re^{i\theta}$. The condition (Eq. 6.53) is always satisfied when $F(s)$ is a rational function, $F(s) = P(s)/Q(s)$, and the degree of $P(s)$ is less than that of $Q(s)$. Let us apply the method of residues to the previous examples.

CASE 1: Distinct Roots

Example 6.5   Consider the Laplace transform of a function $f(t)$:

$$\overline{f}(s) = \frac{1}{s^2 - 5s + 6}. \tag{6.54}$$

The function can be written as

$$\overline{f}(s) = \frac{1}{(s - 3)(s - 2)} \tag{6.55}$$

with simple poles: $a_1 = 2$ and $a_2 = 3$.

The method of residues gives

$$\rho_1(t) = \lim_{s \to a_1} \left[(s - a_1) \frac{P(s)}{Q(s)} \right] e^{a_1 t} \tag{6.56}$$

or

$$\rho_1(t) = \lim_{s \to 2}\left[(s-2)\frac{1}{(s-3)(s-2)}\right]e^{2t}$$

$$\rho_1(t) = \lim_{s \to 2}\left[\frac{1}{s-3}\right]e^{2t} = -e^{2t}. \tag{6.57}$$

Similarly, the residue at $a_2 = 3$ is

$$\rho_2(t) = \lim_{s \to a_2}\left[(s-a_2)\frac{P(s)}{Q(s)}\right]e^{a_2t} \tag{6.58}$$

or

$$\rho_2(t) = \lim_{s \to 3}\left[(s-3)\frac{1}{(s-3)(s-2)}\right]e^{3t}$$

$$\rho_2(t) = \lim_{s \to 3}\left[\frac{1}{s-2}\right]e^{3t} = e^{3t}. \tag{6.59}$$

As a result,

$$f(t) = \mathcal{L}^{-1}\{F(s)\} = \rho_1(t) + \rho_2(t) \tag{6.60}$$

and

$$f(t) = e^{3t} - e^{2t}. \tag{6.61}$$

Example 6.6   Consider the Laplace transform of a function $f(t)$:

$$\overline{f}(s) = \frac{s}{3s^3 + 16s^2 + 23s + 6} \tag{6.62}$$

or

$$\overline{f}(s) = \frac{s}{3(s+\frac{1}{3})(s+2)(s+3)} \tag{6.63}$$

with $a_1 = -\frac{1}{3}$, $a_2 = -2$, and $a_2 = -3$.

The residues are

$$\rho_1(t) = \lim_{s \to a_1} \left[(s-a_1) \frac{P(s)}{Q(s)} \right] e^{a_1 t}$$

$$\rho_1(t) = \lim_{s \to -1/3} \left[\left(s + \frac{1}{3} \right) \frac{s}{3 \left(s + \frac{1}{3} \right) (s+2)(s+3)} \right] e^{-1/3t}$$

$$\rho_1(t) = \lim_{s \to -1/3} \left[\frac{s}{3(s+2)(s+3)} \right] e^{-1/3t} \qquad (6.64)$$

$$\rho_1(t) = -\frac{1}{40} e^{-1/3t}$$

$$\rho_2(t) = \lim_{s \to a_2} \left[(s-a_2) \frac{P(s)}{Q(s)} \right] e^{a_2 t}$$

$$\rho_2(t) = \lim_{s \to -2} \left[(s+2) \frac{s}{3 \left(s + \frac{1}{3} \right)(s+2)(s+3)} \right] e^{-2t}$$

$$\rho_2(t) = \lim_{s \to -2} \left[\frac{s}{3 \left(s + \frac{1}{3} \right)(s+3)} \right] e^{-2t} \qquad (6.65)$$

$$\rho_2(t) = \frac{2}{5} e^{-2t}$$

and

$$\rho_3(t) = \lim_{s \to a_3} \left[(s-a_3) \frac{P(s)}{Q(s)} \right] e^{a_3 t}$$

$$\rho_3(t) = \lim_{s \to -3} \left[(s+3) \frac{s}{3 \left(s + \frac{1}{3} \right)(s+2)(s+3)} \right] e^{-3t}$$

$$\rho_3(t) = \lim_{s \to -3} \left[\frac{s}{3 \left(s + \frac{1}{3} \right)(s+2)} \right] e^{-3t} \qquad (6.66)$$

$$\rho_3(t) = -\frac{3}{8} e^{-3t}.$$

The function $f(t)$ is

$$f(t) = -\frac{1}{40} e^{-1/3t} + \frac{2}{5} e^{-2t} - \frac{3}{8} e^{-3t}. \qquad (6.67)$$

Example 6.7   Consider a Laplace transform with complex roots:

$$\bar{f}(s) = \frac{s+1}{s^2 - 4s + 8}. \qquad (6.68)$$

The roots of the denominator are $a_1 = 2 - 2i$ and $a_2 = 2 + 2i$. The function $\bar{f}(s)$ is written as

$$\bar{f}(s) = \frac{s+1}{[s-(2-2i)][s-(2+2i)]}. \qquad (6.69)$$

The residues are

$$
\begin{aligned}
\rho_1(t) &= \lim_{s \to a_1} \left[(s - a_1) \frac{P(s)}{Q(s)} \right] e^{a_1 t} \\
\rho_1(t) &= \lim_{s \to 2-2i} \left[[s-(2-2i)] \frac{(s+1)}{[s-(2-2i)][s-(2+2i)]} \right] e^{(2-2i)t} \\
\rho_1(t) &= \lim_{s \to 2-2i} \left[\frac{(s+1)}{[s-(2+2i)]} \right] e^{(2-2i)t} \\
\rho_1(t) &= \left(\frac{1}{2} + \frac{3}{4}i \right) e^{(2-2i)t}
\end{aligned}
\qquad (6.70)
$$

and

$$
\begin{aligned}
\rho_2(t) &= \lim_{s \to a_2} \left[(s - a_2) \frac{P(s)}{Q(s)} \right] e^{a_2 t} \\
\rho_2(t) &= \lim_{s \to 2+2i} \left[[s-(2+2i)] \frac{(s+1)}{[s-(2-2i)][s-(2+2i)]} \right] e^{(2+2i)t} \\
\rho_2(t) &= \lim_{s \to 2+2i} \left[\frac{(s+1)}{[s-(2-2i)]} \right] e^{(2+2i)t} \\
\rho_2(t) &= \left(\frac{1}{2} - \frac{3}{4}i \right) e^{(2+2i)t}.
\end{aligned}
\qquad (6.71)
$$

Therefore,

$$
\begin{aligned}
f(t) &= \mathcal{L}^{-1}\{F(s)\} = \rho_1(t) + \rho_2(t) \\
f(t) &= \left(\frac{1}{2} + \frac{3}{4}i \right) e^{(2-2i)t} + \left(\frac{1}{2} - \frac{3}{4}i \right) e^{(2+2i)t} \\
f(t) &= \frac{1}{2} e^{2t} \left[\left(1 + \frac{3}{2}i \right) e^{(-2i)t} + \left(1 - \frac{3}{2}i \right) e^{(2i)t} \right].
\end{aligned}
\qquad (6.72)
$$

After using Euler's identity, the function $f(t)$ becomes

$$f(t) = \frac{1}{2}e^{2t}[2\cos(2t) + 3\sin(2t)]. \tag{6.73}$$

It is worth noting that, when $\overline{f}(s)$ is written as a quotient of polynomials, application of the residue method bears some resemblance to the Heaviside expansion procedure when $\overline{f}(s)$ contains only simple poles. The coefficient c_n in

$$f(t) = \mathcal{L}^{-1}\left\{\frac{c_1}{r_1(s)}\right\} + \mathcal{L}^{-1}\left\{\frac{c_2}{r_2(s)}\right\} + \cdots + \mathcal{L}^{-1}\left\{\frac{c_N}{r_N(s)}\right\} \tag{6.74}$$

is equal to $\rho_n(t)e^{-a_n t}$. Also, because $r_n(s)$ is in the form of $(s - a_n)$ (see Section 6.1), Equation (6.74) becomes

$$
\begin{aligned}
f(t) &= \mathcal{L}^{-1}\left\{\frac{c_1}{r_1(s)}\right\} + \mathcal{L}^{-1}\left\{\frac{c_2}{r_2(s)}\right\} + \cdots + \mathcal{L}^{-1}\left\{\frac{c_N}{r_N(s)}\right\} \\
f(t) &= c_1 e^{a_1 t} + c_2 e^{a_2 t} + \cdots + c_N e^{a_N t} \\
f(t) &= \rho_1(t)e^{-a_1 t}e^{a_1 t} + \rho_2(t)e^{-a_2 t}e^{a_2 t} + \cdots + \rho_N(t)e^{-a_N t}e^{a_N t} \\
f(t) &= \rho_1(t) + \rho_2(t) + \cdots + \rho_N(t).
\end{aligned}
\tag{6.75}
$$

CASE 2: Multiple Roots

Example 6.8   The Laplace transform of a function $f(t)$ is given by

$$\overline{f}(s) = \frac{1}{(s+3)^3(s+1)}. \tag{6.76}$$

The function $\overline{f}(s)$ contains a pole of order 3 (or *triple pole*) at $a_1 = -3$ and a simple pole $a_2 = -1$. Although $f(t)$ is found by adding the residues of $\overline{f}(s)$ at the poles, expressions for $\rho_n(t)$ depend on whether we are dealing with multiple or simple poles.

When $a_2 = -1$, we have

$$\rho_2(t) = \lim_{s \to a_2} \left[(s - a_2) \frac{P(s)}{Q(s)} \right] e^{a_2 t}$$

$$\rho_2(t) = \lim_{s \to -1} \left[(s+1) \frac{1}{(s+3)^3 (s+1)} \right] e^{-t}$$

$$\rho_2(t) = \lim_{s \to -1} \left[\frac{1}{(s+3)^3} \right] e^{-t}$$

$$\rho_2(t) = \frac{1}{8} e^{-t}. \tag{6.77}$$

The triple pole $a_1 = -3$ leads to

$$\rho_1(t) = e^{a_1 t} \left\{ A_1 + t A_2 + \frac{t^2}{2!} A_3 \right\}. \tag{6.78}$$

Application of Equation (6.52) results in

$$A_1 = \lim_{s \to -3} \frac{1}{(3-1)!} \frac{d^{3-1}}{ds^{3-1}} \left\{ (s+3)^3 \frac{1}{(s+3)^3 (s+1)} \right\}$$

$$A_1 = \lim_{s \to -3} \frac{1}{2} \frac{d^2}{ds^2} \left\{ \frac{1}{(s+1)} \right\}$$

$$A_1 = \frac{1}{2} \lim_{s \to -3} \frac{2}{(1+s)^3}$$

$$A_1 = -\frac{1}{8}, \tag{6.79}$$

$$A_2 = \lim_{s \to -3} \frac{1}{(3-2)!} \frac{d^{3-2}}{ds^{3-2}} \left\{ (s+3)^3 \frac{1}{(s+3)^3 (s+1)} \right\}$$

$$A_2 = \lim_{s \to -3} \frac{d}{ds} \left\{ \frac{1}{(s+1)} \right\}$$

$$A_2 = -\lim_{s \to -3} \frac{1}{(1+s)^2}$$

$$A_2 = -\frac{1}{4}, \tag{6.80}$$

and

$$A_3 = \lim_{s \to -3} \frac{1}{(3-3)!} \frac{d^{3-3}}{ds^{3-3}} \left\{ (s+3)^3 \frac{1}{(s+3)^3(s+1)} \right\}$$

$$A_3 = \lim_{s \to -3} \frac{1}{s+1} \tag{6.81}$$

$$A_3 = -\frac{1}{2}.$$

As a result,

$$\rho_1(t) = e^{-3t} \left\{ -\frac{1}{8} - \frac{1}{4}t - \frac{1}{2}\frac{t^2}{2} \right\}$$

$$\rho_1(t) = e^{-3t} \left\{ -\frac{1}{8} - \frac{t}{4} - \frac{t^2}{4} \right\} \tag{6.82}$$

$$\rho_1(t) = -\frac{1}{8}e^{-3t} - \frac{1}{4}te^{-3t} - \frac{1}{4}t^2e^{-3t}.$$

The function $f(t)$ is

$$f(t) = \rho_1(t) + \rho_2(t) \tag{6.83}$$

or

$$f(t) = -\frac{1}{8}e^{-3t} - \frac{1}{4}te^{-3t} - \frac{1}{4}t^2e^{-3t} + \frac{1}{8}e^{-t}. \tag{6.84}$$

Again, a close observation of the strategy used when applying the Heaviside expansion method shows that the coefficients c_1, c_2, and c_3 in

$$\bar{f}(s) = \frac{c_1}{(s+3)} + \frac{c_2}{(s+3)^2} + \frac{c_3}{(s+3)^3} + \frac{b}{(s+1)} \tag{6.85}$$

are equal to (and computed in a similar manner as) A_1, A_2, and A_3. Just as before, $b = \rho_2(t)e^t$. The function $\bar{f}(s)$ takes the form

$$\bar{f}(s) = \frac{A_1}{(s+3)} + \frac{A_2}{(s+3)^2} + \frac{A_3}{(s+3)^3} + \frac{\rho_2(t)e^t}{(s+1)}. \tag{6.86}$$

Consequently,

$$f(t) = A_1 e^{-3t} + t A_2 e^{-3t} + \frac{1}{2} t^2 A_3 e^{-3t} + p_2(t) e^t e^{-t}$$

$$f(t) = e^{-3t} \left(A_1 + t A_2 + \frac{1}{2} t^2 A_3 \right) + p_2(t). \tag{6.87}$$

In conclusion, both procedures are equivalent when $\bar{f}(s)$ is expressed as a quotient of polynomials $(P(s)/Q(s))$. In general, for cases where the degree of $P(s)$ is less than that of $Q(s)$, the Heaviside expansion formula [2] reduces to

$$f(t) = \mathcal{L}^{-1}\{\bar{f}(s)\} = \sum_{n=1}^{N} \frac{P(a_n)}{\left.\dfrac{dQ}{ds}\right|_{s=a_n}} e^{a_n t}, \tag{6.88}$$

where $\bar{f}(s)$ has N simple poles. A systematic proof, given in Spiegel [2], also shows that the steps used when applying the Heaviside expansion inevitably leads to the applications of Equations (6.48) and (6.49). However, significant time is saved with direct application of the residue theorem instead of partial fraction decomposition.

6.3 CONTINUOUS FERMENTATION

The dynamic equation of a bioreactor, after linearization about a stable steady-state point, is

$$\frac{d\tilde{C}_x}{dt} = -0.54\tilde{C}_s$$

$$\frac{d\tilde{C}_s}{dt} = 1.82\tilde{C}_x - 3.15\tilde{C}_s, \tag{6.89}$$

where the subscripts x and s represent the cell and substrate, respectively (see Section 5.5). The equilibrium point, written in terms of deviation variables, is $(\tilde{C}_{xe}, \tilde{C}_{se})$. The linearized system response to a step change of size 1.0 g/L in the initial (or steady-state) substrate concentration led to

$$\bar{C}_x = \frac{-0.54}{(s+0.35)(s+2.80)} \tag{6.90}$$

and

$$\bar{C}_s = \frac{-s}{(s+0.35)(s+2.80)},$$ (6.91)

where the poles are $a_1 = -0.35$ and $a_2 = -2.80$. To calculate $\tilde{C}_x$, we first compute the residues:

$$\rho_1(t) = \lim_{s \to a_1}\left[(s - a_1)\frac{P(s)}{Q(s)}\right]e^{a_1 t}$$

$$\rho_1(t) = \lim_{s \to -0.35}\left[(s + 0.35)\frac{-0.54}{(s+0.35)(s+2.80)}\right]e^{-0.35t}$$ (6.92)

$$\rho_1(t) = \lim_{s \to -0.35}\left[\frac{-0.54}{s+2.80}\right]e^{-0.35t}$$

$$\rho_1(t) = -0.22e^{-0.35t}$$

and

$$\rho_2(t) = \lim_{s \to a_2}\left[(s - a_2)\frac{P(s)}{Q(s)}\right]e^{a_2 t}$$

$$\rho_2(t) = \lim_{s \to -2.80}\left[(s + 2.80)\frac{-0.54}{(s+0.35)(s+2.80)}\right]e^{-2.80t}$$ (6.93)

$$\rho_2(t) = \lim_{s \to -2.80}\left[\frac{-0.54}{s+0.35}\right]e^{-2.80t}$$

$$\rho_2(t) = 0.22e^{-2.80t}.$$

The concentration $\tilde{C}_x(t)$ is

$$\tilde{C}_x(t) = \rho_1(t) + \rho_2(t)$$ (6.94)

or

$$\tilde{C}_x(t) = -0.22e^{-0.35t} + 0.22e^{-2.80t}.$$ (6.95)

A similar procedure applied for the substrate concentration in deviation form gives

$$\tilde{C}_s(t) = 1.14e^{-2.8t} - 0.14e^{-0.35t}.$$ (6.96)

6.4 DEGRADATION OF PLASMID DNA

The transfer of intact plasmid DNA (pDNA) to their target sites is essential to the success of gene delivery/therapy. Plasmids are small extrachromosomal double-stranded, closed-circular DNA molecules that are encountered in a number of bacterial species. Studies are often devoted to understand the degradation of pDNA in the bloodstream following intravenous administration. For example, Houk et al. investigated the kinetic processes to explain the clearance of pDNA in a rat plasma model [3]. The process follows first-order kinetics with time-invariant rate constants:

$$SC \xrightarrow{k_s} OC \xrightarrow{k_0} L \xrightarrow{k_l}, \tag{6.97}$$

where SC (ng/μL), OC (ng/μL), and L (ng/μL) represent the concentrations of supercoiled, open circular, and linear pDNA, respectively. The degradation rate constants (per minute) are $k_s = 0.6$, $k_o = 0.03$, and $k_l = 0.06$. Laplace transform techniques can be used to describe the dynamics.

The degradation of pDNA in the plasma is assumed to follow pseudo-first-order kinetics [3]:

$$\frac{dSC}{dt} = -k_s SC, \tag{6.98}$$

$$\frac{dOC}{dt} = k_s SC - k_o OC, \tag{6.99}$$

and

$$\frac{dL}{dt} = k_o OC - k_l L. \tag{6.100}$$

The model can be solved using Laplace transform methods. The initial conditions are $SC(t = 0) = 10.8$, $OC(t = 0) = 0$, and $L(t = 0) = 0$.

Applying the Laplace transforms, we have

$$s\overline{SC} - SC_0 = -k_s \overline{SC}, \tag{6.101}$$

$$s\overline{OC} = k_s \overline{SC} - k_o \overline{OC}, \tag{6.102}$$

and

$$s\overline{L} = k_o \overline{OC} - k_l \overline{L}, \tag{6.103}$$

where $\overline{SC}, \overline{OC},$ and $\overline{L}$ are the transforms of $SC, OC,$ and $L,$ respectively. The equations are rearranged to

$$
\begin{aligned}
(s+k_s)\overline{SC} + 0.\overline{OC} + 0.\overline{L} &= SC_0 \\
-k_s\overline{SC} + (s+k_o)\overline{OC} + 0.\overline{L} &= 0 \\
0.\overline{SC} - k_o\overline{OC} + (s+k_l)\overline{L} &= 0.
\end{aligned}
\tag{6.104}
$$

Using the Cramer's rule, we have

$$
\Delta = \begin{vmatrix} (s+k_s) & 0 & 0 \\ -k_s & (s+k_o) & 0 \\ 0 & -k_o & (s+k_l) \end{vmatrix} = \begin{vmatrix} 0.6+s & 0 & 0 \\ -0.6 & 0.03+s & 0 \\ 0 & -0.03 & 0.06+s \end{vmatrix}
\tag{6.105}
$$

$$
\Delta = 0.00108 + 0.0558s + 0.69s^2 + s^3 = (0.03+s)(0.06+s)(0.6+s)
$$

$$
\Delta_{\overline{SC}} = \begin{vmatrix} SC_0 & 0 & 0 \\ 0 & (s+k_o) & 0 \\ 0 & -k_o & (s+k_l) \end{vmatrix} = \begin{vmatrix} 10.8 & 0 & 0 \\ 0 & 0.03+s & 0 \\ 0 & -0.03 & 0.06+s \end{vmatrix}
\tag{6.106}
$$

$$
\Delta_{\overline{SC}} = 10.8(0.03+s)(0.06+s)
$$

$$
\Delta_{\overline{OC}} = \begin{vmatrix} (s+k_s) & SC_0 & 0 \\ -k_s & 0 & 0 \\ 0 & 0 & (s+k_l) \end{vmatrix} = \begin{vmatrix} 0.6+s & 10.8 & 0 \\ -0.6 & 0 & 0 \\ 0 & 0 & 0.06+s \end{vmatrix}
\tag{6.107}
$$

$$
\Delta_{\overline{OC}} = 6.48(0.06+s)
$$

$$
\Delta_{\overline{L}} = \begin{vmatrix} (s+k_s) & 0 & SC_0 \\ -k_s & (s+k_o) & 0 \\ 0 & -k_o & 0 \end{vmatrix} = \begin{vmatrix} 0.6+s & 0 & 10.8 \\ -0.6 & 0.03+s & 0 \\ 0 & -0.03 & 0 \end{vmatrix}
\tag{6.108}
$$

$$
\Delta_{\overline{L}} = 0.1944.
$$

As a result, the Laplace transforms of the concentrations are

$$
\overline{SC} = \frac{\Delta_{\overline{SC}}}{\Delta} = \frac{0.01944 + 0.972s + 10.8s^2}{0.00108 + 0.0558s + 0.69s^2 + s^3} = \frac{10.8(0.03+s)(0.06+s)}{(0.03+s)(0.06+s)(0.6+s)}
$$

$$
\overline{SC} = \frac{10.8}{0.6+s},
\tag{6.109}
$$

$$
\overline{OC} = \frac{\Delta_{\overline{OC}}}{\Delta} = \frac{0.3888 + 6.48s}{0.00108 + 0.0558s + 0.69s^2 + s^3} = \frac{6.48(0.06+s)}{(0.03+s)(0.06+s)(0.6+s)}
$$

$$
\overline{OC} = \frac{6.48}{(0.03+s)(0.6+s)},
\tag{6.110}
$$

and

$$\bar{L} = \frac{\Delta_{\bar{L}}}{\Delta} = \frac{0.1944}{0.00108 + 0.0558s + 0.69s^2 + s^3} = \frac{0.1944}{(0.03 + s)(0.06 + s)(0.6 + s)}. \quad (6.111)$$

To calculate the concentration $SC(t)$, we apply the residue theorem. The residue for the root $a_1 = -0.6$ is

$$\rho_1(t) = \lim_{s \to a_1} \left[(s - a_1) \frac{P(s)}{Q(s)} \right] e^{a_1 t}$$

$$\rho_1(t) = \lim_{s \to -0.6} \left[(s + 0.6) \frac{10.8}{0.6 + s} \right] e^{-0.6t} \quad (6.112)$$

$$\rho_1(t) = 10.8e^{-0.6t}.$$

Therefore, $SC(t)$ is

$$SC(t) = 10.8e^{-0.6t}. \quad (6.113)$$

A similar procedure applied to calculate $OC(t)$ and $L(t)$ gives

$$OC(t) = 11.37(e^{-0.03t} - e^{-0.6t}) \quad (6.114)$$

and

$$L(t) = 0.63e^{-0.6t} - 12.0e^{-0.06t} + 11.34e^{-0.03t}. \quad (6.115)$$

6.5 CONSTANT-RATE INTRAVENOUS INFUSION

In Chapter 2, the following equation was obtained for a constant rate of drug infusion (k_0) in a one-compartment model:

$$\frac{dC_p}{dt} = \frac{k_0}{V} - k_{el}C_p, \quad (6.116)$$

where C_p is the plasma drug concentration, k_{el} is the elimination rate constant, and V is the volume of distribution. The initial concentration is $C_p(0) = 0$. After applying the Laplace transform to both sides of Equation (6.116), we obtain

$$s\bar{C}_p = \frac{k_0}{V} \frac{1}{s} - k_{el}\bar{C}_p. \quad (6.117)$$

As a result,

$$\bar{C}_p = \frac{k_0}{V} \frac{1}{s(s+k_{\mathrm{el}})} \tag{6.118}$$

with the poles $a_1 = 0$ and $a_2 = -k_{\mathrm{el}}$.

The residues are the following:

$$\rho_1(t) = \lim_{s \to a_1} \left[(s-a_1) \frac{P(s)}{Q(s)} \right] e^{a_1 t}$$

$$\rho_1(t) = \lim_{s \to 0} \left[(s+0) \frac{k_0}{V} \frac{1}{s(s+k_{\mathrm{el}})} \right] e^{0.t}$$

$$\rho_1(t) = \frac{k_0}{V} \lim_{s \to 0} \frac{1}{(s+k_{\mathrm{el}})} \tag{6.119}$$

$$\rho_1(t) = \frac{k_0}{k_{\mathrm{el}} V}$$

and

$$\rho_2(t) = \lim_{s \to a_2} \left[(s-a_2) \frac{P(s)}{Q(s)} \right] e^{a_2 t}$$

$$\rho_2(t) = \lim_{s \to -k_{\mathrm{el}}} \left[(s+k_{\mathrm{el}}) \frac{k_0}{V} \frac{1}{s(s+k_{\mathrm{el}})} \right] e^{-k_{\mathrm{el}} t}$$

$$\rho_2(t) = \frac{k_0}{V} \lim_{s \to -k_{\mathrm{el}}} \left(\frac{1}{s} \right) e^{-k_{\mathrm{el}} t} \tag{6.120}$$

$$\rho_2(t) = -\frac{k_0}{k_{\mathrm{el}} V} e^{-k_{\mathrm{el}} t}.$$

Finally, the drug concentration is given by

$$C_p(t) = \frac{k_0}{k_{\mathrm{el}} V} \left(1 - e^{-k_{\mathrm{el}} t} \right). \tag{6.121}$$

6.6   TRANSDERMAL DRUG-DELIVERY SYSTEMS

Transdermal drug release involves the delivery of drugs through intact skin to administer a therapeutic dose. Key advantages of using the transdermal route are a decrease in the fluctuations of plasma drug concentration and the ability to keep a constant delivery rate for a long period of time. Mathematical analyses for transdermal drug release usually consider a homogeneous model

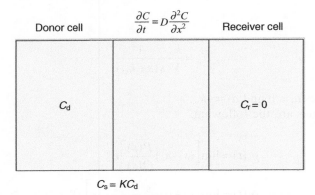

Donor cell $\qquad \dfrac{\partial C}{\partial t} = D\dfrac{\partial^2 C}{\partial x^2} \qquad$ Receiver cell

C_d

$C_r = 0$

$C_s = KC_d$

Figure 6.1. Homogeneous membrane model of transdermal drug delivery.

for the skin and passive diffusion through the stratum corneum as the rate-limiting step.

In vitro test systems, such as two-chamber Franz diffusion cells, are used to investigate drug penetration into human skin (Fig. 6.1). Each half-cell contains a volume of solution and offers an effective diffusional area. The membrane is clamped between the donor and receiver cells and the system is usually maintained at 37°C by using a temperature-controlled water bath and a pump to circulate water through the device. An aliquot of drug solution is added to the donor compartment at the beginning of the experiment. At specific time intervals, the receiver cell content is removed for analysis and replaced with the same volume of fresh solution.

An equation for diffusion in a plane sheet (i.e., Fick's second law) is applied to describe permeation through the membrane:

$$\frac{\partial C}{\partial t} = D\frac{\partial^2 C}{\partial x^2}, \tag{6.122}$$

satisfying the boundary conditions

$$C = KC_d, \quad x = 0, \quad t \geq 0 \tag{6.123}$$

$$C = 0, \qquad x = h, \quad t \geq 0 \tag{6.124}$$

and the initial condition

$$C = 0, \quad 0 < x < h, \quad t = 0. \tag{6.125}$$

Equation (6.122) states the accumulation of drug in the membrane is due to diffusion. The parameter D represents an effective drug diffusion coefficient in the stratum corneum; C is the concentration; x is a distance from the surface of the skin and extends from 0 to the skin thickness; C_d is the donor concentration; and K is the donor/stratum corneum partition coefficient. Equation (6.123) is an equilibrium condition and Equation (6.124) denotes a perfect sink condition assumed for a highly diluted reservoir. Since the drug is being removed, regularly, by blood circulation after diffusing through the skin, a perfect sink condition is applicable in the context of clinical research.

In many applications, equations and variables are often normalized to increase the usefulness of the solutions. The variables can be normalized to give

$$U = \frac{C}{KC_d}; \quad \tau = \frac{D}{h^2}t; \quad X = \frac{x}{h}. \tag{6.126}$$

Equation (6.122) becomes

$$\frac{\partial(UKC_d)}{\partial\left(\dfrac{h^2\tau}{D}\right)} = D\frac{\partial^2(UKC_d)}{\partial(Xh)^2}$$

$$\frac{D\partial(U)}{\partial(h^2\tau)} = D\frac{\partial^2(U)}{\partial(Xh)^2} \tag{6.127}$$

$$\frac{\partial U}{\partial \tau} = \frac{\partial^2 U}{\partial X^2}.$$

The boundary conditions are

$$U = 1, \quad X = 0, \quad \tau \geq 0 \tag{6.128}$$

$$U = 0, \quad X = 1, \quad \tau \geq 0, \tag{6.129}$$

and the initial condition is

$$U = 0, \quad 0 < X < 1, \quad \tau = 0. \tag{6.130}$$

The Laplace transform is applied to Equation (6.127),

$$\mathcal{L}\left\{\frac{\partial U(X,\tau)}{\partial \tau}\right\} = \mathcal{L}\left\{\frac{\partial^2 U(X,\tau)}{\partial X^2}\right\}, \tag{6.131}$$

to give

$$s\bar{U}(X,s) - U(X,0) = \frac{d^2\bar{U}(X,s)}{dX^2} \tag{6.132}$$

or

$$\frac{d^2\bar{U}(X,s)}{dX^2} = s\bar{U}(X,s). \qquad (6.133)$$

The solution to Equation (6.133) is

$$\bar{U}(X,s) = c_1 e^{-X\sqrt{s}} + c_2 e^{X\sqrt{s}}. \qquad (6.134)$$

Note that

$$\exp\left(X\sqrt{s}\right) = \cosh\left(\sqrt{s}X\right) + \sinh\left(\sqrt{s}X\right) \qquad (6.135)$$

and

$$\exp\left(-X\sqrt{s}\right) = \cosh\left(\sqrt{s}X\right) - \sinh\left(\sqrt{s}X\right). \qquad (6.136)$$

Consequently,

$$\bar{U}(X,s) = c_1\left[\cosh\left(\sqrt{s}X\right) - \sinh\left(\sqrt{s}X\right)\right] + c_2\left[\cosh\left(\sqrt{s}X\right) + \sinh\left(\sqrt{s}X\right)\right] \qquad (6.137)$$

or

$$\bar{U}(X,s) = (c_1 + c_2)\cosh\left(\sqrt{s}X\right) + (-c_1 + c_2)\sinh\left(\sqrt{s}X\right). \qquad (6.138)$$

The Laplace transform can be further simplified to

$$\bar{U}(X,s) = k_1 \cosh\left(\sqrt{s}X\right) + k_2 \sinh\left(\sqrt{s}X\right), \qquad (6.139)$$

where $k_1 = c_1 + c_2$ and $k_2 = c_2 - c_1$.

The Laplace transform of the boundary conditions yields

$$\bar{U}(0,s) = \frac{1}{s} \qquad (6.140)$$

and

$$\bar{U}(1,s) = 0. \qquad (6.141)$$

Applying the boundary conditions to Equation (6.139) gives

$$\frac{1}{s} = k_1 \cosh\left(\sqrt{s}\cdot 0\right) + k_2 \sinh\left(\sqrt{s}\cdot 0\right) \qquad (6.142)$$

and

$$0 = k_1 \cosh\left(\sqrt{s} \cdot 1\right) + k_2 \sinh\left(\sqrt{s} \cdot 1\right). \tag{6.143}$$

Since $\sinh(0) = 0$ and $\cosh(0) = 1$, Equation (6.142) is used to yield $k_1 = 1/s$. As a result,

$$0 = \frac{1}{s}\cosh\left(\sqrt{s} \cdot 1\right) + k_2 \sinh\left(\sqrt{s} \cdot 1\right)$$

$$k_2 = -\frac{1}{s}\frac{\cosh\left(\sqrt{s}\right)}{\sinh\left(\sqrt{s}\right)}. \tag{6.144}$$

Equation (6.139) becomes

$$\bar{U}(X,s) = \frac{1}{s}\cosh\left(\sqrt{s}X\right) - \frac{1}{s}\frac{\cosh\left(\sqrt{s}\right)}{\sinh\left(\sqrt{s}\right)}\sinh\left(\sqrt{s}X\right)$$

$$\bar{U}(X,s) = \frac{1}{s}\left\{\cosh\left(\sqrt{s}X\right) - \frac{\cosh\left(\sqrt{s}\right)}{\sinh\left(\sqrt{s}\right)}\sinh\left(\sqrt{s}X\right)\right\} \tag{6.145}$$

$$\bar{U}(X,s) = \frac{1}{s}\left\{\frac{\sinh\left(\sqrt{s}\right)\cosh\left(\sqrt{s}X\right) - \cosh\left(\sqrt{s}\right)\sinh\left(\sqrt{s}X\right)}{\sinh\left(\sqrt{s}\right)}\right\}.$$

Equation (6.145) can be simplified further if we use the formula

$$\sinh(a-b) = \sinh(a)\cosh(b) - \cosh(a)\sinh(b) \tag{6.146}$$

to give

$$\bar{U}(X,s) = \frac{1}{s}\left\{\frac{\sinh\left(\sqrt{s} - X\sqrt{s}\right)}{\sinh\left(\sqrt{s}\right)}\right\}$$

$$\bar{U}(X,s) = \frac{1}{s}\left\{\frac{\sinh\left(\sqrt{s}(1-X)\right)}{\sinh\left(\sqrt{s}\right)}\right\}. \tag{6.147}$$

To find the poles of $\bar{U}(X,s)$, we need to set $Q(s) = s\sinh\left(\sqrt{s}\right) = 0$. The poles are $a_1 = 0$ and the values of s are such that $\sinh\left(\sqrt{s}\right) = 0$. The solution of $\sinh\left(\sqrt{s}\right) = 0$ is $s = -n^2\pi^2$ with $n = 1, 2, 3 \ldots$ obtained after using $\sqrt{s} = n\pi i$.

The concentration in the membrane is

$$U(X,\tau) = \mathcal{L}^{-1}\left\{\bar{U}(X,s)\right\} = \sum_{i=1}^{\infty} \rho_n(X,\tau). \tag{6.148}$$

The upper limit is infinity, in this case, because a countless number of poles can be found.

The residue at $a_1 = 0$ is

$$\rho_1(X, \tau) = \lim_{s \to a_1}\left[(s - a_1)\frac{P(X, s)}{Q(s)}\right]e^{a_1\tau}$$

$$\rho_1(X, \tau) = \lim_{s \to 0}\left[(s - 0)\frac{\sinh\left[\sqrt{s}(1 - X)\right]}{s\sinh\left(\sqrt{s}\right)}\right]e^{0\tau} \tag{6.149}$$

$$\rho_1(X, \tau) = \lim_{s \to 0}\left[\frac{\sinh\left[\sqrt{s}(1 - X)\right]}{\sinh\left(\sqrt{s}\right)}\right].$$

L'Hôpital's rule is used because of the indeterminate form $\frac{0}{0}$. After taking the derivative of the numerator and denominator, we have

$$\rho_1(X, \tau) = \lim_{s \to 0}\left[\frac{\dfrac{(1 - X)\cosh\left[\sqrt{s}(1 - X)\right]}{2\sqrt{s}}}{\dfrac{\cosh\left(\sqrt{s}\right)}{2\sqrt{s}}}\right]$$

$$\rho_1(X, \tau) = \lim_{s \to 0}\left[\frac{(1 - X)\cosh\left[\sqrt{s}(1 - X)\right]}{\cosh\left(\sqrt{s}\right)}\right] \tag{6.150}$$

$$\rho_1(X, \tau) = \frac{(1 - X)\cosh(0)}{\cosh(0)}$$

$$\rho_1(X, \tau) = 1 - X.$$

The equivalent form of the residue can be used for the other poles:

$$\rho_n(X, \tau) = \frac{P(X, a_n)}{\left.\dfrac{dQ}{ds}\right|_{s=a_n}}e^{a_n\tau}. \tag{6.151}$$

The derivative of $Q(s)$ is

$$\frac{dQ}{ds} = \frac{1}{2}\sqrt{s}\cosh\left(\sqrt{s}\right) + \sinh\left(\sqrt{s}\right). \tag{6.152}$$

Then, at the poles, we have

$$\left.\frac{dQ}{ds}\right|_{s=a_n} = \frac{1}{2}(n\pi i)\cosh(n\pi i) + \sinh(n\pi i)$$

$$\left.\frac{dQ}{ds}\right|_{s=a_n} = \frac{1}{2}(n\pi i)\cosh(n\pi i) \tag{6.153}$$

and

$$P(X, a_n) = \sinh(n\pi i)\cosh(n\pi Xi) - \cosh(n\pi i)\sinh(n\pi Xi)$$
$$P(X, a_n) = -\cosh(n\pi i)\sinh(n\pi Xi)$$

(6.154)

after making the following substitution: $\sqrt{s} = n\pi i$. Therefore,

$$\frac{P(X, a_n)}{\left.\dfrac{dQ}{ds}\right|_{s=a_n}} = \frac{-\cosh(n\pi i)\sinh(n\pi iX)}{\frac{1}{2}(n\pi i)\cosh(n\pi i)}$$

$$\frac{P(X, a_n)}{\left.\dfrac{dQ}{ds}\right|_{s=a_n}} = -\frac{2\sinh(n\pi iX)}{n\pi i}$$

$$\frac{P(X, a_n)}{\left.\dfrac{dQ}{ds}\right|_{s=a_n}} = -\frac{2i\sin(n\pi X)}{n\pi i}$$

(6.155)

$$\frac{P(X, a_n)}{\left.\dfrac{dQ}{ds}\right|_{s=a_n}} = -\frac{2}{n\pi}\sin(n\pi X).$$

The normalized concentration is

$$U(X, \tau) = \rho_1(X, \tau) + \sum_{n=1}^{\infty} \rho_n(X, \tau)$$

$$U(X, \tau) = 1 - X + \sum_{n=1}^{\infty}\left\{-\frac{2}{n\pi}\sin(n\pi X)e^{-n^2\pi^2\tau}\right\}$$

(6.156)

$$U(X, \tau) = 1 - X - \frac{2}{\pi}\sum_{n=1}^{\infty}\left\{\frac{1}{n}\sin(n\pi X)e^{-n^2\pi^2\tau}\right\}.$$

The concentration is

$$\frac{C(x, t)}{KC_d} = 1 - \frac{x}{h} - \frac{2}{\pi}\sum_{n=1}^{\infty}\left[\frac{1}{n}\sin\left(n\pi\frac{x}{h}\right)e^{-n^2\pi^2(D/h^2)t}\right]$$

(6.157)

$$C(x, t) = KC_d\left\{\frac{h-x}{h} - \frac{2}{\pi}\sum_{n=1}^{\infty}\left[\frac{1}{n}\sin\left(\frac{n\pi x}{h}\right)e^{-n^2\pi^2(D/h^2)t}\right]\right\}.$$

Additional examples of the use of the residue theorem to solve differential equations can be found in Loney [1].

6.7 SUMMARY

The Heaviside expansion formula and the residue theorem were used to invert Laplace transforms and to find a solution in the time domain. Several examples show that the two procedures give equivalent results. The residue theorem was implemented to provide the substrate and cell concentration profiles in a continuous bioreactor. Equations for the levels of supercoiled, open circular, and linear pDNA were provided to explain the clearance of pDNA in a rat plasma model. The plasma drug concentration during a constant-rate intravenous infusion was calculated using a one-compartment model. A transdermal drug-delivery system was analyzed to derive the drug concentration as a function of time and skin depth. This controlled release example shows an application of the residue method to solve partial differential equations.

PROBLEMS

6.1. Evaluate $\mathcal{L}^{-1}\left\{\dfrac{s+2}{s^2-4s+14}\right\}$.

6.2. Evaluate $\mathcal{L}^{-1}\left\{\dfrac{s}{s^2-5s+6}\right\}$.

6.3. Solve $\begin{cases} \dfrac{dx}{dt}=2x-y \\[2mm] \dfrac{dy}{dt}=-x+y \end{cases}$ subject to $x(0)=2$ and $y(0)=1$.

6.4. Plot the concentrations of supercoiled, open circular, and linear pDNA using the model discussed in Section 6.4.

6.5. The differential equation describing a sinusoidal drug-delivery system is given by Burnette [4]:

$$\frac{dC}{dt}=-k_eC+\frac{R_0}{V}(1+\sin\omega t)$$
$$C(0)=0,$$

where C, V, and k_{el} are the drug concentration, volume of distribution, and elimination rate constant, respectively. The rate of infusion is k_0. Evaluate $C(t)$.

6.6. A constant rate of drug infusion (k_0) in a one-compartment model is described by

$$\frac{dC}{dt} = -k_{el}C + \frac{k_0}{V}$$
$$C(0) = C_0.$$

Determine $C(t)$.

6.7. The Laplace transform of a plasma drug concentration is given by

$$\bar{C}(s) = \frac{F \times \text{Dose} \times k_a}{V(s+k_a)(s+k_{el})},$$

where F is the bioavailability, k_a is the absorption rate constant, and k_{el} is the elimination rate constant.

(a) Find $C(t)$.

(b) Evaluate $C(t \to \infty)$.

(c) Derive the area under the concentration curve using Laplace transforms.

6.8. A one-compartment pharmacokinetic model with a ramp input is described by

$$\frac{dC}{dt} = -k_{el}C + \frac{m \times t}{V}$$
$$C(0) = 0.$$

Find $C(t)$.

6.9. The concentration profile derived in Section 6.6 is

$$C(x,t) = KC_d \left\{ \frac{h-x}{h} - \frac{2}{\pi} \sum_{n=1}^{\infty} \left[\frac{1}{n} \sin\left(\frac{n\pi x}{h}\right) e^{-n^2\pi^2(D/h^2)t} \right] \right\}.$$

(a) Evaluate the drug-delivery rate.

(b) Find the steady-state delivery rate.

(c) Derive the cumulative amount of drug released.

6.10. The Laplace transform of the amount of drug in the central compartment of a two-compartment model is given by

$$\bar{m}(s) = \frac{F \times \text{Dose} \times k_a(s+k_{21})}{(s+k_a)(s+\alpha)(s+\beta)}.$$

Find $m(t)$.

REFERENCES

1. Loney NW. *Applied Mathematical Methods for Chemical Engineers*. Boca Raton, FL: CRC, 2007.
2. Spiegel MR. *Schaum's Outline of Theory and Problems of Laplace Transforms*. New York: McGraw Hill, 1965.
3. Houk BE, Hochhaus G, Hughes JA. Kinetic modeling of plasmid DNA degradation in rat plasma. *AAPS PharmSci* 1999; 1:E9.
4. Burnette RR. Fundamental pharmacokinetic limits on the utility of using a sinusoidal drug delivery system to enhance therapy. *Journal of Pharmacokinetics and Pharmacodynamics* 1992; 20:477–500.

CHAPTER 7

TRANSFER FUNCTIONS

Continuous system transfer functions describe the input–output relationships in a process and offer a way to analyze its transient behavior without solving the governing linear differential equation. Procedures to derive these expressions include the applications of Laplace transforms. The final mathematical forms contain algebraic or transcendental functions of the Laplace variable s. Methodologies for calculating and analyzing transfer functions are essential for controller design.

7.1 INPUT–OUTPUT MODELS

In Chapter 2, input, output, and state variables were defined. The outputs were functions of the state and input variables. Many control applications involve models that only show the effects of stimuli (inputs) on process responses (outputs). A single-input and single-output (SISO) process in shown is Figure 7.1. These systems are usually described by differential equations (e.g., $d^2y/dt^2 + 2dy/dt + y = 3u$ and $d^2y/dt^2 + 2dy/dt + y = 3u + du/dt$, where u and y are the input and output variables, respectively). Given that most processes exhibit nonlinear dynamics, linearized differential equations are first written, using techniques described in Chapter 3, to allow the use of

Control of Biological and Drug-Delivery Systems for Chemical, Biomedical, and Pharmaceutical Engineering, First Edition. Laurent Simon.
© 2013 John Wiley & Sons, Inc. Published 2013 by John Wiley & Sons, Inc.

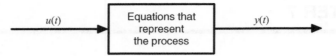

Figure 7.1. A single-input and single-output (SISO) process.

transfer functions. An example of multiple-input and multiple-output systems is given:

$$\frac{dy_1}{dt} = y_1 + y_2 + 2u_1 + 3u_2$$
$$\frac{dy_2}{dt} = 3y_1 - y_2 + 5u_1 + 3u_2,$$

(7.1)

where u_1 and u_2 are two input variables and y_1 and y_2 are output variables.

7.2 DERIVATION OF TRANSFER FUNCTIONS

Consider a linear input–output model defined by

$$a_2 \frac{d^2 y}{dt^2} + a_1 \frac{dy}{dt} + a_0 y = bu,$$

(7.2)

where the coefficients a_0, a_1, a_2, and b are independent of time. Equation (7.2) describes a *linear time invariant* (LTI) system. The transfer function is obtained by first taking the Laplace transform of both sides and solving for the ratio $\bar{y}(s)/\bar{u}(s)$. The functions $\bar{y}(s)$ and $\bar{u}(s)$ are the Laplace transforms of the deviation variables $y(t)$ and $u(t)$, respectively.

Applying Laplace transforms to both sides of Equation (7.2) yields

$$a_2 \left[s^2 \bar{y}(s) - sy(0) - y'(0) \right] + a_1 \left[s\bar{y} - y(0) \right] + a_0 \bar{y}(s) = b\bar{u}(s).$$

(7.3)

Since deviation variables are used, the following equation is obtained:

$$a_2 s^2 \bar{y}(s) + a_1 s\bar{y}(s) + a_0 \bar{y}(s) = b\bar{u}(s).$$

(7.4)

Therefore, the transfer function is

$$G(s) = \frac{\bar{y}(s)}{\bar{u}(s)} = \frac{b}{a_2 s^2 + a_1 s + a_0}$$

(7.5)

When we consider that the transform of the unit impulse function is 1 ($\delta(t = 0) = 1$ and $\delta(t > 0) = 0$), the transfer function $G(s)$ can be defined as the Laplace transform of the response (i.e., $\bar{y}(s)$) to a unit impulse (i.e., $\bar{u}(s) = 1$).

Example 7.1 The procedure, outlined here, to find the transfer function of a SISO system can also be applied to other systems involving more input and/or output variables. Consider the energy balance for a tank oil heater in Figure 7.2:

$$\rho V c_p \frac{dT}{dt} = F \rho c_p (T_{in} - T) + UA(T_{st} - T), \tag{7.6}$$

where ρ (kg/m^3) is the fluid density; V (m^3) is the liquid volume in the tank; c_p (J/kg·K) is the specific heat of the fluid; F (m^3/s) is the volumetric flow rate of an outlet stream; T_{in} (K) and T (K) are the inlet and outlet stream temperatures, respectively; T_{st} (K) is the steam temperature; U (W/m^2·K) is the overall heat transfer coefficient; and A (m^2) is the heat transfer surface area. Equation (7.6) takes into account assumptions such as a uniform temperature in the tank and a constant U.

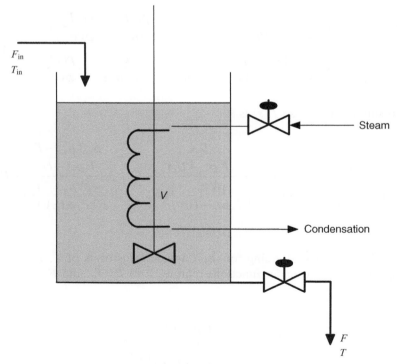

Figure 7.2. Diagram of a tank oil heater.

The input variables are F, T_{in}, and T_{st}, and the output variable is the outlet stream temperature T. Equation (7.6) should be linearized before applying Laplace transform methodologies. Algebraic manipulation of Equation (7.6) gives

$$\frac{dT}{dt} = \frac{F\rho c_p}{\rho V c_p}(T_{in} - T) + \frac{UA}{\rho V c_p}(T_{st} - T)$$

$$\frac{dT}{dt} = \frac{F\rho c_p}{\rho V c_p}T_{in} - \frac{F\rho c_p}{\rho V c_p}T + \frac{UA}{\rho V c_p}T_{st} - \frac{UA}{\rho V c_p}T. \tag{7.7}$$

Using the techniques outlined in Chapter 3, the above equation is linearized:

$$\frac{d\tilde{T}}{dt} = \frac{F_s\rho c_p}{\rho V c_p}\tilde{T}_{in} - \left(\frac{F_s\rho c_p}{\rho V c_p} + \frac{UA}{\rho V c_p}\right)\tilde{T} + \frac{UA}{\rho V c_p}\tilde{T}_{st} + \left(\frac{\rho c_p T_{in,s}}{\rho V c_p} - \frac{\rho c_p T_s}{\rho V c_p}\right)\tilde{F}, \quad (7.8)$$

where the subscript s is used to represent a steady-state point and $\tilde{x} = x - x_s$ is the deviation of a variable x from the steady-state x_s. After using Laplace transforms, we have

$$s\bar{T} = \frac{F_s\rho c_p}{\rho V c_p}\bar{T}_{in} - \left(\frac{F_s\rho c_p}{\rho V c_p} + \frac{UA}{\rho V c_p}\right)\bar{T} + \frac{UA}{\rho V c_p}\bar{T}_{st} + \left(\frac{\rho c_p T_{in,s}}{\rho V c_p} - \frac{\rho c_p T_s}{\rho V c_p}\right)\bar{F}$$

$$\left(s + \frac{F_s\rho c_p}{\rho V c_p} + \frac{UA}{\rho V c_p}\right)\bar{T} = \frac{F_s\rho c_p}{\rho V c_p}\bar{T}_{in} + \frac{UA}{\rho V c_p}\bar{T}_{st} + \left(\frac{\rho c_p T_{in,s}}{\rho V c_p} - \frac{\rho c_p T_s}{\rho V c_p}\right)\bar{F}. \tag{7.9}$$

As a result, the response $\bar{T}(s)$ is

$$\bar{T} = \frac{\dfrac{F_s\rho c_p}{F_s\rho c_p + UA}}{\left(\dfrac{\rho V c_p}{F_s\rho c_p + UA}\right)s + 1}\bar{T}_{in} + \frac{\dfrac{UA}{F_s\rho c_p + UA}}{\left(\dfrac{\rho V c_p}{F_s\rho c_p + UA}\right)s + 1}\bar{T}_{st} + \frac{\dfrac{\rho c_p(T_{in,s} - T_s)}{F_s\rho c_p + UA}}{\left(\dfrac{\rho V c_p}{F_s\rho c_p + UA}\right)s + 1}\bar{F},$$

$$\tag{7.10}$$

with $\bar{T}$, $\bar{T}_{in}$, $\bar{T}_{st}$ and $\bar{F}$ standing for the Laplace transforms of $\tilde{T}$, $\tilde{T}_{in}$, $\tilde{T}_{st}$, and $\tilde{F}$, respectively. The transfer functions relating $\bar{T}$ to $\bar{T}_{in}$, $\bar{T}_{st}$, and $\bar{F}$ are

$$G_{11}(s) = \frac{\dfrac{F_s\rho c_p}{F_s\rho c_p + UA}}{\left(\dfrac{\rho V c_p}{F_s\rho c_p + UA}\right)s + 1}, \tag{7.11}$$

$$G_{12}(s) = \frac{\dfrac{UA}{F_s\rho c_p + UA}}{\left(\dfrac{\rho V c_p}{F_s\rho c_p + UA}\right)s+1},$$ (7.12)

and

$$G_{13}(s) = \frac{\dfrac{\rho c_p\left(T_{in,s} - T_s\right)}{F_s\rho c_p + UA}}{\left(\dfrac{\rho V c_p}{F_s\rho c_p + UA}\right)s+1}.$$ (7.13)

A block diagram, representing the relationships, is shown in Figure 7.3.

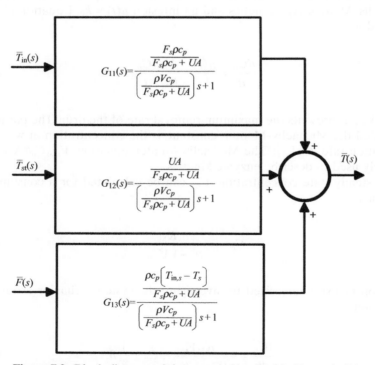

Figure 7.3. Block diagram of the system described in Example 7.1.

7.3 ONE- AND TWO-COMPARTMENT MODELS: MICHAELIS–MENTEN KINETICS

A constant-rate infusion is recommended for certain drugs to maintain the plasma drug concentration within a prescribed range. Transfer functions can be derived to study the effects of changes in the delivery rate on the drug concentrations in one- and two-compartment models.

7.3.1 A One-Compartment Model with Michaelis–Menten Elimination Kinetics

The equation for a constant-rate drug infusion (k_0) in a one-compartment model is given by

$$\frac{dC_p}{dt} = \frac{k_0}{V} - k_{el}C_p, \tag{7.14}$$

where C_p is the plasma drug concentration, k_{el} is the elimination rate constant, and V is the volume of distribution. The input variable is k_0 and the output is C_p with the initial concentration defined as $C_p(0) = 0$. If we consider Michaelis–Menten-type kinetics and an infusion $u(t) = k_0$, Equation (7.14) is changed to

$$\frac{dC_p}{dt} = \frac{u(t)}{V} - \frac{V_{max}C_p}{K_M + C_p}, \tag{7.15}$$

where V_{max} represents the maximum removal rate of the drug. The parameter K_M, called the Michaelis–Menten constant, is the concentration at which the elimination rate is $\frac{1}{2}V_{max}$. The Michaelis–Menten equation, $V_{max}C_p/(K_M + C_p)$, is mainly used to describe enzyme kinetics.

The steady-state concentration, C_{ps}, can be calculated for a constant-rate infusion, u_s:

$$C_{ps} = -\frac{u_s K_M}{u_s - V V_{max}}. \tag{7.16}$$

Equation (7.15) is linearized to investigate the system behavior around the steady state:

$$\frac{d\tilde{C}_p}{dt} = -\frac{K_M V_{max}}{\left(C_{ps} + K_M\right)^2}\tilde{C}_p + \frac{1}{V}\tilde{u}, \tag{7.17}$$

with $\tilde{C}_p = C_p - C_{ps}$ and $\tilde{u} = u - u_s$. The initial concentration $\tilde{C}_p(t=0)$ is zero. To determine the transfer function, the Laplace operator is applied to both sides of Equation (7.17) to give

$$s\bar{C}_p = -\frac{K_M V_{max}}{\left(C_{ps} + K_M\right)^2}\bar{C}_p + \frac{1}{V}\bar{u}$$

$$\left(s + \frac{K_M V_{max}}{\left(C_{ps} + K_M\right)^2}\right)\bar{C}_p = \frac{1}{V}\bar{u} \tag{7.18}$$

$$G(s) = \frac{\bar{C}_p}{\bar{u}} = \frac{\dfrac{1}{V}}{s + \dfrac{K_M V_{max}}{\left(C_{ps} + K_M\right)^2}}.$$

Consequently, the transfer function is

$$G(s) = \frac{\dfrac{1}{V}}{s + \dfrac{K_M V_{max}}{\left(C_{ps} + K_M\right)^2}}. \tag{7.19}$$

7.3.2 A Two-Compartment Model with Michaelis–Menten Elimination Kinetics

A two-compartment model is described by the following equations:

$$\frac{dC_I V_I}{dt} = -\left(k_1 + \frac{V_{max}}{K_M + C_I}\right)C_I V_I + k_2 C_{II} V_{II} + u(t) \tag{7.20}$$

and

$$\frac{dC_{II} V_{II}}{dt} = k_1 C_I V_I - k_2 C_{II} V_{II}. \tag{7.21}$$

The first-order mass transfer rate constants are represented by k_1 and k_2 (per hour); k_E is the elimination rate constant (per hour); V_I and V_{II} are the distribution volumes in the central and peripheral compartments, respectively; and C_I and C_{II} are the drug concentrations in the two compartments. If we use the mass (m) instead of the concentration (i.e., $m_I = C_I V_I$ and $m_{II} = C_{II} V_{II}$), Equations (7.20) and (7.21) become

$$\frac{dm_I}{dt} = -\left(k_1 + \frac{v_{max}}{k_M + m_I}\right)m_I + k_2 m_{II} + u(t) \tag{7.22}$$

and

$$\frac{dm_{II}}{dt} = k_1 m_I - k_2 m_{II}, \tag{7.23}$$

where $v_{\max} = V_I V_{\max}$ and $k_M = V_I K_M$.

The steady-state masses m_{Is} and m_{IIs} can be calculated for a constant-rate infusion, u_s:

$$m_{Is} = -\frac{k_M u_s}{u_s - v_{\max}} \tag{7.24}$$

and

$$m_{IIs} = -\frac{k_1 k_M u_s}{k_2 (u_s - v_{\max})}. \tag{7.25}$$

In terms of deviation variables, Equations (7.22) and (7.23) become

$$\frac{d\tilde{m}_I}{dt} = -\left(k_1 + \frac{v_{\max}}{(k_M + m_I)^2} \right) \tilde{m}_I + k_2 \tilde{m}_{II} + \tilde{u}(t) \tag{7.26}$$

and

$$\frac{d\tilde{m}_{II}}{dt} = k_1 \tilde{m}_I - k_2 \tilde{m}_{II}, \tag{7.27}$$

with $\tilde{m}_I = m_I - \tilde{m}_{Is}$, $\tilde{m}_{II} = m_{II} - \tilde{m}_{IIs}$, and $\tilde{u} = u - u_s$. The initial masses $\tilde{m}_I(t = 0)$ and $\tilde{m}_{II}(t = 0)$ are set to zero. To simplify Equation (7.26), we use

$$k_{E1} = \frac{v_{\max}}{(k_M + m_I)^2}. \tag{7.28}$$

Equations (7.26) and (7.27) are transformed to

$$s\bar{m}_I(s) = -(k_1 + k_{E1})\bar{m}_I(s) + k_2 \bar{m}_{II}(s) + \bar{U}(s) \tag{7.29}$$

and

$$s\bar{m}_{II}(s) = k_1 \bar{m}_I(s) - k_2 \bar{m}_{II}(s). \tag{7.30}$$

From Equation (7.30),

$$\bar{m}_{II}(s) = \frac{k_1}{(s+k_2)}\bar{m}_I(s). \tag{7.31}$$

Plugging Equation (7.31) into Equation (7.29) and rearranging the terms yield the transfer functions relating

$$\begin{bmatrix} \bar{m}_I(s) \\ \bar{m}_{II}(s) \end{bmatrix}$$

to $\bar{U}(s)$:

$$\frac{\bar{m}_I(s)}{\bar{U}(s)} = \frac{1}{\left[s + (k_1 + k_{E1}) - \dfrac{k_1 k_2}{(s+k_2)} \right]} = \frac{(s+k_2)}{(s+k_2)\left[s + (k_1 + k_{E1}) - \dfrac{k_1 k_2}{(s+k_2)} \right]}$$

$$= \frac{(s+k_2)}{\left[s(s+k_2) + (k_1 + k_{E1})(s+k_2) - k_1 k_2 \right]} = \frac{(s+k_2)}{\left[s^2 + (k_1 + k_2 + k_{E1})s + k_{E1}k_2 \right]} \tag{7.32}$$

and

$$\frac{\bar{m}_{II}(s)}{\bar{U}(s)} = \frac{k_1}{(s+k_2)\left[s + (k_1 + k_{E1}) - \dfrac{k_1 k_2}{(s+k_2)} \right]} = \frac{k_1}{\left[s^2 + (k_1 + k_2 + k_{E1})s + k_{E1}k_2 \right]}. \tag{7.33}$$

The transfer functions are

$$G_{11}(s) = \frac{\bar{m}_I(s)}{\bar{U}(s)} = \frac{(s+k_2)}{\left[s^2 + (k_1 + k_2 + k_{E1})s + k_{E1}k_2 \right]} \tag{7.34}$$

and

$$G_{12} = \frac{\bar{m}_{II}(s)}{\bar{U}(s)} = \frac{k_1}{\left[s^2 + (k_1 + k_2 + k_{E1})s + k_{E1}k_2 \right]}. \tag{7.35}$$

7.4 CONTROLLED-RELEASE SYSTEMS

The following dimensionless equation was obtained in Section 6.6 for the homogeneous membrane model of transdermal drug delivery:

$$\frac{\partial U}{\partial \tau} = \frac{\partial^2 U}{\partial X^2} \tag{7.36}$$

with the following boundary conditions:

$$U = 1, \quad X = 0, \quad \tau \geq 0 \tag{7.37}$$

$$U = 0, \quad X = 1, \quad t \geq 0, \tag{7.38}$$

and initial condition,

$$U = 0, \quad 0 < X < 1, \quad \tau = 0. \tag{7.39}$$

The Laplace transform of the dimensionless concentration was

$$\bar{U}(X, s) = \frac{1}{s} \left\{ \frac{\sinh\left[\sqrt{s}(1 - X)\right]}{\sinh(\sqrt{s})} \right\}. \tag{7.40}$$

To derive the transfer function between $\bar{U}(X, s)$ and $\bar{U}(0, s)$, $1/s$ in Equation (7.40) is replaced by $\bar{U}(0, s)$:

$$\bar{U}(X, s) = \bar{U}(0, s) \left\{ \frac{\sinh\left[\sqrt{s}(1 - X)\right]}{\sinh(\sqrt{s})} \right\}. \tag{7.41}$$

Consequently, the transfer function is

$$G(X, s) = \frac{\sinh\left[\sqrt{s}(1 - X)\right]}{\sinh(\sqrt{s})}. \tag{7.42}$$

7.5 SUMMARY

Transfer functions were derived for systems involving the transport of drugs into the body or through a biological membrane. The methodologies applied to develop these functions use Laplace transforms and are suitable for both ordinary and partial differential equations (PDEs). Block diagrams, which include the transfer functions, can be drawn to show the relationships between input and output variables.

PROBLEMS

7.1. What is the transfer function of the following system?

$$\frac{d^2 y}{dt^2} + \frac{dy}{dt} + 3y = 2u + \frac{du}{dt},$$

where u and y are the input and output, respectively. Note: $u(0) = y(0) = dy(0)/dt(0) = 0$.

7.2. What is the transfer function of the following system with the time delay t_0?

$$2\frac{dy}{dt} + 3y = u(t - t_0)\psi(t - t_0),$$

where u is the input, y is the output, and $\psi(t - t_0)$ is the unit step function. *Note*: $u(0) = y(0) = 0$.

7.3. A system is described by

$$\begin{bmatrix} \dfrac{dx_1}{dt} \\ \dfrac{dx_2}{dt} \end{bmatrix} = \begin{bmatrix} 0 & 1 \\ -1 & -0.5 \end{bmatrix} \begin{bmatrix} x_1 \\ x_2 \end{bmatrix} + \begin{bmatrix} 0 \\ 2 \end{bmatrix} u,$$

where u is the input. Find the transfer functions relating x_1 and x_2 to u. *Note*: $u(0) = x_1(0) = x_2(0) = 0$.

7.4. The component balances of reactant A and product B are given by

$$\frac{dC_A}{dt} = \frac{F_i}{V}(C_{Ai} - C_A) - k_1 C_A + k_2 C_B$$

$$\frac{dC_B}{dt} = -\frac{F_i}{V}C_B + k_1 C_A - k_2 C_B,$$

with $F_i(0) = C_{Ai}(0) = C_A(0) = C_B(0) = 0$. The inputs are F_i and C_{Ai}.
(a) Linearize the system around the equilibrium point: $(F_{ie}, C_{Aie}, C_{Ae}, C_{Be})$.
(b) Using the linearized system, determine the transfer functions relating the state variables C_A and C_B to F_i and C_{Ai}.

7.5. The heat balance in a stirred-tank heater is given by (see Example 7.1)

$$\rho V c_p \frac{dT}{dt} = F\rho c_p(T_{in} - T) + UA(T_{st} - T).$$

The input variables are T_{in} and T_{st}; the output variable is the outlet stream temperature T; F is kept constant. Write the transfer functions relating $\overline{T}$ to $\overline{T}_{in}$ and $\overline{T}_{st}$.

7.6. The component balances in a two-compartment model with a constant-rate infusion in the central compartment are (see Section 2.3.2)

$$\frac{dC_1}{dt} = \frac{k_0}{V_1} - (k_{el} + k_{12})C_1 + k_{21}C_2\frac{V_2}{V_1}$$

and

$$\frac{dC_2}{dt} = k_{12}C_1\frac{V_1}{V_2} - k_{21}C_2.$$

(a) Find the equilibrium concentrations C_{1s} and C_{2s}. The input variable is k_0.

(b) Write the transfer functions relating C_1 and C_2 to k_0.

7.7. The mass balances in a two-compartment model with a constant-rate infusion in the central compartment are

$$\frac{dm_I}{dt} = -(k_1 + k_{1e})m_I + k_2 m_{II} + u(t)$$

and

$$\frac{dm_{II}}{dt} = k_1 m_I - k_2 m_{II} - k_{2e} m_{II}.$$

The initial masses $m_I(t = 0)$ and $m_{II}(t = 0)$ are set to zero. Write the transfer functions relating m_I and m_{II} to u.

7.8. The component balances in two well-mixed tanks connected in series are

$$V_1\frac{dc_1}{dt} = F_0(c_0 - c_1)$$

and

$$V_2\frac{dc_2}{dt} = F_0(c_1 - c_2),$$

where F_0, V_1, and V_2 remain unchanged. The initial concentrations are $c_1(t = 0) = 0$ and $c_2(t = 0) = 0$.

(a) Write the transfer function relating c_1 and c_0.

(b) Write the transfer function relating c_2 and c_1.

(c) Write the transfer function relating c_2 and c_0.

7.9. Consider the PDE used to explain diffusion through a semi-infinite membrane:

$$\frac{\partial C(x,t)}{\partial t} = \frac{\partial^2 C(x,t)}{\partial x^2}, \quad x > 0, \quad t > 0,$$
$$C(0,t) = u(t), \quad C(x,0) = 0.$$

(a) Derive the transfer function relating $C(x,t)$ to u.

(b) Derive the transfer function relating $C(1,t)$ to u.

7.10. Consider the PDE used to explain diffusion through a semi-infinite membrane:

$$\frac{\partial C(x,t)}{\partial t} = D\frac{\partial^2 C(x,t)}{\partial x^2}, \quad x > 0, \quad t > 0,$$
$$C(0,t) = u(t), \quad C(x,0) = 0.$$

The parameter D is the diffusion coefficient of the permeant through the membrane. Derive the transfer function relating J to u, where

$$J = -D\frac{\partial C(x,t)}{\partial x}.$$

CHAPTER 8

DYNAMIC BEHAVIORS OF TYPICAL PLANTS

The dynamic behavior of a process plays an important role in the design of a controller. An understanding of how the plant responds to changes in the disturbances, or manipulated variables, will help an engineer devise a fitting control scheme. In this chapter, we will focus on typical plants, such as the models studied in Chapters 2, 5, and 7, and their characteristic features.

8.1 FIRST-, SECOND- AND HIGHER-ORDER SYSTEMS

8.1.1 First-Order Systems

A first-order differential equation describes the output of a first-order system:

$$a_1 \frac{dy}{dt} + a_0 y = bu(t), \qquad (8.1)$$

where $u(t)$ is the input and a_0, a_1, and b represent time-invariant coefficients. Equation (8.1) can be written, in standard form, as

$$\tau_p \frac{dy}{dt} + y = K_p u(t) \qquad (8.2)$$

Control of Biological and Drug-Delivery Systems for Chemical, Biomedical, and Pharmaceutical Engineering, First Edition. Laurent Simon.
© 2013 John Wiley & Sons, Inc. Published 2013 by John Wiley & Sons, Inc.

with

$$\tau_p = \frac{a_1}{a_0}; \quad K_p = \frac{b}{a_0}. \tag{8.3}$$

The constants τ_p and K_p are the *time constant* and *steady-state gain* (also called *static gain* or the *gain*) of the process, respectively. Assuming deviation variables, the standard form transfer function of the system described by Equation (8.2) is

$$G(s) = \frac{\bar{y}(s)}{\bar{u}(s)} = \frac{K_p}{\tau_p s + 1}. \tag{8.4}$$

If $a_0 = 0$, Equation (8.1) becomes

$$a_1 \frac{dy}{dt} = bu(t). \tag{8.5}$$

In this case, the transfer function is

$$G(s) = \frac{\bar{y}(s)}{\bar{u}(s)} = \frac{K_{1p}}{s}, \tag{8.6}$$

with $K_{1p} = b/a_1$. The process defined by Equation (8.5), or the transfer function given by Equation (8.6), is a *pure integrator*.

The transfer function represented by Equation (8.4) is that of a system that has the capacity to collect mass, momentum, or energy. Examples of such processes include a water reservoir and an energy storage unit.

8.1.2 Second-Order Systems

A second-order differential equation is

$$a_2 \frac{d^2 y}{dt^2} + a_1 \frac{dy}{dt} + a_0 y = bu(t). \tag{8.7}$$

Equation (8.7) is often used to model a mass–spring–damper system derived using Newton's second law (Fig. 8.1). In this representation, a_2 is the mass of an object attached to the spring and damper; a_1 is the coefficient of viscous friction (i.e., damping coefficient); a_0 is the spring constant; $bu(t)$ is an external force applied to the system; and y is the displacement. The coefficient b may stand for an area and $u(t)$ is an applied pressure. A damper is any device, or effect, that can help reduce vibrations, or oscillations, in a system. This definition and the spring–mass–damper analogy will help the reader better understand the influence of the model parameters on the transient behavior of the process.

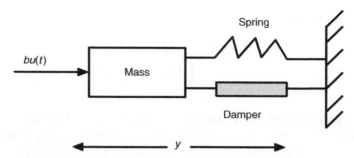

Figure 8.1. The mass–spring–damper system.

The standard form of Equation (8.7) is

$$\tau^2 \frac{d^2 y}{dt^2} + 2\varsigma\tau \frac{dy}{dt} + y = K_p u(t), \tag{8.8}$$

where $\tau = a_2/a_0$, $2\varsigma\tau = a_1/a_0$, and $K_p = b/a_0$. The transfer function of the system described by Equation (8.8) is

$$G(s) = \frac{\overline{y}(s)}{\overline{u}(s)} = \frac{K_p}{\tau^2 s^2 + 2\varsigma\tau s + 1}. \tag{8.9}$$

Just as in the case of a first-order system, the variables are written in deviation forms. The constant τ is the *characteristic time*, ς is the *damping ratio* (or *damping factor*), and K_p is the steady-state gain. Note that ς increases with the damping coefficient in the spring–mass–damper system (a_1). As a result, small values of ς lead to large oscillations.

8.1.3 Higher-Order Systems

8.1.3.1 Combination of First-Order Systems Some higher-order systems are obtained by combining first-order systems. The transfer function is given by

$$G(s) = \frac{\overline{y}(s)}{\overline{u}(s)} = \frac{K_1 K_2 \cdots K_n}{(\tau_1 s + 1)(\tau_2 s + 1) \cdots (\tau_n s + 1)} = \prod_{i=1}^{n} \frac{K_i}{(\tau_i s + 1)}, \tag{8.10}$$

where K_i and τ_i are the gain and time constant of process i. If an overall gain K is used, such that

$$K = \prod_{i=1}^{n} K_i, \tag{8.11}$$

the transfer function is

$$G(s) = \frac{K}{\prod_{i=1}^{n}(\tau_i s + 1)}. \tag{8.12}$$

8.1.3.2 Systems with Time Delays

A time delay (or *dead time*) is usually associated with the motion of material through a process. Plug flow motion through a pipe is represented by t_d, defined as the length of the pipe divided by the fluid velocity [1]. To represent the transfer function of such systems, recall from Chapter 5 the definition of a translation function $g(t)$:

$$g(t) = y(t - t_d)\psi(t - t_d), \tag{8.13}$$

where

$$\psi(t - t_0) = \begin{cases} 0 & t < t_0 \\ 1 & t_0 \geq t_0. \end{cases} \tag{8.14}$$

Equation (8.13) shows a response $y(t)$ delayed by t_d time units. The Laplace transform of $g(t)$ was found to be

$$\mathcal{L}\{g(t)\} = \overline{g}(s) = e^{-st_d}\overline{y}(s). \tag{8.15}$$

The transfer function between $\overline{u}(s)$ and the delayed response $\overline{g}(s)$ for a process with transfer function $G_p(s)$ is

$$G(s) = \frac{\overline{g}(s)}{\overline{u}(s)} = \frac{\overline{g}(s)}{\overline{y}(s)}\frac{\overline{y}(s)}{\overline{u}(s)} = e^{-st_d}G_p(s) \tag{8.16}$$

or

$$G(s) = G_p(s)e^{-st_d}. \tag{8.17}$$

If $G_p(s)$ represents, for example, a first-order process, we have

$$G(s) = \frac{K_p}{\tau s + 1}e^{-st_d}. \tag{8.18}$$

Padé approximations are widely used to write the time delay as a rational function. A first-order approximation yields

$$e^{-st_d} \approx \frac{1 - \dfrac{t_d}{2}s}{1 + \dfrac{t_d}{2}s}, \tag{8.19}$$

while a second-order approximation is

$$e^{-st_d} \approx \frac{t_d^2 s^2 - 6t_d s + 12}{t_d^2 s^2 + 6t_d s + 12}. \tag{8.20}$$

The nth-order Padé approximation is given by

$$e^{-st_d} \approx \frac{\sum_{k=0}^{n} \frac{(2n-k)!}{k!(n-k)!}(-st_d)^k}{\sum_{k=0}^{n} \frac{(2n-k)!}{k!(n-k)!}(st_d)^k}. \tag{8.21}$$

8.2 REDUCED-ORDER MODELS

Higher-order dynamics can be estimated with lower-order models. Several methods, using Laplace transforms, are available in the literature [2–4].

8.2.1 Removal of Insignificant Poles

One of the simplest approaches to construct reduced-order models is to remove insignificant poles from the transfer function. These poles have negative real parts and are much larger than the remaining poles [3]. Caution should be taken to maintain the steady-state gain because it affects the ultimate response of the system.

Example 8.1   Consider the following transfer function:

$$G(s) = \frac{\bar{y}(s)}{\bar{u}(s)} = \frac{K}{(s+20)(s^2+3s+2)}. \tag{8.22}$$

The poles of $G(s)$ are -1, -2, and -20. To eliminate the effects of the pole at -20 on the dynamic response, we first write Equation (8.22) as

$$G(s) = \frac{K}{20\left(\dfrac{s}{20}+1\right)(s^2+3s+2)}. \tag{8.23}$$

The fraction $s/20$ is neglected so that

$$G(s) \approx \frac{K}{20(s^2+3s+2)}. \tag{8.24}$$

This approach guarantees that the steady-state response is not affected by the approximation. For example, if $\bar{u}(s) = 1/s$, then $\bar{y}(s)$, calculated from Equation (8.23), is

$$\bar{y}(s) = \frac{K}{20\left(\dfrac{s}{20}+1\right)(s^2+3s+2)}\frac{1}{s}, \tag{8.25}$$

while the expression from Equation (8.24) is

$$\bar{y}(s) = \frac{K}{20(s^2+3s+2)}\frac{1}{s}. \tag{8.26}$$

Application of the final value theorem yields

$$\lim_{s \to 0}[sy(s)] = \frac{K}{40} \tag{8.27}$$

for the two processes. The responses are shown in Figure 8.2 for comparison.

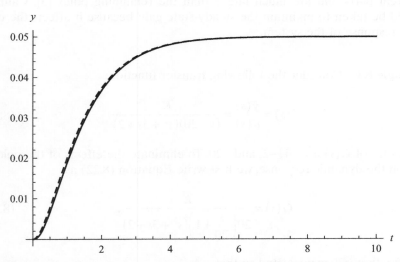

Figure 8.2. The system response to a step change in the input variable when the original (solid line) and approximated transfer functions (dashed line) are used. The gain K is set to 2.0.

8.2.2 Approximation with Lower-Order Functions and a Time Delay

Processes, consisting of a combination of first-order systems, such as the one represented by Equation (8.12), can be estimated by a second-order model if these systems are dominated by two time constants, τ_1 and τ_2:

$$G(s) \approx \frac{Ke^{-t_d s}}{(\tau_1 s + 1)(\tau_2 s + 1)} \tag{8.28}$$

where

$$t_d = \sum_{i=3}^{n} \tau_i. \tag{8.29}$$

A similar idea can be applied to generate first-order-plus-dead-time approximations [5].

Example 8.2 The transfer function

$$G(s) = \frac{4}{(0.1s + 1)(0.2s + 1)(1.0s + 1)} \tag{8.30}$$

can be approximated by

$$G(s) \approx \frac{4e^{-0.3s}}{1.0s + 1}. \tag{8.31}$$

The two responses are plotted in Figure 8.3.

8.3 TRANSCENDENTAL TRANSFER FUNCTIONS

In addition to the time delay, other transcendental functions are encountered in the modeling and control of distributed-parameter systems (i.e., processes described by partial differential equations). One of the major differences between rational and transcendental functions is that the latter usually contains an infinite number of poles. In the case of a heat flow in a rod [6],

$$C_p \rho \frac{\partial z(x, t)}{\partial t} = K_0 \frac{\partial^2 z(x, t)}{\partial x^2}, \quad x \in (0, L), \quad t \geq 0, \tag{8.32}$$

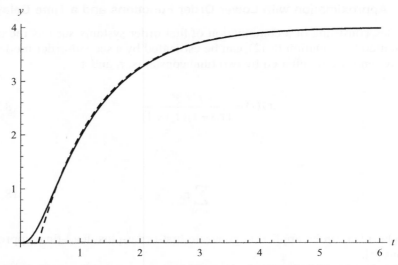

Figure 8.3. The system response to a step change in the input variable when the original (solid line) and approximated transfer functions (dashed line) are used.

with the following boundary conditions:

$$\frac{\partial z(0, t)}{\partial x} = 0 \tag{8.33}$$

and

$$K_0 \frac{\partial z(L, t)}{\partial x} = u(t), \tag{8.34}$$

it can be shown, following the procedure outlined in Chapter 7, that the transfer function is

$$G(x_0, s) = \frac{\overline{z}(x_0, s)}{\overline{u}(s)} = \frac{\alpha \cosh\left(\dfrac{x_0\sqrt{s}}{\alpha}\right)}{K_0\sqrt{s}\sinh\left(\dfrac{L\sqrt{s}}{\alpha}\right)}. \tag{8.35}$$

The rod length is L, the thermal conductivity is K_0, and the specific heat capacity is C_p. In Equation (8.35), the thermal diffusivity is

$$\alpha = \sqrt{\frac{K_0}{C_p\rho}}.$$

Equation (8.35) represents the ratio of the temperature z measured at a point x_0 along the rod to the rate of heat flow u into the rod. Both variables are represented by their Laplace transforms $\bar{z}(x_0, s)$ and $\bar{u}(s)$. The poles of $G(x_0, s)$ are $s = 0$ and $s = -(n\pi\alpha/L)^2$ with $n = 1, 2, 3 \ldots$.

8.4   TIME RESPONSES OF SYSTEMS WITH RATIONAL TRANSFER FUNCTIONS

8.4.1   Dynamic Response of First-Order Systems

The response of a first-order lag system with transfer functions described by Equation (8.4),

$$\frac{\bar{y}(s)}{\bar{u}(s)} = \frac{K_p}{\tau_p s + 1}, \tag{8.36}$$

is

$$\bar{y}(s) = \left(\frac{K_p}{\tau_p s + 1}\right)\frac{U}{s} \tag{8.37}$$

after introducing a step change in $\bar{u}(s)$ of magnitude U. The response is

$$y(t) = UK_p\left(1 - e^{-t/\tau_p}\right). \tag{8.38}$$

A plot of $y(t)/UK_p$ versus t/τ_p is shown in Figure 8.4. The maximum rate of change occurs immediately after applying the step change. A steady-state response is obtained by applying the final value theorem:

$$\lim_{t\to\infty} y(t) = UK_p\left(1 - e^{-t/\tau_p}\right) = UK_p. \tag{8.39}$$

For a unit step change, the ultimate value of $y(t)$ is the steady state or static gain K_p. Table 8.1 shows that the response reaches 63.2% of its final value after one time constant. In four time constants, the system is assumed to reach a steady state.

The Laplace transform of the response to a *ramp input* $u(t) = at, t \geq 0$ is given by

$$\bar{y}(s) = \left(\frac{K_p}{\tau_p s + 1}\right)\frac{a}{s^2}. \tag{8.40}$$

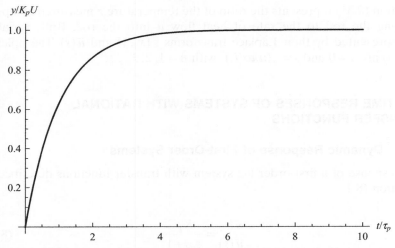

Figure 8.4. Normalized response of a first-order lag to a step change in the input variable.

TABLE 8.1 Normalized First-Order Response

t/τ_p	y/K_pU
0.0	0.000
1.0	0.632
2.0	0.865
3.0	0.950
4.0	0.982
5.0	0.993
6.0	0.998
7.0	0.999
8.0	1.000
9.0	1.000
10.0	1.000

Using the Laplace transform inversion techniques in Chapter 6, the output variable is

$$y(t) = aK_p\left(t - \tau_p + e^{-t/\tau_p}\tau_p\right). \tag{8.41}$$

The response to a ramp input becomes (Fig. 8.5)

$$y(t) = aK_p\left(t - \tau_p\right) \tag{8.42}$$

over a long period of time.

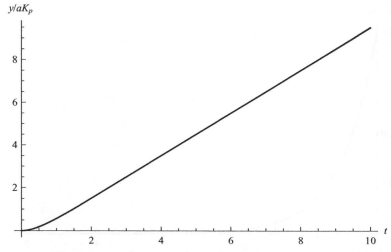

Figure 8.5. Normalized ramp response of a first-order lag ($\tau_p = 0.5$).

The *impulse function* is another typical test signal:

$$u(t) = A\delta(t), \tag{8.43}$$

where A is the area. The Laplace transform is $\bar{u}(s) = A$. The Laplace transform of the response is

$$\bar{y}(s) = \left(\frac{K_p}{\tau_p s + 1} \right) A. \tag{8.44}$$

As a result,

$$y(t) = \frac{AK_p}{\tau_p} e^{-t/\tau_p} \tag{8.45}$$

and

$$\lim_{t \to \infty} y(t) = \frac{AK_p}{\tau_p} e^{-t/\tau_p} = 0 \tag{8.46}$$

as illustrated in Figure 8.6.

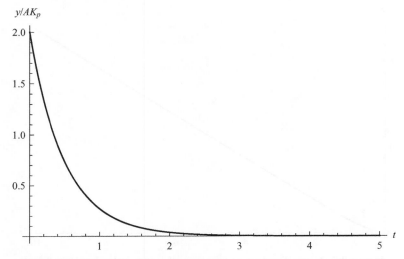

Figure 8.6. Normalized impulse response of a first-order lag ($\tau_p = 0.5$).

8.4.2 Dynamic Response of Second-Order Systems

The relationship between input and output variables using Laplace transforms is

$$\overline{y}(s) = \frac{K_p}{\tau^2 s^2 + 2\varsigma\tau s + 1}\overline{u}(s). \tag{8.47}$$

A step change of size U leads to

$$\overline{y}(s) = \frac{K_p}{\tau^2 s^2 + 2\varsigma\tau s + 1}\frac{U}{s}. \tag{8.48}$$

To invert $\overline{y}(s)$, the simple poles are

$$a_1 = \frac{-\varsigma - \sqrt{-1 + \varsigma^2}}{\tau}, a_2 = \frac{-\varsigma + \sqrt{-1 + \varsigma^2}}{\tau}$$

(when $a_1 \neq a_2$) and $a_3 = 0$. Equation (8.48) can also be written as

$$\overline{y}(s) = \frac{K_p}{\tau^2 (s - a_1)(s - a_2)}\frac{U}{s}. \tag{8.49}$$

The residue theorem yields

$$\rho_1(t) = \lim_{s \to a_1}\left[(s-a_1)\frac{P(s)}{Q(s)}\right]e^{a_1 t}$$

$$\rho_1(t) = \lim_{s \to a_1}\left[(s-a_1)\frac{UK_p}{s\tau^2(s-a_1)(s-a_2)}\right]e^{a_1 t} \qquad (8.50)$$

$$\rho_1(t) = \frac{UK_p}{a_1\tau^2(a_1-a_2)}e^{a_1 t}.$$

Similarly,

$$\rho_2(t) = \lim_{s \to a_2}\left[(s-a_2)\frac{P(s)}{Q(s)}\right]e^{a_2 t}$$

$$\rho_2(t) = \lim_{s \to a_2}\left[(s-a_2)\frac{UK_p}{s\tau^2(s-a_1)(s-a_2)}\right]e^{a_2 t} \qquad (8.51)$$

$$\rho_2(t) = \frac{UK_p}{a_2\tau^2(a_2-a_1)}e^{a_2 t}$$

and

$$\rho_3(t) = \lim_{s \to a_3}\left[(s-a_3)\frac{P(s)}{Q(s)}\right]e^{a_3 t}$$

$$\rho_3(t) = \lim_{s \to 0}\left[s\frac{UK_p}{s\tau^2(s-a_1)(s-a_2)}\right]e^{0\times t} \qquad (8.52)$$

$$\rho_3(t) = \frac{UK_p}{\tau^2 a_1 a_2}.$$

As a result,

$$y(t) = \mathcal{L}^{-1}\{\bar{y}(s)\} = \rho_1(t)+\rho_2(t)+\rho_3(t) \qquad (8.53)$$

or

$$y(t) = \frac{UK_p}{a_1\tau^2(a_1-a_2)}e^{a_1 t} + \frac{UK_p}{a_2\tau^2(a_2-a_1)}e^{a_2 t} + \frac{UK_p}{\tau^2 a_1 a_2}. \qquad (8.54)$$

After replacing the expressions for a_1 and a_2 into Equation (8.54), we have

$$y(t) = UK_p\left[\frac{e^{\left(\left(-\varsigma-\sqrt{-1+\varsigma^2}\right)/\tau\right)t}}{2\left(-1+\varsigma^2+\varsigma\sqrt{-1+\varsigma^2}\right)} + \frac{e^{\left(\left(-\varsigma+\sqrt{-1+\varsigma^2}\right)/\tau\right)t}}{2\left(-1+\varsigma^2-\varsigma\sqrt{-1+\varsigma^2}\right)} + 1\right]. \qquad (8.55)$$

Applying the identity, $e^x = \cosh(x) + \sinh(x)$, Equation (8.55) becomes

$$y(t) = UK_p\left\{1 - e^{-\varsigma t/\tau}\left[\frac{(-1+\varsigma^2)\cosh\left(\frac{t\sqrt{-1+\varsigma^2}}{\tau}\right) + \varsigma\sqrt{-1+\varsigma^2}\sinh\left(\frac{t\sqrt{-1+\varsigma^2}}{\tau}\right)}{-1+\varsigma^2}\right]\right\}$$

$$y(t) = UK_p\left\{1 - e^{-\varsigma t/\tau}\left[\cosh\left(\frac{t\sqrt{-1+\varsigma^2}}{\tau}\right) + \varsigma\frac{\sqrt{-1+\varsigma^2}}{-1+\varsigma^2}\sinh\left(\frac{t\sqrt{-1+\varsigma^2}}{\tau}\right)\right]\right\},$$

(8.56)

where $\cosh(\bullet)$ and $\sinh(\bullet)$ are hyperbolic trigonometric functions. The two following cases are considered:

(a) If $\varsigma > 1$ (*overdamped response*), the two distinct poles, a_1 and a_2, are real and $y(t)$ can be further simplified:

$$y(t) = UK_p\left\{1 - e^{-\varsigma t/\tau}\left[\cosh\left(\frac{t\sqrt{-1+\varsigma^2}}{\tau}\right) + \frac{\varsigma}{\sqrt{-1+\varsigma^2}}\sinh\left(\frac{t\sqrt{-1+\varsigma^2}}{\tau}\right)\right]\right\}.$$

(8.57)

(b) If $0 \leq \varsigma < 1$ (*underdamped response*), the two distinct poles, a_1 and a_2, are complex and $y(t)$ is

$$y(t) = UK_p\left\{1 - e^{-\varsigma t/\tau}\left[\cosh\left(i\frac{t\sqrt{1-\varsigma^2}}{\tau}\right) + \varsigma\frac{i\sqrt{1-\varsigma^2}}{-1+\varsigma^2}\sinh\left(i\frac{t\sqrt{1-\varsigma^2}}{\tau}\right)\right]\right\}$$

$$y(t) = UK_p\left\{1 - e^{-\varsigma t/\tau}\left[\cos\left(\frac{t\sqrt{1-\varsigma^2}}{\tau}\right) + \varsigma\frac{\sqrt{1-\varsigma^2}}{1-\varsigma^2}\sin\left(\frac{t\sqrt{1-\varsigma^2}}{\tau}\right)\right]\right\}$$ (8.58)

$$y(t) = UK_p\left\{1 - e^{-\varsigma t/\tau}\left[\cos\left(\frac{t\sqrt{1-\varsigma^2}}{\tau}\right) + \frac{\varsigma}{\sqrt{1-\varsigma^2}}\sin\left(\frac{t\sqrt{1-\varsigma^2}}{\tau}\right)\right]\right\}.$$

Alternatively, $y(t)$ can be written as

$$y(t) = UK_p\left\{1 - \frac{e^{-\varsigma t/\tau}}{\sqrt{1-\varsigma^2}}\left[\sqrt{1-\varsigma^2}\cos\left(\frac{t\sqrt{1-\varsigma^2}}{\tau}\right) + \varsigma\sin\left(\frac{t\sqrt{1-\varsigma^2}}{\tau}\right)\right]\right\}$$

$$y(t) = UK_p\left\{1 - \frac{e^{-\varsigma t/\tau}}{\sqrt{1-\varsigma^2}}\sin(\omega t + \phi)\right\},$$

(8.59)

where

$$\omega = \frac{\sqrt{1-\varsigma^2}}{\tau} \text{ and } \phi = \tan^{-1}\left(\frac{\sqrt{1-\varsigma^2}}{\varsigma}\right) = \cos^{-1}\varsigma.$$

Proof of $\sin(\omega t + \phi) = \sqrt{1-\varsigma^2}\cos\left(\frac{t\sqrt{1-\varsigma^2}}{\tau}\right) + \varsigma\sin\left(\frac{t\sqrt{1-\varsigma^2}}{\tau}\right).$

The final expression in Equation (8.59) is obtained by considering a right triangle where $\sqrt{1-\varsigma^2}$ and ς are the respective lengths of the sides opposite and adjacent to angle ϕ. We have

$$\cos\phi = \frac{\varsigma}{\varsigma^2 + \left(\sqrt{1-\varsigma^2}\right)^2} = \varsigma$$

$$\sin\phi = \frac{\sqrt{1-\varsigma^2}}{\varsigma^2 + \left(\sqrt{1-\varsigma^2}\right)^2} = \sqrt{1-\varsigma^2}. \tag{8.60}$$

Then,

$$\sqrt{1-\varsigma^2}\cos\left(\frac{t\sqrt{1-\varsigma^2}}{\tau}\right) + \varsigma\sin\left(\frac{t\sqrt{1-\varsigma^2}}{\tau}\right)$$

$$= \sin\phi\cos\left(\frac{t\sqrt{1-\varsigma^2}}{\tau}\right) + \cos\phi\sin\left(\frac{t\sqrt{1-\varsigma^2}}{\tau}\right) \tag{8.61}$$

$$= \sin\left(\frac{t\sqrt{1-\varsigma^2}}{\tau} + \phi\right) = \sin(\omega t + \phi).$$

End of proof.

If $\varsigma = 1$ (*critically damped response*), that is, $a_1 = a_2$, $\bar{y}(s)$ contains a pole of order 2 (or a *multiple pole*) at $-1/\tau$ and a simple pole $a_3 = 0$. At $a_3 = 0$, we have

$$\rho_3(t) = \lim_{s \to a_3}\left[(s-a_3)\frac{P(s)}{Q(s)}\right]e^{a_3 t}$$

$$\rho_3(t) = \lim_{s \to 0}\left[s\frac{UK_p}{s\tau^2(s+1/\tau)^2}\right]e^{0 \times t} \tag{8.62}$$

$$\rho_3(t) = UK_p.$$

The multiple pole leads to

$$\rho_1(t) = e^{a_1 t}\{A_1 + tA_2\} \tag{8.63}$$

with

$$A_1 = \lim_{s \to a_1} \frac{1}{(2-1)!} \frac{d^{2-1}}{ds^{2-1}} \left\{ (s-a_1)^2 \frac{UK_p}{s\tau^2(s-a_1)^2} \right\}$$

$$A_1 = UK_p \lim_{s \to -1/\tau} \frac{d}{ds} \left(\frac{1}{s\tau^2} \right)$$

$$A_1 = \frac{UK_p}{\tau^2} \lim_{s \to -1/\tau} \left(-\frac{1}{s^2} \right)$$

$$A_1 = -UK_p$$

(8.64)

and

$$A_2 = \lim_{s \to a_1} \frac{1}{(2-2)!} \frac{d^{2-2}}{ds^{2-2}} \left\{ (s-a_1)^2 \frac{UK_p}{s\tau^2(s-a_1)^2} \right\}$$

$$A_2 = UK_p \lim_{s \to -1/\tau} \left(\frac{1}{s\tau^2} \right)$$

$$A_2 = -\frac{UK_p}{\tau}.$$

(8.65)

As a result,

$$\rho_1(t) = -UK_p e^{-t/\tau} \left(1 + \frac{t}{\tau} \right).$$

(8.66)

The response is

$$y(t) = \rho_1(t) + \rho_3(t)$$

(8.67)

or

$$y(t) = -UK_p e^{-t/\tau} \left(1 + \frac{t}{\tau} \right) + UK_p$$

$$y(t) = UK_p \left[1 - \left(1 + \frac{t}{\tau} \right) e^{-t/\tau} \right].$$

(8.68)

Let us study the effects of the damping ratio on the response. A damper is used to reduce a system vibration. As a result, we expect that an increase in ς would lead to a decrease in the oscillations (Fig. 8.7). Several observations can be made concerning the responses displayed in Figure 8.7:

1. Oscillations are obtained for values of ς less than 1.
2. The response is *sluggish* when $\varsigma > 1$.
3. The response for a critically damped system ($\varsigma = 1$) is faster than any overdamped system.

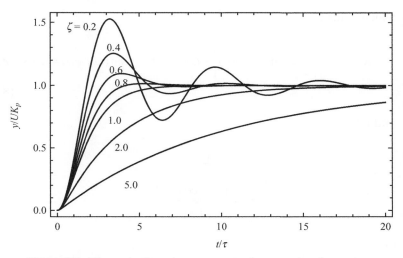

Figure 8.7. Dimensionless step response of a second-order system.

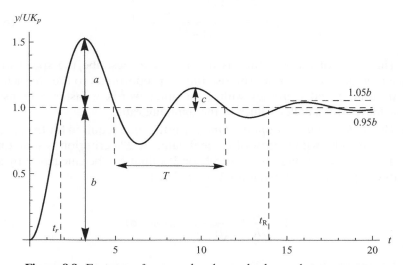

Figure 8.8. Features of a second-order underdamped step response.

Some characteristics of the underdamped response are defined below (Fig. 8.8):

1. The *rise time*, t_r, is the time when the process output first reaches the steady-state value. The rise time is computed by writing

$$y(t_r) = UK_p \left\{ 1 - \frac{e^{-\varsigma t_r / \tau}}{\sqrt{1-\varsigma^2}} \sin(\omega t_r + \phi) \right\} = y(\infty) = UK_p, \qquad (8.69)$$

which leads to

$$UK_p\left\{1-\frac{e-\varsigma t_r/\tau}{\sqrt{1-\varsigma^2}}\sin\left(\omega t_r+\phi\right)\right\}-UK_p=0 \qquad \sin\left(\omega t_r+\phi\right)=0=\sin\pi,$$

(8.70)

and finally,

$$t_r = \frac{\pi-\phi}{\omega}$$
$$t_r = \frac{\tau\left(\pi-\cos^{-1}\varsigma\right)}{\sqrt{1-\varsigma^2}}$$

(8.71)

or

$$t_r = \frac{\tau\left[\pi-\tan^{-1}\left(\frac{\sqrt{1-\varsigma^2}}{\varsigma}\right)\right]}{\sqrt{1-\varsigma^2}}.$$

(8.72)

The value of the rise time is often used to describe the speed of an underdamped response. The rise time is proportional to the characteristic time τ and increases with ς. In other words, a faster response is achieved for smaller values of the damping ratio.

2. The *settling time* (or *response time*), t_R, is the time required for the process output to be within ±5% of its final value. A 2% criterion is sometimes used for the settling time. The 5% or 2% limit can be calculated by first developing an expression for $1 - y(t)/UK_p$:

$$1-\frac{y(t)}{UK_p}=\frac{e^{-\varsigma t/\tau}}{\sqrt{1-\varsigma^2}}\sin\left(\omega t+\phi\right).$$

(8.73)

For the 5% limit,

$$\frac{e^{-\varsigma t_R/\tau}}{\sqrt{1-\varsigma^2}}=0.05$$
$$e^{-\varsigma t_R/\tau}=0.05\sqrt{1-\varsigma^2}$$
$$-\frac{\varsigma t_R}{\tau}=\ln\left(0.05\sqrt{1-\varsigma^2}\right)$$

(8.74)

and

$$t_R = -\frac{\tau}{\varsigma}\ln\left(0.05\sqrt{1-\varsigma^2}\right). \tag{8.75}$$

The settling time increases in proportion to the characteristic time τ. On the other hand, as ς increases, a smaller t_R is obtained. With a rise in the damping factor, the system reaches its final value in a shorter time.

3. The *peak overshoot* or *overshoot* (*OS*) is the ratio of the maximum amount by which the response exceeds the steady-state value to the ultimate value of the response: *a/b*. The *percent overshoot* is $100 \times OS$ and the time required to reach this maximum value is the *peak time* (t_p). An expression for t_p is obtained by first setting the derivative of the underdamped response to $dy/dt = 0$ and then solving for the time. Using Equation (8.59), we have

$$\frac{dy(t)}{dt} = -\frac{UK_p e^{-\varsigma t/\tau}[\tau\omega\cos(\omega t+\phi)-\varsigma\sin(\omega t+\phi)]}{\sqrt{1-\tau^2}\,\tau} = 0. \tag{8.76}$$

As a result,

$$\tau\omega\cos(\omega t+\phi)-\varsigma\sin(\omega t+\phi)=0 \tag{8.77}$$

or

$$\tau\omega\cos(\omega t+\phi)-\varsigma\sin(\omega t+\phi)=0$$
$$\tan(\omega t+\phi) = \frac{\tau\omega}{\varsigma} = \tau\frac{\sqrt{1-\zeta^2}}{\tau\varsigma} = \frac{\sqrt{1-\zeta^2}}{\varsigma}. \tag{8.78}$$

Since

$$\phi = \tan^{-1}\left(\frac{\sqrt{1-\zeta^2}}{\varsigma}\right),$$

we have

$$\tan(\phi) = \frac{\sqrt{1-\zeta^2}}{\varsigma} \tag{8.79}$$

and, consequently,

$$\tan(\omega t+\phi) = \tan(\phi), \tag{8.80}$$

which means that

$$\omega t = n\pi \quad n = 0, 1, 2, \dots. \tag{8.81}$$

The peak time corresponds to $n = 1$. Therefore,

$$t_\mathrm{p} = \frac{\pi}{\omega} = \frac{\pi\tau}{\sqrt{1-\zeta^2}}. \tag{8.82}$$

To calculate the overshoot, we first notice that

$$OS = \frac{a}{b} = \frac{y(t_\mathrm{p}) - y(\infty)}{y(\infty)}. \tag{8.83}$$

Since

$$y(t_\mathrm{p}) = UK_p\left\{1 - e^{-\varsigma/\tau\left(\pi\tau/\sqrt{1-\zeta^2}\right)}\left[\cos\left(\frac{\left(\frac{\pi\tau}{\sqrt{1-\zeta^2}}\right)\sqrt{1-\varsigma^2}}{\tau}\right) + \frac{\varsigma}{\sqrt{1-\varsigma^2}}\sin\left(\frac{\left(\frac{\pi\tau}{\sqrt{1-\zeta^2}}\right)\sqrt{1-\varsigma^2}}{\tau}\right)\right]\right\}$$

$$y(t_\mathrm{p}) = UK_p\left\{1 - e^{-\varsigma/\tau\left(\pi\tau/\sqrt{1-\zeta^2}\right)}\left[\cos(\pi) + \frac{\varsigma}{\sqrt{1-\varsigma^2}}\sin(\pi)\right]\right\}$$

$$y(t_\mathrm{p}) = UK_p\left(1 + e^{-\pi\varsigma/\sqrt{1-\varsigma^2}}\right), \tag{8.84}$$

the overshoot is

$$OS = \frac{UK_p\left[1 + e^{-\pi\varsigma/\sqrt{1-\zeta^2}}\right] - UK_p}{UK_p}, \tag{8.85}$$

which results in

$$OS = e^{-\pi\varsigma/\sqrt{1-\zeta^2}}. \tag{8.86}$$

As expected, from the definition of the damping factor, the overshoot decreases with increasing values of ς.

4. The *decay ratio* (DR) is the ratio of the amplitudes of the second oscillation to the height of the first peak:

$$DR = \frac{c}{a} = \frac{y(t_{p2}) - y(\infty)}{y(t_p) - y(\infty)},$$ (8.87)

where t_{p2} is the time the second peak occurs (see Eq. 8.81):

$$t_{p2} = \frac{3\pi}{\omega} = \frac{3\pi\tau}{\sqrt{1-\zeta^2}}.$$ (8.88)

As a result, the response at this time is given by

$$y(t_{p2}) = UK_p \left\{ 1 - e^{-\zeta/\tau\left(3\pi\tau/\sqrt{1-\zeta^2}\right)} \left[\cos(3\pi) + \frac{\zeta}{\sqrt{1-\zeta^2}} \sin(3\pi) \right] \right\}$$

$$y(t_{p2}) = UK_p \left(1 + e^{-3\pi\zeta/\sqrt{1-\zeta^2}} \right).$$ (8.89)

The decay ratio is

$$DR = \frac{UK_p \left[1 + e^{-3\pi\zeta/\sqrt{1-\zeta^2}} \right] - UK_p}{UK_p \left[1 + e^{-\pi\zeta/\sqrt{1-\zeta^2}} \right] - UK_p},$$ (8.90)

which simplifies to

$$DR = e^{-2\pi\zeta/\sqrt{1-\zeta^2}} = OS^2.$$ (8.91)

5. The *period of oscillation* (T) can be calculated as $t_{p2} - t_p$ (i.e., the time elapsed between two successive peaks):

$$T = \frac{3\pi\tau}{\sqrt{1-\zeta^2}} - \frac{\pi\tau}{\sqrt{1-\zeta^2}}$$

$$T = \frac{2\pi\tau}{\sqrt{1-\zeta^2}}.$$ (8.92)

The *frequency*, expressed in radian/time, is

$$\omega = \frac{2\pi}{T} = \frac{\sqrt{1-\zeta^2}}{\tau}. \tag{8.93}$$

The *cyclical frequency* (cycles/time) is

$$f = \frac{1}{T} = \frac{\sqrt{1-\zeta^2}}{2\pi\tau}. \tag{8.94}$$

If the time is measured in seconds, the unit of f is cycles/second or hertz. The *natural period of oscillation* is defined by setting $\varsigma = 0$:

$$T_n = 2\pi\tau, \tag{8.95}$$

and the corresponding *natural frequency* is

$$\omega_n = \frac{1}{\tau}. \tag{8.96}$$

In the absence of damping, sustained (or continuous) oscillations with a constant amplitude are observed. Note that for these systems, the *simple poles* are located on the imaginary axis.

Example 8.3   An underdamped second-order system has an overshoot of 10% after a step change in the input. Calculate the decay ratio and damping factor

Solution   Application of Equation (8.91) yields

$$DR = \left(\frac{10}{100}\right)^2 = \frac{1}{100}. \tag{8.97}$$

The damping factor is calculated by solving Equation (8.86):

$$\frac{10}{100} = e^{-\pi\varsigma/\sqrt{1-\zeta^2}}. \tag{8.98}$$

Therefore,

$$\zeta = \frac{\ln(10)}{\sqrt{\pi^2 + \ln^2(10)}} = 0.591. \tag{8.99}$$

Example 8.4 The period of oscillation of an underdamped second-order system is 100 seconds after a step change in the input. Calculate the frequency and the characteristic time if $\zeta = 0.4$.

Solution The frequency is obtained by applying Equation (8.93):

$$\omega = \frac{2\pi}{T} = \frac{2\pi}{100} = 0.0628 \text{ rad/s.} \tag{8.100}$$

From Equation (8.93), the characteristic time is

$$\tau = \frac{\sqrt{1-\zeta^2}}{\omega} = \frac{\sqrt{1-0.4^2}}{0.0628} = 14.59 \text{ seconds.} \tag{8.101}$$

8.4.3 Dynamic Response of General Linear Systems

Transfer functions of physical systems can be written as

$$G(s) = \frac{N(s)}{D(s)} = K \frac{(s-z_1)(s-z_2)\cdots(s-z_m)}{(s-p_1)(s-p_2)\cdots(s-p_m)} \tag{8.102}$$

where z_i and p_i denote the system *zeros* and *poles*, respectively, and K is the steady-state gain. The zeros are the solution of the equation $N(s) = 0$; the roots of the equation $D(s) = 0$ are the poles. Equation (8.102) is often written using the *gain–time constant* form:

$$G(s) = K' \frac{(\tau_a s+1)(\tau_b s+1)\cdots}{(\tau_1 s+1)(\tau_2 s+1)\cdots}. \tag{8.103}$$

The responses of these systems to changes in the input are found by the methods outlined in Chapter 6 and are influenced by both the zeros and the poles. Two cases are chosen to illustrate the effects of the zeros on the process dynamics: *lead–lag element* and systems with two *competing dynamic effects* [1, 5].

8.4.3.1 *Response of a Lead–Lag Element* The lead–lag element is a transfer function of the form

$$G(s) = K \frac{\tau_a s+1}{\tau_1 s+1}. \tag{8.104}$$

The characteristics of this transfer function will be explored in the chapter dealing with frequency response (e.g., phase lag). In this section, the response of the lead–lag element to a step change is explored. Equation (8.104) can be written as

$$\bar{y}(s) = K \frac{\tau_a s + 1}{\tau_1 s + 1} \bar{u}(s). \tag{8.105}$$

For a step change of size U, the response becomes

$$Y(s) = K \frac{\tau_a s + 1}{\tau_1 s + 1} \frac{U}{s}. \tag{8.106}$$

After inversion, $y(t)$ is given by

$$y(t) = KU \left[1 + e^{-t/\tau_1} \left(\frac{\tau_a}{\tau_1} - 1 \right) \right]. \tag{8.107}$$

To study the effects of τ_a on the response, a plot of $y(t)/KU$ is shown in Figure 8.9 for $\tau_1 = 3$ and $\tau_a = 7, 5, 3, 1, -1, -3, -5$. Some key observations are summarized below:

(a) The response depends on whether $0 < \tau_1 < \tau_a$, $0 < \tau_a < \tau_1$, or $\tau_a < 0 < \tau_1$.
(b) When $\tau_1 = \tau_a$, the normalized response is 1 and the transfer function is K. The constant transfer function is obtained after canceling the numerator and denominator (i.e., *pole–zero cancellation*).

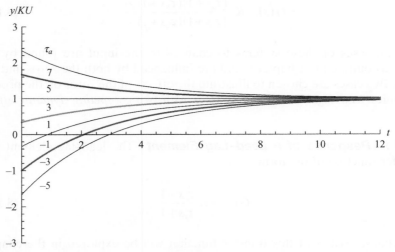

Figure 8.9. Step response of a lead–lag element: $\tau_1 = 3$ and $\tau_a = 7, 5, 3, 1, -1, -3, -5$.

(c) The initial response is $\tau_a/\tau_1\, KU$ (normalized response is τ_a/τ_1), although the transfer function was obtained using the following differential equation:

$$\tau_1 \frac{dy}{dt} + y = K\left(\tau_a \frac{dy}{dt} + u\right), \tag{8.108}$$

with $y(0) = 0$ and $u = 0$. The jump, or instantaneous response, occurs because the numerator and the denominator have the same order (e.g., first order). This illustration helps explain why, in the case of most physical processes, the order of the denominator (n) is greater than that of the numerator (m). Therefore, the condition $n \geq m$ is always applied to make sure that the system is *physically realizable*.

8.4.3.2 Systems Exhibiting Competing Dynamic Effects Processes, such as the one shown in Figure 8.10, can be described by a transfer function of the form

$$G(s) = \frac{K_1}{\tau_1 s + 1} + \frac{K_2}{\tau_2 s + 1}. \tag{8.109}$$

Competing dynamic effects may be observed due to the different time scales (i.e., τ_1 and τ_2). In such systems, the direction of the short-term response is

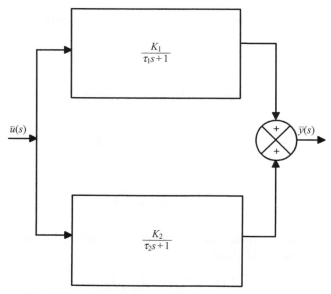

Figure 8.10. Block diagram of two parallel first-order processes.

opposite to that of the long-term behavior. Physical examples are distillation columns, tubular catalytic reactors, and multiple-stage membrane processes, such as continuous ultrafiltration of whey [7].

The transfer function can be represented by

$$G(s) = \frac{(K_1\tau_2 + K_2\tau_1)s + K_1 + K_2}{(\tau_1 s + 1)(\tau_2 s + 1)}$$

$$G(s) = \frac{(K_1 + K_2)\left[\left(\dfrac{K_1\tau_2 + K_2\tau_1}{K_1 + K_2}\right)s + 1\right]}{(\tau_1 s + 1)(\tau_2 s + 1)}. \tag{8.110}$$

After the substitutions $K = K_1 + K_2$ and $\tau = (K_1\tau_2 + K_2\tau_1)/(K_1 + K_2)$, the following expression is derived:

$$G(s) = \frac{K(\tau s + 1)}{(\tau_1 s + 1)(\tau_2 s + 1)}. \tag{8.111}$$

The response of such systems to a step change of size U in the input variable is

$$y(t) = KU\left(1 + \frac{\tau - \tau_1}{\tau_1 - \tau_2}e^{-t/\tau_1} + \frac{\tau - \tau_2}{\tau_2 - \tau_1}e^{-t/\tau_2}\right). \tag{8.112}$$

The steady-state response is

$$y(t \to \infty) = KU. \tag{8.113}$$

Let us analyze some characteristics of the dynamic response.

CASE 1: $y(t \to \infty)/U = K > 0$

Note that the response is positive. The initial slope of $y(t)/M$ is

$$d\left(\frac{y(t)}{U}\right)\Big/dt\Big|_{t=0} = K\left(-\frac{1}{\tau_1}\frac{\tau - \tau_1}{\tau_1 - \tau_2} - \frac{1}{\tau_2}\frac{\tau - \tau_2}{\tau_2 - \tau_1}\right) \tag{8.114}$$

or

$$d\left(\frac{y(t)}{U}\right)\Big/dt\Big|_{t=0} = \frac{\tau K}{\tau_1\tau_2} \tag{8.115}$$

after some routine algebraic manipulations. Substituting the expression for τ and K in Equation (8.115) leads to

$$d\left(\frac{y(t)}{U}\right)\Big/dt\Big|_{t=0} = \frac{K_1}{\tau_1} + \frac{K_2}{\tau_2}. \tag{8.116}$$

If

$$\frac{K_1}{\tau_1} + \frac{K_2}{\tau_2} < 0,$$

the slope is negative at $t = 0$. As a result, the system exhibits an *inverse response*, initially, as the step response takes on negative values, at short times, before approaching a positive steady-state value, K.

CASE 2: $y(t \rightarrow \infty)/U = K < 0$

If

$$\frac{K_1}{\tau_1} + \frac{K_2}{\tau_2} > 0,$$

the slope is positive at $t = 0$ and an inverse response is observed. From the results of Cases 1 and 2, an opposite behavior is expected when

$$\left(\frac{K_1}{\tau_1} + \frac{K_2}{\tau_2}\right)\Big/ K < 0 \text{ or } \frac{1}{\tau_1\tau_2}\left(\frac{K_1\tau_2 + K_2\tau_1}{K}\right) = \frac{\tau}{\tau_1\tau_2} < 0.$$

Because the time scales τ_1 and τ_2 are positive, τ *should be less than zero; that is, the transfer function* $G(s)$ *must have a positive zero for the system to show an inverse response for a step change in the input variable.* Figure 8.11 shows

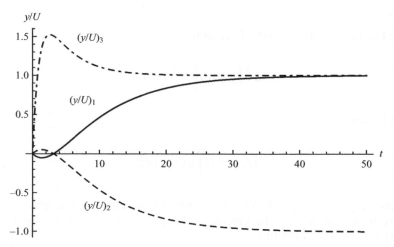

Figure 8.11. Step response of a second-order system with different time constants and a single zero: $(y/U)_1$: $K_1 = 2, K_2 = -1, \tau_1 = 8, \tau_2 = 3, \tau = -2$; $(y/U)_2$: $K_1 = -2, K_2 = 1, \tau_1 = 8, \tau_2 = 3, \tau = -2$; $(y/U)_3$: $K_1 = -\frac{4}{3}, K_2 = \frac{7}{3}, \tau_1 = 4, \tau_2 = 1, \tau = 8$.

the system described by Equation (8.112) for different time constants and gains. In addition, large values of τ can cause some overshoot as well.

8.5   TIME RESPONSES OF SYSTEMS WITH TRANSCENDENTAL TRANSFER FUNCTIONS

Considering the transfer function in Section 8.3, the response is

$$\bar{z}(x_0, s) = \frac{\alpha \cosh\left(\dfrac{x_0\sqrt{s}}{\alpha}\right)}{K_0\sqrt{s}\sinh\left(\dfrac{L\sqrt{s}}{\alpha}\right)}\bar{u}(s) \tag{8.117}$$

and, in particular, the temperature at the end of the rod for a step change of size U in the input variable is

$$\bar{z}(L, s) = G(L, s)\frac{U}{s} \tag{8.118}$$

with

$$G(L, s) = \frac{\alpha \cosh\left(\dfrac{L\sqrt{s}}{\alpha}\right)}{K_0\sqrt{s}\sinh\left(\dfrac{L\sqrt{s}}{\alpha}\right)}. \tag{8.119}$$

Using the Laplace transform of integrals

$$\mathcal{L}\left\{\int_0^t f(t)\,dt\right\} = \frac{1}{s}\bar{f}(s),$$

$z(L,t)$ becomes

$$z(L, t) = U\int_0^t g(L, \tau)\,d\tau. \tag{8.120}$$

The first step in finding $z(L,t)$ is to compute the inverse Laplace transform of $G(L,s)$ to obtain $g(L,t)$. To calculate the poles of $G(L,s)$, we first write $G(L,s) = P(s)/Q(s)$, such that

$$P(s) = \alpha \cosh\left(\frac{L\sqrt{s}}{\alpha}\right)$$

and

$$Q(s) = K_0 \sqrt{s} \sinh\left(\frac{L\sqrt{s}}{\alpha}\right),$$

then set $Q(s)$ equal to zero. The residue at $a_1 = 0$ is

$$\rho_0(t) = \lim_{s \to a_1}\left[(s - a_1)\frac{P(s)}{Q(s)}\right]e^{a_1 t}$$

$$\rho_0(t) = \lim_{s \to 0}\left[\frac{\alpha s \cosh\left(\dfrac{L\sqrt{s}}{\alpha}\right)}{K_0 \sqrt{s}\sinh\left(\dfrac{L\sqrt{s}}{\alpha}\right)}\right]e^{0t} \tag{8.121}$$

$$\rho_0(t) = \lim_{s \to 0}\left[\frac{\alpha s \cosh\left(\dfrac{L\sqrt{s}}{\alpha}\right)}{K_0 \sqrt{s}\sinh\left(\dfrac{L\sqrt{s}}{\alpha}\right)}\right].$$

An indeterminate form is obtained. L'Hospital's rule gives

$$\rho_0(t) = \frac{\alpha^2}{LK_0}. \tag{8.122}$$

The other residues at $a_n = -(n\pi\alpha/L)^2$ with $n = 1, 2, 3, \ldots$ are given by

$$\rho_n(t) = \frac{P(a_n)}{\left.\dfrac{dQ}{ds}\right|_{s=a_n}}e^{a_n t}, \tag{8.123}$$

where

$$\frac{dQ}{ds} = \frac{LK_0 \cosh\left(\dfrac{L\sqrt{s}}{\alpha}\right)}{2\alpha} + \frac{K_0 \sinh\left(\dfrac{L\sqrt{s}}{\alpha}\right)}{2\sqrt{s}}. \tag{8.124}$$

Consequently,

$$\left.\frac{dQ}{ds}\right|_{s=a_n} = \frac{LK_0 \cosh(n\pi i)}{2\alpha} \tag{8.125}$$

and

$$P(a_n) = \alpha \cosh(n\pi i). \tag{8.126}$$

The residues, given by Equation (8.123), simplify to

$$\rho_n(t) = \frac{2\alpha^2}{LK_0} e^{a_n t}. \tag{8.127}$$

Finally,

$$g(L,t) = \rho_0 + \sum_{n=1}^{\infty} \rho_n(t)$$

$$g(L,t) = \frac{\alpha^2}{LK_0} + \sum_{n=1}^{\infty} \left\{ \frac{2\alpha^2}{LK_0} e^{a_n t} \right\} \tag{8.128}$$

$$g(L,t) = \frac{\alpha^2}{LK_0} + \sum_{n=1}^{\infty} \left\{ \frac{2\alpha^2}{LK_0} e^{-(n\pi\alpha/L)^2 t} \right\}.$$

The temperature profile is

$$z(L,t) = U \int_0^t \left\{ \frac{\alpha^2}{LK_0} + \sum_{n=1}^{\infty} \left[\frac{2\alpha^2}{LK_0} e^{-(n\pi\alpha/L)^2 \tau} \right] \right\} d\tau$$

$$z(L,t) = U \left[\frac{\alpha^2}{LK_0} t + \frac{2L}{\pi^2 K_0} \sum_{n=1}^{\infty} \left(\frac{1 - e^{-(n\pi\alpha/L)^2 t}}{n^2} \right) \right]. \tag{8.129}$$

Other processes involving transcendental functions can be handled in a similar manner. Note that this procedure offers a simple way to approximate the transfer functions. Indeed, the Laplace transform of $g(L,t)$ from Equation (8.128) is

$$G_e(L,s) = \frac{\alpha^2}{LK_0 s} + \frac{2\alpha^2}{LK_0} \sum_{n=1}^{\infty} \left\{ \frac{1}{s + (n\pi\alpha/L)^2} \right\}$$

$$G_e(L,s) = \frac{\alpha^2}{LK_0 s} + \frac{2L\alpha^2}{K_0} \sum_{n=1}^{\infty} \left\{ \frac{1}{L^2 s + (n\pi\alpha)^2} \right\}. \tag{8.130}$$

8.6 BONE REGENERATION

Several theories have been proposed to explain the underlying mechanisms directing bone morphogenesis. While considerable efforts are devoted to the identification of degradable biocomposites that bind to bone tissue and accelerate healing, the modeling of fracture healing is also essential. These studies may help develop physical therapy programs.

A first-order model showing the relationship between mechanical stress stimulus (r) and fracture strength (c) during bone healing is proposed [8]:

$$T\frac{dc(t)}{dt} + c(t) = Kr(t), \tag{8.131}$$

where T and K stand for the time constant and gain, respectively. The variable r represents the proportion of stress on the healing bone to the pressure applied to the intact bone; c compares the percent of the ultimate bending strength of the healing bone to that of intact bone.

The transfer function is

$$G(s) = \frac{K}{Ts+1} \tag{8.132}$$

and the step response of size U is

$$c(t) = UK(1 - e^{-t/T}). \tag{8.133}$$

8.7 NITRIC OXIDE TRANSPORT TO PULMONARY ARTERIOLES

Gases, such as oxygen and nitric oxide (NO), play essential roles in the human body. Nitric oxide serves as a neurotransmitter that allows signals to be carried from one neuron to the next and a potent vasodilator that causes the relaxation of smooth muscles in the walls of vessels [9]. The full NO transport mechanism involves diffusion, convection, reaction, and complex interactions between the gas and hemoglobin.

Jeh and Georg present a dynamic model of NO inhalation and delivery to the pulmonary circulation [10]:

$$G_T(s) = \frac{\overline{c}_{sm}(s)}{\overline{c}_{feed}(s)} = G_{inh}(s)G_{art}(s), \tag{8.134}$$

where $G_T(s)$ is the overall transfer function relating the mean nitric oxide (NO) concentration in the smooth muscle cells of the arterioles ($c_{sm}(t)$) to the feed concentration of the inhaled gas ($c_{feed}(t)$). The Laplace transformed variables are represented by the overbars. The other transfer functions are

$$G_{inh}(s) = \frac{\overline{c}_{alv}(s)}{\overline{c}_{feed}(s)} \quad \text{and} \quad G_{art}(s) = \frac{\overline{c}_{sm}(s)}{\overline{c}_{alv}(s)}.$$

Note that the NO gas concentration in the alveolar region around the arteriole is $c_{alv}(t)$.

A first-order system is used for $G_T(s)$:

$$G_T(s) = \frac{K_p}{\tau s + 1}, \tag{8.135}$$

which allows the calculation of the overall response of $c_{sm}(t)$ (nM) to a step change of magnitude U (ppm) in $c_{feed}(t)$:

$$c_{sm}(t) = K_p(1 - e^{-t/\tau}). \tag{8.136}$$

The model easily predicts the steady-state concentration and the time it takes to reach this value after a step change in the inhaled gas concentration.

8.8   TRANSDERMAL DRUG DELIVERY

The transdermal drug-delivery system was described in Chapter 6 by the following equation:

$$\frac{\partial C}{\partial t} = D \frac{\partial^2 C}{\partial x^2}, \tag{8.137}$$

with the conditions $C = KC_d, x = 0, t \geq 0; C = 0, x = h, t \geq 0$ and $C = 0, 0 < x = h$, $t = 0$, where D is the drug diffusion coefficient in the stratum corneum. Note that C is the concentration; x is the distance; C_d is the donor concentration and K is the donor/stratum corneum partition coefficient. After the normalizations $U = C/KC_d$, $\tau = Dt/h^2$, and $X = x/h$, the following transfer function between the input $\bar{U}(0, s)$ and output $\bar{U}(X, s)$,

$$\bar{U}(X, s) = \left\{ \frac{\sinh\left[\sqrt{s}(1 - X)\right]}{\sinh(\sqrt{s})} \right\} \bar{U}(0, s), \tag{8.138}$$

is obtained. The solution, derived in Chapter 6 for $\bar{U}(0, s) = 1/s$, can be generalized for a step change of magnitude M:

$$U(X, \tau) = M \left\{ 1 - X - \frac{2}{\pi} \sum_{n=1}^{\infty} \left[\frac{1}{n} \sin(n\pi X) e^{-n^2\pi^2\tau} \right] \right\}. \tag{8.139}$$

8.9   SUMMARY

The dynamics of a range of physical processes are considered. Transfer functions describing these systems may include both rational and transcendental functions. Important transient characteristics, such as time constant, settling time, and damping factor are identified for lower-order systems. Lead–lag transfer functions and inverse response behaviors are analyzed following step changes in the input variable. The study of the transient responses for transcendental transfer functions is facilitated by the use of the residue theorem.

PROBLEMS

8.1. Derive the transfer function

$$G(x_0, s) = \frac{\overline{z}(x_0, s)}{\overline{u}(s)} = \frac{\alpha \cosh\left(\dfrac{x_0 \sqrt{s}}{\alpha}\right)}{K_0 \sqrt{s} \sinh\left(\dfrac{L\sqrt{s}}{\alpha}\right)} \overline{u}(s)$$

for the heated rod problem described in Section 8.3. Note that

$$z(x, 0) = 0.$$

8.2. Derive the transfer function

$$G(x_0, s) = \frac{\overline{z}(x_0, s)}{\overline{u}(s)}$$

for the heated rod in Section 8.3 using the following boundary conditions:

$$z(0, t) = 0$$

and

$$z(L, t) = u(t).$$

Note that

$$z(x, 0) = 0.$$

8.3. The energy balance for a thermocouple junction is given by

$$mC_p \frac{dT}{dt} = hA(T_0 - T),$$

where

m: mass of thermocouple junction
C_p: specific heat of the junction material
h: heat transfer coefficient
A: heat transfer area in the film
T_0: fluid temperature
T: junction temperature.

(a) Derive the transfer function relating T_0 to T.

(b) What is the time constant?

(c) What is the gain?

8.4. The following irreversible reaction occurs in a continuous stirred-tank reaction (CSTR):

$$A \rightarrow B,$$

where the reaction rate per unit volume is $r = kC_A$. The CSTR has a single feed stream and a single outlet stream; C_{A0} is the concentration of A in the feed. Assume that the volume remains constant.

(a) Derive the state equation dC_A/dt.

(b) Derive the gain and time constant of the process with input C_{A0} and output C_A.

8.5. The transfer function relating P_A (alveolar pressure) to P_{A0} (airway pressure) in a simplified lung mechanics model is given by

$$\frac{P_A(s)}{P_{A0}(s)} = \frac{1}{LCs^2 + RCs + 1},$$

where R and C account for resistance and compliance, respectively; L is the inductance. Identify the characteristic time, the damping ratio, and the steady-state gain.

8.6. Derive the ramp response of the second-order system:

$$G(s) = \frac{\bar{y}(s)}{\bar{u}(s)} = \frac{K_p}{\tau^2 s^2 + 2\varsigma\tau s + 1}.$$

Consider the critically damped case.

8.7. Derive the impulse response of the second-order system:

$$G(s) = \frac{\bar{y}(s)}{\bar{u}(s)} = \frac{K_p}{\tau^2 s^2 + 2\varsigma\tau s + 1}.$$

Consider the critically damped case.

8.8. Consider the following transfer function:

$$G(s) = \frac{\bar{y}(s)}{\bar{u}(s)} = \frac{5}{(s+30)(s^2 + 4s + 3)}.$$

(a) Approximate the transfer function by a second-order system so that the steady-state response remains the same.

(b) Plot the responses of the original and estimated systems for a unit step change in u.

8.9. Approximate the transfer function

$$G(s) = \frac{\overline{y}(s)}{\overline{u}(s)} = \frac{5}{(0.3s+1)(0.2s+1)(2.0s+1)}$$

by a first-order plus time-delay system. Plot the responses of the original and estimated systems for a step change in u of size 2.

8.10. Using the first-order Padé approximation, write the following transfer function as a quotient of two polynomials:

$$G(s) = \frac{\overline{y}(s)}{\overline{u}(s)} = \frac{e^{-3s}}{8s+1}.$$

Compare the first- and second-order Padé approximations by plotting the response $y(t)$ to a unit step change in $u(t)$.

REFERENCES

1. Seborg DE, Edgar TF, Mellichamp DA. *Process Dynamics and Control.* Hoboken, NJ: Wiley, 2004.
2. Papadourakis A, Doherty M, Douglas J. Approximate dynamic models for chemical process systems. *Industrial and Engineering Chemistry Research* 1989; 28:546–552.
3. Nagrath IJ, Gopal M. *Control Systems Engineering.* Anshan: Tunbridge Wells, 2008.
4. Simon L, Goyal A. Dynamics and control of percutaneous drug absorption in the presence of epidermal turnover. *Journal of Pharmaceutical Sciences* 2009; 98:187–204.
5. Chau PC. *Process Control: a First Course with MATLAB.* Cambridge: Cambridge University Press, 2002.
6. Curtain R, Morris K. Transfer functions of distributed parameter systems: a tutorial. *Automatica* 2009; 45:1101–1116.
7. Yee K, Alexiadis A, Bao J, Wiley DE. Effects of multiple-stage membrane process designs on the achievable performance of automatic control. *Journal of Membrane Science* 2008; 320:280–291.
8. Wang X, Zhang X, Li Z, Yu X. A first order system model of fracture healing. *Journal of Zhejiang University Science B* 2005; 6:926–930.
9. Truskey GA, Yuan F, Katz DF. *Transport Phenomena in Biological Systems.* Upper Saddle River, NJ: Pearson/Prentice Hall, 2004.
10. Jeh HS, Georg SC. Dynamic modeling and simulation of nitric oxide gas delivery to pulmonary arterioles. *Annals of Biomedical Engineering* 2002; 30:946–960.

CHAPTER 9

CLOSED-LOOP RESPONSES WITH P, PI, AND PID CONTROLLERS

The topics introduced in previous chapters (e.g., stability and dynamics) were designed to assist in the analysis and design of control systems. For example, in Chapter 1, specific control mechanisms that governed pupil light reflex or the regulation of human arterial blood pressure were outlined. The ability to analyze the behavior of these processes is a necessary step toward the development of devices intended to partially replace certain biological functions in case of failure. For blood glucose control by insulin in a nondiabetic person, a feedback control system that involves pancreatic beta cells is able to maintain glucose at a desired level. However, for a type 1 diabetic patient, the design of an external mechanism is critical, and its success depends largely on a basic understanding of the physiological process.

The analytical tools, discussed in Chapter 8, allow the control engineer to study the transient behavior of processes. Such methods make it possible to investigate and quantify how fast a bioreactor responds to changes in key limiting nutrients or airflow into a sparger. Proper pH control also requires an understanding of the effects of acid, or base, addition on the pH of the liquid medium. Failure to capture the relationships between input and output variables may lead to inadequate control strategies, reduce cell growth, and decrease product formation. In this chapter, the focus is placed on three commonly used feedback controllers: proportional (P), proportional–integral (PI), and proportional–integral–derivative (PID).

Control of Biological and Drug-Delivery Systems for Chemical, Biomedical, and Pharmaceutical Engineering, First Edition. Laurent Simon.
© 2013 John Wiley & Sons, Inc. Published 2013 by John Wiley & Sons, Inc.

9.1 BLOCK DIAGRAM OF CLOSED-LOOP SYSTEMS

The relationship between the input and output of a system is shown in a block diagram (Fig. 9.1). The transfer function that describes the connection is usually written in the interior of the block. In this context, each device in the process plant (e.g., sensor, controller) needs to be included in the block diagram of the feedback closed-loop system. The transient behaviors of these hardware elements will be considered when designing the controller. For example, the response time of a dissolved oxygen (DO) probe, or how fast it takes to measure the concentrations of key nutrients in real time, is an important factor that may influence controller performance.

Just like *open-loop systems* (i.e., no controller), arrows are drawn to show the direction in which information propagates. Signals are also added or subtracted by including summing points in the block diagram (Fig. 9.2). These rules will be applied in analyzing the closed-loop system. The transmission path between the comparator and $\bar{y}(s)$ makes up the *forward path*: G_c, G_f, and G_p; the *feedback path* is between $\bar{y}(s)$ and the comparator. Block G_m

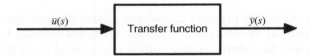

Figure 9.1. Block diagram of a single-input single-output system.

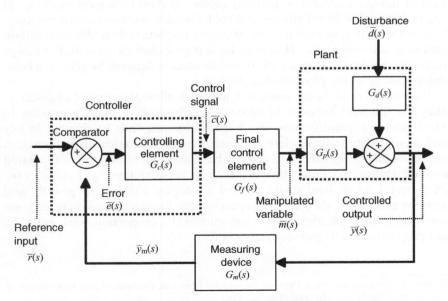

Figure 9.2. Block diagram of a feedback closed-loop system.

is located on the feedback path. Figure 9.2 is an example of a *negative-feedback* control system: $\bar{e}(s) = \bar{r}(s) - \bar{y}_m(s)$. In a positive-feedback control system, $\bar{e}(s) = \bar{r}(s) + \bar{y}_m(s)$.

A close examination of Figure 9.2 helps address two problems often encountered in feedback control systems: (1) How does the controlled variable $\bar{y}(s)$ respond to a change in the reference input $\bar{r}(s)$? and (2) How does $\bar{y}(s)$ respond to a change in the load (or disturbance) $\bar{d}(s)$? These questions can be answered by deriving two *closed-loop transfer functions* G_{sp} and G_{load} such that

$$\bar{y}(s) = G_{sp}(s)\bar{r}(s) + G_{load}(s)\bar{d}(s). \tag{9.1}$$

From the block diagram, the following equations can be written for the various blocks:

(a) *The Process*:

$$\bar{y}(s) = G_P(s)\bar{m}(s) + G_d(s)\bar{d}(s) \tag{9.2}$$

(b) *The Final Control Element*:

$$\bar{m}(s) = G_f(s)\bar{c}(s) \tag{9.3}$$

(c) *The Controlling Element*:

$$\bar{c}(s) = G_c(s)\bar{e}(s) \tag{9.4}$$

(d) *The Comparator*:

$$\bar{e}(s) = \bar{r}(s) - \bar{y}_m(s) = \bar{r}(s) - G_m(s)\bar{y}(s) \tag{9.5}$$

Incorporating Equation (9.3) into Equation (9.2) yields

$$\bar{y}(s) = G_P(s)G_f(s)\bar{c}(s) + G_d(s)\bar{d}(s). \tag{9.6}$$

Further substitution using Equations (9.4) and (9.5) gives

$$\bar{y}(s) = G_P(s)G_f(s)\{G_c(s)[\bar{r}(s) - G_m(s)\bar{y}(s)]\} + G_d(s)\bar{d}(s) \tag{9.7}$$

or

$$[1 + G_P(s)G_f(s)G_c(s)G_m(s)]\bar{y}(s) = G_P(s)G_f(s)G_c(s)\bar{r}(s) + G_d(s)\bar{d}(s), \tag{9.8}$$

and finally,

$$\bar{y}(s) = \frac{G_P(s)G_f(s)G_c(s)}{1+G_P(s)G_f(s)G_c(s)G_m(s)}\bar{r}(s) + \frac{G_d(s)}{1+G_P(s)G_f(s)G_c(s)G_m(s)}\bar{d}(s).$$

(9.9)

As a result,

$$G_{sp}(s) = \frac{G_p(s)G_f(s)G_c(s)}{1+G_p(s)G_f(s)G_c(s)G_m(s)}$$

(9.10)

and

$$G_{load}(s) = \frac{G_d(s)}{1+G_p(s)G_f(s)G_c(s)G_m(s)}$$

(9.11)

by reference to Equation (9.1). Equation (9.9) represents the process *closed-loop response*. When $\bar{d}(s) = 0$, we have

$$\bar{y}(s) = G_{sp}(s)\bar{r}(s).$$

(9.12)

In this case, the plant only responds to a change in the reference point. The poles of $G_{sp}(s)$ and $G_{load}(s)$ are obtained by solving the equation

$$1+G_P(s)G_f(s)G_c(s)G_m(s) = 0.$$

(9.13)

Equation (9.13) is the *characteristic equation* of the feedback system. The term $G_p(s)G_f(s)G_c(s)G_m(s)$ is the *open-loop transfer function* (G_{OL}):

$$G_{OL}(s) = G_p(s)G_f(s)G_c(s)G_m(s).$$

(9.14)

The characteristic equation becomes $1 + G_{OL}(s) = 0$.

The purpose of the controller, when $\bar{d}(s) = 0$, is to follow the set point trajectory as close as possible (also called *set point tracking*). This case is referred to as the *servo* problem. The feedback controller may also be designed to maintain the controlled variable at a desired set point (i.e., $\bar{r}(s) = 0$) even when the load changes (also called *disturbance rejection*):

$$\bar{y}(s) = G_{load}(s)\bar{d}(s).$$

(9.15)

9.2 PROPORTIONAL CONTROL

This chapter focuses on PID control because of its broad acceptance in process industries. These systems are composed of three elements (or *modes*), each processing the receiving error signal in a different manner. Their popularity is due, in part, to their commercial availability, proven track record, and flexibility of the control algorithm (i.e., three tuning parameters). In the absence of the integral and derivative modes, a *P-only controller* is obtained.

Differences among controllers stem from the ways the error signal is processed. A closed-loop system may be designed to regulate a given process by simply switching a device between two states: *on* and *off*. In the liquid level example given in Chapter 1 (Fig. 1.12), the controller computed a signal based on the error ($h - h_{sp}$) and a *control law*. Afterward, the output from the controller (level controller [LC]) adjusted the valve in such a way that the liquid level (h) remained as close as possible to the desired reference point. If $h < h_{sp}$, an *on–off* valve will stay closed until the tank level reaches h_{sp} or is within acceptable bounds. The valve will stay open if $h > h_{sp}$. The variable electric signal from a P-only controller would increase or decrease the valve opening, thereby increasing or reducing the outlet flow (F_0) in an attempt to keep the liquid at h_{sp}. *The control action is proportional to the error between the set point and the measurement*:

$$c(t) = K_c e(t) + c_0, \tag{9.16}$$

where K_c is the controller gain, $c(t)$ is the controller output, and $e(t)$ the error signal defined as $r(t) - y_m(t)$. The corresponding Laplace transforms of these variables, in deviation form, are shown in Fig. 9.2 with the overbars. Equation (9.16) contains a fixed value, c_0, which is the *controller bias* (or *null value*). This constant is the desired controller output when the error is zero. Recall from Chapter 3, when defining deviation variables, the system was considered to be at steady state at time $t = 0$. We were interested in finding out the dynamic behavior of the process variable y if the manipulated variable u changed suddenly from its steady-state condition. In that context, the bias represented the value of c required to keep the process variable, in manual mode, at the steady-state condition (i.e., $t < 0$, before any change in u). Note that the deviation actuating signal $\tilde{c}$ is defined as $\tilde{c}(t) = c(t) - c_0$. Hence, Equation (9.16) becomes

$$\tilde{c}(t) = K_c e(t). \tag{9.17}$$

The transfer function for the P-only controller is obtained by taking the Laplace transform of both sides of Equation (9.17):

$$G_c(s) = K_c. \tag{9.18}$$

From Equation (9.17), it can be concluded that the larger the error, the greater the corrective action taken by the controller. Instead of the gain K_c, many manufacturers often use the value of the *proportional band* (*PB*) to describe the controller. The parameter *PB* is defined as the error that would cause the controller output to change from its lowest to its highest value and is given by

$$PB = \frac{100}{K_c'}, \tag{9.19}$$

where *PB* is expressed as a percentage and K_c' is a scaled gain:

$$K_c' = K_c \frac{VAL}{\Delta c}, \tag{9.20}$$

with *VAL* defined as the span of the sensor measuring $c(t)$; Δc is the range of the controller output.

For the liquid control system discussed above, consider, for example, a controller output pressure that changes 10 psig or $10/(15 - 3) = 83\%$ of the full range for a variation of 3 in. in the liquid level. Assume a level sensor span of 24 in. and a controller output between 0% and 100%, then $K_c = 83\%/3$ in. $= 27.7\%$/in. The dimensionless gain is

$$K_c' = \frac{27.7\%}{\text{in.}} \frac{24 \text{ in.}}{100\%} = 6.48$$

and the proportional band is $PB = 100\%/6.48 = 15.4\%$.

A final comment on proportional controllers is that an error (or *offset*) subsides in the system because the control action is proportional to the error. Also, as will be shown in later chapters, this offset is reduced when the gain is increased. Very large K_c values may lead to oscillatory behaviors. For processes that exhibit integrating properties (i.e., $1/s$), the offset is eliminated for a set point change, while a steady-state error is observed after a change in the load. In the industry, P-only controllers are commonly used in applications where it is not critical to eliminate offsets completely (e.g., liquid level).

9.3 PI CONTROL

The integral mode is often used with the P controller to eliminate the steady-state offset:

$$c(t) = K_c e(t) + \frac{K_c}{\tau_I} \int_0^t e(\tau)d\tau + c_0, \tag{9.21}$$

where τ_I is the *integral time* or *reset time*. The signal $\tilde{c}$ is given by

$$\tilde{c}(t) = K_c e(t) + \frac{K_c}{\tau_I} \int_0^t e(\tau) d\tau. \qquad (9.22)$$

Applying Laplace transforms to Equation (9.22) yields

$$\overline{c}(s) = K_c \overline{e}(s) + \frac{K_c}{\tau_I s} \overline{e}(s)$$

$$\overline{c}(s) = K_c \left(1 + \frac{1}{\tau_I s} \right) \overline{e}(s). \qquad (9.23)$$

The Laplace transform is

$$G_c(s) = K_c \left(1 + \frac{1}{\tau_I s} \right). \qquad (9.24)$$

For comparison, note that Equation (9.16) suggests that a finite error remains when using a P controller. The controller output continues to vary to overcome the offset. Depending on the size of K_c, y will reach a value in the neighborhood of (but different from) the reference. At this point, the controller output no longer changes. With the PI algorithm, the integral term (i.e., sum of the error) does not approach zero with the error. Instead, the integral sum continues to vary until $e(t) = 0$, at which time it takes on a constant final value: Int_f. The control signal required to maintain y at its desired set point is

$$c = \frac{K_c}{\tau_I} Int_f + c_0. \qquad (9.25)$$

Some manufacturers use the reset rate $1/\tau_I$ instead of the reset time.

Because the output of the controller changes as long as the error remains different from zero, the integral term may become very large and the control signal will remain at its saturation value (e.g., a fully open actuator). *Integral windup* refers to the integral term continuing to build up while the controller is saturated. In the case of a step change in the set point, large and undesirable overshoots are expected until the error changes sign (e.g., from below to above the new set point). This causes the integral contribution to decrease to a point where the controller and the manipulated variable are no longer saturated. *Antireset windup* strategies, such as not allowing the integral term to accumulate, are implemented to prevent integral windup. PI controllers are extensively used to regulate flow, pressure, level, and other process variables that do not exhibit long time lags [1].

9.4   PID CONTROL

The equation of the PID controller is

$$c(t) = K_c e(t) + \frac{K_c}{\tau_I} \int_0^t e(\tau)\,d\tau + K_c \tau_D \frac{de(t)}{dt} + c_0, \qquad (9.26)$$

where τ_D is the *derivative time constant*. Due to the derivative mode, which measures the rate at which the error is changing at the current time, any variation in $e(t)$ is immediately detected by the controller. The derivative action is often called *anticipatory control* because of its ability to predict the error in the near future. The net result is a reduction in the maximum deviation and the amount of oscillation. However, the use of the derivative makes the controller sensitive to random fluctuations, causing the output signal to become unnecessarily large. The derivative mode is used with the proportional and/or integral mode because it acts only on $de(t)/dt$ and not on the error itself. A constant error would not dictate any adjustment in $c(t)$:

$$c(t) = \tau_D \frac{de(t)}{dt} + c_0 = c_0. \qquad (9.27)$$

The transfer function of the PID controller is

$$G_c(s) = K_c\left(1 + \frac{1}{\tau_I s} + \tau_D s\right) \qquad (9.28)$$

or

$$G_c(s) = K_c\left(\frac{\tau_D \tau_I s^2 + \tau_I s + 1}{\tau_I s}\right). \qquad (9.29)$$

From Chapter 8, it was established that for most physical processes, the order of the denominator (n) is greater than that of the numerator (m) for a *physically realizable* system. As a result, in commercial controllers, Equation (9.28) is often approximated by [2]

$$G_c(s) = K_c\left(1 + \frac{1}{\tau_I s} + \frac{\tau_D s}{\alpha \tau_D s + 1}\right). \qquad (9.30)$$

An alternative implementation is [3]

$$G_c(s) = K_c\left(\frac{\tau_I s + 1}{\tau_I s}\right)\left(\frac{\tau_D s + 1}{\alpha \tau_D s + 1}\right), \qquad (9.31)$$

where α is a small number that usually lies between 0.05 and 0.2. It is possible to replace the derivative $de(t)/dt$ with $d(r - y_m)/dt = -dy_m(t)/dt$ to prevent a possible *derivative kick* (i.e., sudden change in the error):

$$c(t) = K_c e(t) + \frac{K_c}{\tau_I} \int_0^t e(\tau) d\tau - K_c \tau_D \frac{dy_m(t)}{dt} + c_0. \tag{9.32}$$

In the industry, PID controllers are typically used to control slow process variables (e.g., pH, temperature) [1].

9.5 TOTAL SUGAR CONCENTRATION IN A GLUTAMIC ACID PRODUCTION

Tsao et al. [4] designed a strategy to estimate and control glutamic acid concentration during fed-batch operations [4]. *Brevibacterium divaricatum* initially grew at 30°C for 19 hours in flasks with cane molasses, brown sugar, $(NH_4)_2HPO_4$, beef extract, $(NH_4)_2H_2PO_4$, and urea. The cells were later transferred to a 3-dm^3 fermentation vessel containing cane molasses and phosphoric acid. In the experiment, the substrate feeding rate (u) was used to control the total sugar concentration (y) in the bioreactor. A linear system was applied to estimate how y changes with respect to time:

$$\frac{dy(t)}{dt} = ay(t) + bu(t). \tag{9.33}$$

A transfer function ($G_p(s)$) relating u and y can be developed assuming $y(0) = 0$ and $u(0) = 0$. First, Equation (9.33) is transformed to

$$s\overline{y}(s) - y(0) = a\overline{y}(s) + b\overline{u}(s). \tag{9.34}$$

Solving Equation (9.34) for $\overline{y}(s)/\overline{u}(s)$ yields the transfer function

$$\frac{\overline{y}(s)}{\overline{u}(s)} = G_p(s) = \frac{b}{s-a} \tag{9.35}$$

or

$$G_p(s) = \frac{K_p}{\tau_p s + 1} \tag{9.36}$$

in standard form, where $K_p = -b/a$ and $\tau_p = -1/a$; a is a negative number.

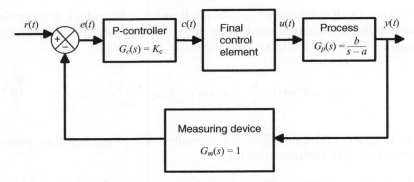

Figure 9.3. Block diagram of the proportional control of total sugar concentration.

The closed-loop response of the process $(G_p(s))$ can be obtained assuming a measuring device transfer function $(G_m(s) = 1)$, a final control element transfer function $(G_f(s) = 1)$, and a P controller $(G_c(s))$. First, the block diagram of the closed-loop system is given in Figure 9.3. Equation (9.10) is used to calculate $G_{sp}(s)$:

$$G_{sp}(s) = \frac{G_p(s)G_f(s)G_c(s)}{1+G_p(s)G_f(s)G_c(s)G_m(s)} = \frac{K_c \dfrac{K_p}{\tau_p s+1}}{1+K_c \dfrac{K_p}{\tau_p s+1}} \tag{9.37}$$

or

$$G_{sp}(s) = \frac{K_p K_c}{\tau_p s+1+K_p K_c} = \frac{K_{p1}}{\tau_{p1}s+1}, \tag{9.38}$$

where

$$K_{p1} = \frac{K_p K_c}{K_p K_c+1} \tag{9.39}$$

and

$$\tau_{p1} = \frac{\tau_p}{K_p K_c+1}. \tag{9.40}$$

For comparison, the Laplace transform of the open-loop response to a step change of size U is given by

$$\bar{y}_{OL}(s) = \frac{K_p}{\tau_p s + 1} \frac{U}{s}, \tag{9.41}$$

and a steady-state value of UK_p is reached at $4\tau_p$. To obtain a steady-state value of $r = 1$, U can be set to $1/K_p$. Note that the speed at which this value is reached depends only on τ_p and is not influenced by the size of U.

The closed-loop response for the servo problem is

$$\bar{y}_{CL}(s) = \frac{K_{p1}}{\tau_{p1} s + 1} \bar{r}(s). \tag{9.42}$$

With $\bar{r}(s) = 1/s$, the ultimate response is K_{p1}. It takes $4\tau_{p1}$ to attain this value. In the absence of a controller, the substrate feeding rate can be set manually in order to obtain a desired total sugar concentration in the bioreactor. Because the time constant depends on the model parameters, the speed of the response cannot be increased by simply fixing the flow rate. Feedback control can be employed to improve the response dynamics. The closed-loop dynamics is faster than the open-loop response because $\tau_{p1} < \tau_p$. The main drawback is that an offset, defined as the difference between the *new set point* and the *ultimate value of the response*, is now measured:

$$\text{Offset} = 1 - K_{p1} = \frac{1}{1 + K_p K_c}. \tag{9.43}$$

This offset approaches 0 as $K_c \rightarrow \infty$.

9.6 TEMPERATURE CONTROL OF FERMENTATIONS

Controllers are designed to maintain large-scale batch bioreactors at desired temperatures because heat transfer can be significant. Cooling water is run through a coil placed in the fermenter to mitigate the effects of the heat generated by the microorganisms. One option is to devise a controller such that the error between the reactor temperature and a reference value is sent to the controller, which then adjusts the cooling water flow rate (Fig. 9.4). An alternative is to manipulate the inlet temperature of the cooling water, T_{Cin}.

From Section 2.2.1, the mass balance equations for the cells and substrate can be written as

$$\frac{dC_x}{dt} = \mu C_x \tag{9.44}$$

and

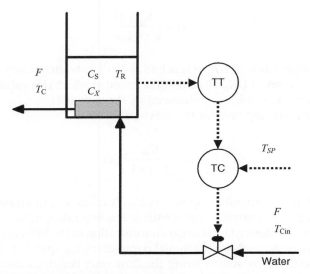

Figure 9.4. Feedback control of the reactor temperature.

$$\frac{dC_s}{dt} = -\frac{\mu C_x}{Y_{x/s}}, \qquad (9.45)$$

respectively, where

$$\mu = \frac{\mu_{max} C_s}{K_s + C_s}. \qquad (9.46)$$

The energy balances for the bioreactor and coil are [5]

$$\frac{dT_R}{dt} = \frac{r_Q}{\rho C_p} - \frac{UA}{V \rho C_p}(T_R - T_C) \qquad (9.47)$$

and

$$\frac{dT_C}{dt} = \frac{F}{V_C}(T_{Cin} - T_C) + \frac{UA}{V_C \rho_C C_{pC}}(T_R - T_C), \qquad (9.48)$$

where T_R (K) is the bioreactor temperature (K), r_Q (kcal/(m³h)) is the rate of heat production due to the presence of biomass, ρ (kg/m³) is the density of the fermentation broth, UA is the heat transfer coefficient (kcal/h·K), V is the bioreactor volume, and C_p (kcal/(kg C)) is the heat capacity of the fermentation broth. T_C (K), V_C (m³), T_{Cin} (K), ρ_C (kg/m³), and C_{pC} (kcal/(kg C)) are the coolant outlet temperature, volume, inlet temperature, density, and heat capacity, respectively.

Manipulation of the flow rate using a PI controller yields the following control law [5]:

$$F(t) = F_0 + K_c e(t) + \frac{K_c}{\tau_I} \int_0^t e(\tau) dt \tag{9.49}$$

and

$$e(t) = T_R(t) - T_{SP}. \tag{9.50}$$

In practice, $F(t)$ can be constrained to take on positive values. Note that this control strategy, although easy to implement, makes Equation (9.48) nonlinear. For simplicity, we assume a P-only controller and a constant flow rate. The manipulated variable is

$$T_{Cin}(t) = T_{Cin,ss} + K_c e(t). \tag{9.51}$$

The output variables are T_R and T_C; r_Q is considered a disturbance.

To develop the open-loop transfer functions, the following deviation variables are used: $\tilde{T}_R(t) = T_R(t) - T_{R,ss}$, $\tilde{T}_C(t) = T_C(t) - T_{C,ss}$, $\tilde{r}_Q(t) = r_Q(t) - r_{Q,ss}$, and $\tilde{T}_{Cin}(t) = T_{Cin}(t) - T_{Cin,ss}$ where the subscript ss denotes a steady-state condition. Because Equations (9.47) and (9.48) are linear, we can write

$$\frac{d\tilde{T}_R}{dt} = \frac{\tilde{r}_Q}{\rho C_p} - \frac{UA}{V \rho C_p} \left(\tilde{T}_R - \tilde{T}_C \right) \tag{9.52}$$

and

$$\frac{d\tilde{T}_C}{dt} = \frac{F}{V_C} \left(\tilde{T}_{Cin} - \tilde{T}_C \right) + \frac{UA}{V_C \rho_C C_{pC}} \left(\tilde{T}_R - \tilde{T}_C \right). \tag{9.53}$$

The transfer function relating $\tilde{T}_R$ to $\tilde{T}_{Cin}$ is

$$G_{RC}(s) = \frac{\overline{T}_R(s)}{\overline{T}_{Cin}(s)} = \frac{\phi_F \phi_U}{s^2 + (\phi_F + \phi_U + \phi_{UC})s + \phi_F \phi_U}, \tag{9.54}$$

where

$$\phi_U = \frac{UA}{V \rho C_p}, \phi_{UC} = \frac{UA}{V_C \rho_C C_{pC}} \text{ and } \phi_F = \frac{F}{V_C}.$$

The derivation of the other three transfer functions is left as an exercise for the reader. Equation (9.54) is rewritten in standard form to generalize the results:

$$G_{RC}(s) = \frac{K_p}{\tau^2 s^2 + 2\varsigma\tau s + 1}, \qquad (9.55)$$

with $K_p = 1$, $\tau = \sqrt{1/\phi_F\phi_U}$, and $2\varsigma\tau = (\phi_F + \phi_U\phi_{UC})/\phi_F\phi_U$. For the servo problem, where $\bar{r}_Q(s) = 0$, the closed-loop transfer function is

$$G_{SP}(s) = \frac{\bar{T}_R(s)}{\bar{T}_{R,SP}(s)} = \frac{G_{RC}K_c}{1 + G_{RC}K_c}, \qquad (9.56)$$

assuming $G_m = G_f = 1$. Algebraic manipulation of Equation (9.56) yields

$$G_{SP}(s) = \frac{K_{p1}}{\tau_1^2 s^2 + 2\varsigma_1\tau_1 s + 1}, \qquad (9.57)$$

where

$$K_{p1} = \frac{K_p K_c}{1 + K_p K_c}, \qquad (9.58)$$

$$\tau_1 = \frac{\tau}{\sqrt{1 + K_p K_c}}, \qquad (9.59)$$

and

$$\varsigma_1 = \frac{\varsigma}{\sqrt{1 + K_p K_c}}. \qquad (9.60)$$

A unit step change in the reference reactor temperature leads to

$$\tilde{T}_R(t \to \infty) = K_{p1}. \qquad (9.61)$$

As a result, the offset is

$$\text{Offset} = 1 - K_{p1} = \frac{1}{1 + K_p K_c}. \qquad (9.62)$$

Just like the first-order process, an offset exists when a P controller is applied to the second-order system. The offset approaches zero as $K_c \to \infty$. Compared to the open-loop response, where the dynamics is described by Equation (9.55), a faster response (i.e., lower damping factor) is obtained for higher values of K_c. However, since the overshoot increases as the damping factor decreases, $\hat{T}_R$ may exceed its ultimate value by a significant margin and undesirable oscillations may occur. When tuning the controller, it is important to achieve a balance among these conflicting characteristics.

9.7 DO CONCENTRATION

A block diagram of the feedback control of DO is shown in Figure 1.2. The optimum DO level in a bioreactor is maintained by manipulating the partial pressure of oxygen in the liquid phase. The oxygen mass balance in the liquid phase of a cell-free cultivation medium is given by [6]

$$\frac{dC_{o2,L}}{dt} = k_L a_B \left(\frac{P}{H} y_{O_2,B} - C_{o2,L} \right) + k_L a_H \left(\frac{P}{H} y_{O_2,H} - C_{o2,L} \right), \qquad (9.63)$$

where $C_{O_2,L}$ (output variable) is the oxygen concentration in the liquid phase (mol/L), $k_L a_B$ is the mass transfer coefficient in the bubble phase (s^{-1}), P is the total pressure (bar), H is Henry's constant (bar L/mol), $y_{O_2,B}$ is the oxygen mole fraction in the bubble phase (manipulated variable), $y_{O_2,H}$ is the oxygen mole fraction in the head phase (disturbance), and $k_L a_H$ is the mass transfer coefficient in the headspace. We assume that the variables are given in deviation form. Therefore, the transfer functions relating $y_{O_2,B}$ and $y_{O_2,H}$ to $C_{O_2,L}$ are

$$G_1(s) = \frac{P}{H} \frac{k_L a_B}{(s + k_L a_B + k_L a_H)} \qquad (9.64)$$

and

$$G_2(s) = \frac{P}{H} \frac{k_L a_H}{(s + k_L a_B + k_L a_H)}, \qquad (9.65)$$

respectively. The transfer function for the measuring device is defined by

$$G_m(s) = \frac{\bar{D}_{O_2}(s)}{\bar{C}_{O_2,L}(s)} = \left(\frac{100}{C_{O_2,\text{sat}}} \right) \frac{k_m}{\tau_m s + 1} e^{-t_d s}. \qquad (9.66)$$

A DO probe is used to measure the level of DO (% D_{O_2}). Experiments show that the dynamics of the probe can be approximated as a first-order system with time delay. Assuming that $G_f = 1$, Figure 9.5 represents the block diagram

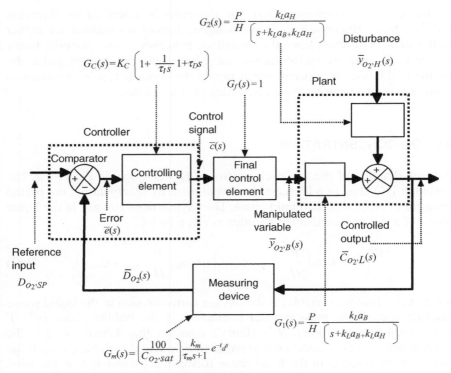

Figure 9.5. Block diagram of dissolved oxygen feedback control.

of the feedback process when a PID controller is used. Calculations, similar to the ones described in Sections 9.5 and 9.6, can be performed to derive the transfer functions and to study the effects of the three modes on the performance of the closed-loop system.

9.8 SUMMARY

This chapter introduced block diagrams of closed-loop systems in the presence of PID controllers. The effects of the three modes on the dynamic response were analyzed. In general, the proportional term leaves a steady-state offset that can be reduced by increasing the gain. The integral mode drives the error to zero but can saturate the controller if some antireset windup strategies are not applied. Derivative action can predict the error in the short term and can decrease the maximum deviation and amount of oscillation. Fermentation control applications are provided to illustrate the operation and performance of closed-loop systems.

PROBLEMS

9.1. A step change of size 2 in the error is fed into a PI controller. If $K_c = 5$ and $\tau_I = 0.7$, sketch the response of the controller.

9.2. The dose–response following the administration of a drug is given by

$$\frac{dx(t)}{dt} = -\frac{1}{T} x(t) + \frac{K}{T} u(t),$$

where $x(t)$ is the state variable that is related to a measuring quantity (e.g., the mean arterial pressure); u is the infusion rate; K and T are parameters that vary with the patient. Derive the closed-loop transfer function G_{sp}, assuming $G_m = G_f = 1$ and $G_c = K_c$.

9.3. The dynamics of a one-compartment model for drug administration is given by (Chapter 2)

$$\frac{dC_p}{dt} = \frac{k_0}{V} - k_{el} C_p$$

where k_0 is the input and C_p is the controlled variable. Derive the closed-loop transfer function G_{sp}, assuming $G_m = G_f = 1$ and

$$G_c = K_c \left(1 + \frac{1}{\tau_I s} \right).$$

9.4. The heat balance in a stirred-tank heater is given by (see Example 7.1)

$$\rho V c_p \frac{dT}{dt} = F \rho c_p (T_{in} - T) + UA (T_{st} - T).$$

The manipulated variable is T_{in} and the input disturbance is T_{st}. The output variable is the outlet stream temperature T; F is kept constant. Derive the closed-loop transfer function G_{sp}, assuming $G_m = G_f = 1$ and

$$G_c = K_c \left(1 + \frac{1}{\tau_I s} + \tau_D s \right).$$

9.5. The heat balance in a stirred-tank heater is given by

$$\rho V c_p \frac{dT}{dt} = F \rho c_p (T_{in} - T) + UA (T_{st} - T).$$

The manipulated variable is T_{in} and the input disturbance is T_{st}. The output variable is the outlet stream temperature T; F is kept constant. Derive the closed-loop transfer function G_{load}, assuming $G_m = G_f = 1$ and $G_c = K_c$.

9.6. In Section 9.6, the transfer function relating $\tilde{T}_R$ to $\tilde{T}_{Cin}$ is

$$G_{RC}(s) = \frac{\bar{T}_R(s)}{\bar{T}_{Cin}(s)} = \frac{\phi_F \phi_U}{s^2 + (\phi_F + \phi_U + \phi_{UC})s + \phi_F \phi_U},$$

where

$$\phi_U = \frac{UA}{V\rho C_p}, \quad \phi_{UC} = \frac{UA}{V_C \rho_C C_{pC}} \quad \text{and} \quad \phi_F = \frac{F}{V_C}.$$

(a) Derive the closed-loop transfer function G_{sp}, assuming $G_m = G_f = 1$ and

$$G_c = K_c \left(1 + \frac{1}{\tau_I s} \right).$$

(b) Calculate the offset for a unit step change in the reference reactor temperature.

9.7. In Section 9.5, the Laplace transform of the open-loop response to a step change of size U is given by

$$\bar{y}_{OL}(s) = \frac{K_p}{\tau_p s + 1} \frac{U}{s}.$$

(a) Derive the closed-loop transfer function G_{sp}, assuming $G_m = G_f = 1$ and

$$G_c = K_c \left(1 + \frac{1}{\tau_I s} + \tau_D s \right).$$

(b) Calculate the offset for a unit step change in the set point.

9.8. Redo Problem 9.7 for

$$G_c = K_c \left(1 + \frac{1}{\tau_I s} \right).$$

9.9. The dynamics of a two-compartment model for drug administration is given by the following equations (Chapter 2):

$$\frac{dC_1}{dt} = \frac{k_0}{V_1} - (k_{el} + k_{12})C_1 + k_{21}C_2 \frac{V_2}{V_1}$$

and

$$\frac{dC_2}{dt} = k_{12}C_1 \frac{V_1}{V_2} - k_{21}C_2.$$

(a) Write the system as

$$d\mathbf{C}^T/dt = \begin{bmatrix} a_{11} & a_{12} \\ a_{21} & a_{22} \end{bmatrix} \mathbf{C}^T + \begin{bmatrix} b_1 \\ b_2 \end{bmatrix} u$$
$$y = \begin{bmatrix} 1 & 0 \end{bmatrix} \mathbf{C}^T,$$

where $C = [C_1\ C_2]$ is the state vector; C_1 is the output y(or controlled variable); and k_0 is the input.

(b) Derive the closed-loop transfer function G_{sp}, assuming $G_m = G_f = 1$ and

$$G_c = K_c \left(1 + \frac{1}{\tau_I s}\right).$$

Assume $C_1(0) = C_2(0) = 0$.

9.10. Redo part (b) of Problem 9.9 for $y = C_2$.

REFERENCES

1. Lipták BG, ed. *Instrument Engineers' Handbook, Third Edition: Process Measurement and Analysis*. Oxford: Butterworth-Heinemann, 1995.
2. Chau PC. *Process Control: a First Course with MATLAB*. Cambridge: Cambridge University Press, 2002.
3. Seborg DE, Edgar TF, Mellichamp DA. *Process Dynamics and Control*. Hoboken, NJ: Wiley, 2004.
4. Tsao J, Chuang H, Wu W. On-line state estimation and control in glutamic acid production. *Bioprocess Engineering* 1991; 7:35–39.
5. Dunn IJ. *Biological Reaction Engineering*. Weinheim: Wiley-VCH, 2003.
6. Simon L, Karim MN. Identification and control of dissolved oxygen in hybridoma cell culture in a shear sensitive environment. *Biotechnology Progress* 2001; 17: 634–642.

CHAPTER 10

FREQUENCY RESPONSE ANALYSIS

Frequency-domain methods are commonly applied to evaluate the effects of noise and to identify pharmacokinetic models and industrial processes directly from input–output measurements. To conduct such analyses, a system is first approximated by the sum of first-order plus time-delay transfer functions. The frequency response data, obtained using pulse testing or a sinusoidal input, can then be fitted to the assumed transfer function over a frequency range of interest by minimizing an error function. For certain applications, this strategy is preferred over modeling in the time domain. One notable advantage of the identification technique is the development of simpler models that are more amenable to simulation and control. This chapter lays the foundation for frequency response methods, which are valuable for designing feedback control systems.

10.1 FREQUENCY RESPONSE FOR LINEAR SYSTEMS

The transfer function of a linear time-invariant system is given by

$$G(s) = \frac{\bar{y}(s)}{\bar{u}(s)}, \qquad (10.1)$$

Control of Biological and Drug-Delivery Systems for Chemical, Biomedical, and Pharmaceutical Engineering, First Edition. Laurent Simon.
© 2013 John Wiley & Sons, Inc. Published 2013 by John Wiley & Sons, Inc.

where $u(t)$ is a sinusoidal input,

$$u(t) = A \sin \omega t, \tag{10.2}$$

with amplitude A and frequency ω. If the system is stable, the ultimate response $y_{ss}(t)$ is given by

$$y_{ss}(t) = A|G(i\omega)| \sin(\omega t + \phi). \tag{10.3}$$

The frequency response is the ultimate response of the system to a sinusoidal input. It is obtained after replacing s with the complex number $i\omega$ in $G(s)$ and by considering long-term behavior. Note that $G(i\omega)$ is a complex number with real and imaginary parts. The notation $|G(i\omega)|$ represents the *magnitude* of $G(i\omega)$; ϕ is the argument, $\arg G(i\omega)$ (also denoted $\angle[G(i\omega)]$). Depending on whether $|G(i\omega)|$ is less than or greater than unity, the system may attenuate or amplify the input signal. The output frequency is similar to that of the input, and its magnitude is equal to the product of the input magnitude A and $|G(i\omega)|$. The phase angle of the input is shifted by $\arg G(i\omega)$. A phase shift corresponds to a displacement of a function along the time axis (Fig. 10.1). Figure 10.1 compares the plots for $\sin(t)$ and $\sin(t+3)$. A similar plot can be generated for the function $\sin(t-3)$, which lags behind $\sin t$ with a phase difference of -3 rad.

Example 10.1 Consider the first-order system:

$$G(s) = \frac{2}{s+3}. \tag{10.4}$$

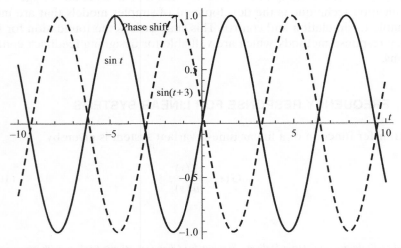

Figure 10.1. The function $\sin(t+3)$ leads with a phase difference of 3 rad relative to $\sin t$.

Find the frequency response of the system when $u(t) = A \sin \omega t$.

Solution   Let $s = i\omega$ and multiply the numerator and denominator by the conjugate $3 - i\omega$:

$$G(i\omega) = \frac{2(3-i\omega)}{(3+i\omega)(3-i\omega)}$$

$$G(i\omega) = \left(\frac{6}{9+\omega^2}\right) + \left(-\frac{2\omega}{9+\omega^2}\right)i, \tag{10.5}$$

which can be represented as a point in the *Argand plane* (or *complex plane*) (Fig. 10.2). The real part is

$$a = \frac{6}{9+\omega^2} \tag{10.6}$$

and the imaginary part is

$$b = -\frac{2\omega}{9+\omega^2}. \tag{10.7}$$

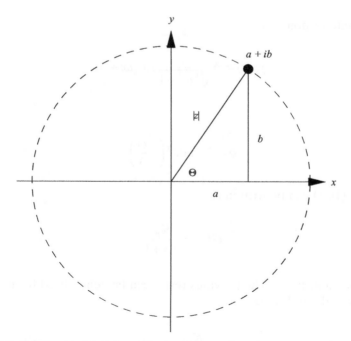

Figure 10.2. Representation of a point in the complex plane: $z = a + ib = |z|(\cos\theta + i\sin\theta) = |z|e^{i\theta}$.

The *magnitude or modulus* of $G(i\omega)$ is then calculated:

$$|G(i\omega)| = \sqrt{\frac{36}{(9+\omega^2)^2} + \frac{4\omega^2}{(9+\omega^2)^2}}$$

$$|G(i\omega)| = 2\sqrt{\frac{1}{9+\omega^2}} \tag{10.8}$$

$$|G(i\omega)| = \frac{\frac{2}{3}}{\sqrt{\frac{1}{9}\omega^2+1}}.$$

The argument (or *phase*) of $G(i\omega)$ is θ (i.e., $\angle[G(i\omega)]$). From Figure 10.2,

$$\tan\theta = \frac{b}{a} = -\frac{\omega}{3}, \tag{10.9}$$

which yields

$$\angle[G(i\omega)] = \theta = \tan^{-1}\left(-\frac{\omega}{3}\right). \tag{10.10}$$

The ultimate response is

$$y_{ss}(t) = A\frac{\frac{2}{3}}{\sqrt{\frac{1}{9}\omega^2+1}}\sin(\omega t + \phi) \tag{10.11}$$

with

$$\phi = \theta = \tan^{-1}\left(-\frac{\omega}{3}\right). \tag{10.12}$$

Equation (10.4) can be written as

$$G(s) = \frac{K_p}{\tau_p s + 1}, \tag{10.13}$$

with $K_p = \frac{2}{3}$ and $\tau_p = \frac{1}{3}$. The previous result can be generalized for first-order processes as the following:

$$y_{ss}(t) = \frac{K_p A}{\sqrt{\tau_p^2\omega^2 + 1}}\sin(\omega t + \phi) \tag{10.14}$$

and

$$\phi = \tan^{-1}\left(-\omega\tau_p\right). \tag{10.15}$$

The ratio of the output to the input signal amplitude is

$$AR = \frac{K_p}{\sqrt{\tau_p^2\omega^2 + 1}} = |G(i\omega)|. \tag{10.16}$$

Equations (10.14) and (10.15) can also be obtained by taking the Laplace transform of the response:

$$\bar{y}(s) = G(s)\bar{u}(s) \tag{10.17}$$

or

$$\bar{y}(s) = \frac{K_p}{\tau_p s + 1}\frac{A\omega}{s^2 + \omega^2}. \tag{10.18}$$

In the time domain, the response is

$$y(t) = \frac{AK_p\left[\sin(t\omega) + \omega\left(e^{-t/\tau_p} - \cos(t\omega)\right)\tau_p\right]}{1 + \omega^2\tau_p^2}. \tag{10.19}$$

Consequently,

$$y_{ss}(t) = \frac{AK_p\left[\sin(t\omega) - \omega\tau_p\cos(t\omega)\right]}{\omega^2\tau_p^2 + 1}, \tag{10.20}$$

which is equivalent to Equation (10.14) after using the following trigonometric identity:

$$a_1\sin\omega + a_2\cos\omega = a_3\sin(\omega + \phi), \tag{10.21}$$

where

$$a_3 = \sqrt{a_1^2 + a_2^2} \tag{10.22}$$

and

$$\phi = \tan^{-1}\left(\frac{a_2}{a_1}\right). \tag{10.23}$$

Example 10.2   Consider the transfer function:

$$G(s) = \frac{K_p}{\tau^2 s^2 + 2\varsigma\tau s + 1}. \tag{10.24}$$

Find the frequency response of the system when $u(t) = A \sin \omega t$.

Solution   After substituting $i\omega$ for s and multiplying by the conjugate, we have

$$G(i\omega) = \left[\frac{(1-\tau^2\omega^2)K_p}{4\varsigma^2\tau^2\omega^2 + (1-\tau^2\omega^2)^2}\right] - i\left[\frac{2\varsigma\tau\omega K_p}{4\varsigma^2\tau^2\omega^2 + (1-\tau^2\omega^2)^2}\right]. \tag{10.25}$$

The modulus is

$$|G(i\omega)| = \frac{K_p}{\sqrt{4\varsigma^2\tau^2\omega^2 + (1-\tau^2\omega^2)^2}}. \tag{10.26}$$

The argument is calculated:

$$\begin{aligned} \theta &= \tan^{-1}\left(-\frac{2\varsigma\tau\omega K_p}{(1-\tau^2\omega^2)K_p}\right) \\ \theta &= \tan^{-1}\left(-\frac{2\varsigma\tau\omega}{1-\tau^2\omega^2}\right). \end{aligned} \tag{10.27}$$

The ultimate response is

$$y_{ss}(t) = A|G(i\omega)|\sin(\omega t + \phi), \tag{10.28}$$

with

$$\phi = \tan^{-1}\left(-\frac{2\varsigma\tau\omega}{1-\tau^2\omega^2}\right). \tag{10.29}$$

The frequency response provides the asymptotic behavior of a stable system to periodic input changes. The time of interest is larger than four times the longest time constant [1]. Although in previous chapters the focus was placed on step changes, the behavior and control of a system to sinusoidal inputs may be important to help minimize variations in product quality. The two previous examples show that the amplitude of the output is influenced by the frequency of the input variable and that the signal tends to lag behind periodic changes introduced in the process. Based on the results obtained so far, *the output is not affected by rapid changes in the input variable* (i.e., high frequency ω). For example, from Equations (10.16) and (10.26), the amplitude ratio AR approaches 0 as $\omega \to \infty$. At low frequencies (i.e., slow changes in the input variable), AR converges to the process steady-state gain (K_p).

Example 10.3 A first-order process has a steady-state gain of 1.0°F-min/lb and a time constant of 6 minutes. At time $t = 0$, the input (in deviation form) is characterized by a sine function with an amplitude of 2.0 lb/min and a period of 3 minutes. These fluctuations may represent noises in the input signal. Calculate the frequency response.

Solution   The transfer function is

$$G(s) = \frac{1}{6s+1} \tag{10.30}$$

and the periodic input is

$$u(t) = 2.0 \sin \frac{2\pi}{3} t.$$

In this case, the frequency ω is $2\pi/3$ rad/min and

$$\bar{u}(s) = \frac{\dfrac{4\pi}{3}}{s^2 + \dfrac{4\pi^2}{9}}. \tag{10.31}$$

The frequency response is obtained:

$$
\begin{aligned}
y_{ss}(t) &= \frac{K_p A}{\sqrt{\tau_p^2 \omega^2 + 1}} \sin(\omega t + \phi) \\
y_{ss}(t) &= \frac{2 \sin(\omega t - \tan^{-1}(4\pi))}{\sqrt{1 + 16\pi^2}}
\end{aligned}
\tag{10.32}
$$

or

$$y_{ss}(t) = \frac{2 \sin(\omega t - 1.49)}{\sqrt{1 + 16\pi^2}}. \tag{10.33}$$

Figure 10.3 shows the input $u(t)$, transient response $y(t)$, and frequency response $y_{ss}(t)$. Figure 10.4 only shows $y(t)$ and $y_{ss}(t)$, to allow direct comparison. The response can be approximated by its steady-state behavior during the initial 24 minutes $(4\tau_p)$. The input sine function reaches its first peak at 0.75

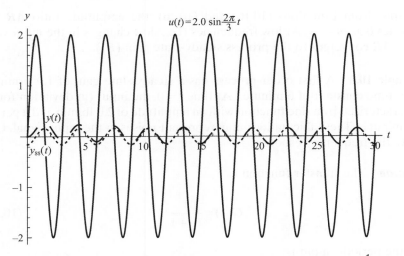

Figure 10.3. Input, transient, and frequency responses of $G(s) = \dfrac{1}{6s+1}$.

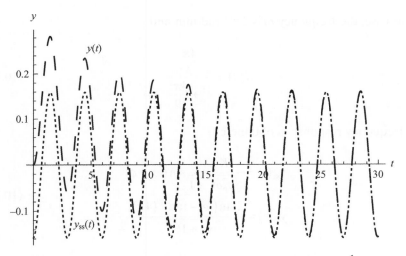

Figure 10.4. Transient and frequency responses of $G(s) = \dfrac{1}{6s+1}$.

minute, while the frequency response reaches its first peak 0.71 minute, later. Therefore, the phase lag is $-(2\pi/3) \times 0.71 = -1.49$ rad as predicted by Equation (10.33). Note that the input variable changes around its nominal value by a function of the form $U(t) = U_{ss} + 2.0\sin(2\pi/3)t$, where $u(t) = U(t) - U_{ss}$. This perturbation causes $y(t)$ to deviate from its desired nominal value with an amplitude of $K_p A / \sqrt{\tau_p^2 \omega^2 + 1} = 0.159°F$ and a period of 3 minutes. The amplitude ratio is $K_p / \sqrt{\tau_p^2 \omega^2 + 1} = 0.079°F\text{-min/lb}$. Depending on process specifications, corrective action may be necessary to reduce such perturbations.

10.2 BODE DIAGRAMS

Bode diagrams represent frequency responses. The amplitude ratio and phase shifts are plotted versus the frequency. Basic definitions and properties of the logarithm to the base 10 (i.e., $\log(.)$) functions are given below:

- $z_1 = |z_1| e^{i\theta_1} = |z_1|(\cos\theta_1 + i\sin\theta_1)$, $z_2 = |z_2| e^{i\theta_2} = |z_2|(\cos\theta_2 + i\sin\theta_2)$
- $z_1 z_2 = |z_1||z_2| e^{i(\theta_1 + \theta_2)} = |z_1||z_2|[\cos(\theta_1 + \theta_2) + i\sin(\theta_1 + \theta_2)]$
- $\dfrac{z_1}{z_2} = \dfrac{|z_1|}{|z_2|} e^{i(\theta_1 - \theta_2)} = \dfrac{|z_1|}{|z_2|}[\cos(\theta_1 - \theta_2) + i\sin(\theta_1 - \theta_2)]$
- $z^n = |z|^n e^{in\theta} = |z|^n (\cos n\theta + i\sin n\theta)$
- $\log|z_1 z_2| = \log|z_1| + \log|z_2|$
- $\angle(z_1 z_2) = \theta_1 + \theta_2$
- $\log\left|\dfrac{z_1}{z_2}\right| = \log|z_1| - \log|z_2|$
- $\angle\left(\dfrac{z_1}{z_2}\right) = \theta_1 - \theta_2$
- Magnitude of $G(i\omega)$ in decibels $= 20\log|G(i\omega)|$.

Example 10.4 Derive the equations for the logarithm of the amplitude ratio and the phase shift for the following first-order system:

$$G(s) = \frac{K_p}{\tau_p s + 1}. \tag{10.34}$$

Solution From Equation (10.16),

$$
\begin{aligned}
\log(AR) &= \log(K_p) - \log(\tau_p^2 \omega^2 + 1)^{1/2} \\
\log(AR) &= \log(K_p) - \frac{1}{2}\log(\tau_p^2 \omega^2 + 1).
\end{aligned}
\tag{10.35}
$$

The phase shift is given by $\angle G(j\omega) = \tan^{-1}(-\omega\tau_p)$. The Bode diagram for $K_p = 1$ and $\tau_p = 1$ is given in Figure 10.5 using semilog plots. Figure 10.6 shows the sinusoidal input and frequency response of the process. The amplitude ratio

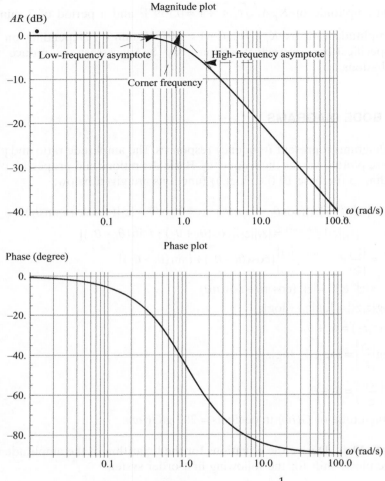

Figure 10.5. Bode plot for $G(s) = \dfrac{1}{s+1}$.

remains fairly constant at low frequency ω (*low-frequency asymptote* [LFA]) and varies linearly with the frequency for higher values of ω (*high-frequency asymptote* [HFA]). The equations for LFA and HFA are given by $\log(AR) = 0$ and $\log(AR) = -\log\tau_p\omega$, respectively. The two lines meet at the *corner frequency*: $\omega = 1/\tau_p$.

Example 10.5 Derive the frequency response for the following pure dead-time process:

$$G(s) = e^{-t_d s}. \tag{10.36}$$

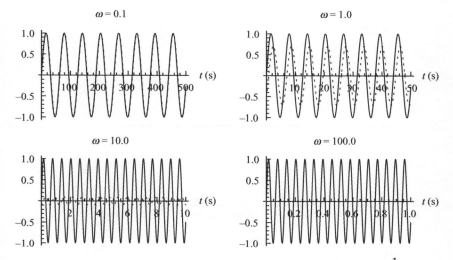

Figure 10.6. Sinusoidal input and frequency response for $G(s) = \dfrac{1}{s+1}$.

Solution   The transfer function when $s = i\omega$ is

$$
\begin{aligned}
G(i\omega) &= e^{-t_{\mathrm{d}} i\omega} = \cos(\omega t_{\mathrm{d}}) - i\sin(\omega t_{\mathrm{d}}) \\
G(i\omega) &= \cos(-\omega t_{\mathrm{d}}) + i\sin(-\omega t_{\mathrm{d}}).
\end{aligned}
\tag{10.37}
$$

The amplitude ratio is given by (Fig. 10.7)

$$
|G(i\omega)| = 1,
\tag{10.38}
$$

and the phase shift is

$$
\angle G(i\omega) = -t_{\mathrm{d}}\omega.
\tag{10.39}
$$

10.3 NYQUIST PLOTS

The Nyquist diagram uses a single polar plot to describe the frequency response. Figure 10.2 shows that a complex number (i.e., $z = a + ib$) can be represented by placing the imaginary and real parts of z in a complex plane. The magnitude, defined by $\sqrt{a^2 + b^2}$, is the length of the vector from the origin to the point, and θ is the angle of the vector with the real axis, which is shown by the x-axis in Figure 10.2. Such a plot can be used to represent the transfer function $G(i\omega)$ with magnitude $|G(i\omega)|$ and $\theta = \angle G(i\omega)$. The Nyquist diagram is generated by varying the frequency from 0 to ∞ and can be drawn by taking

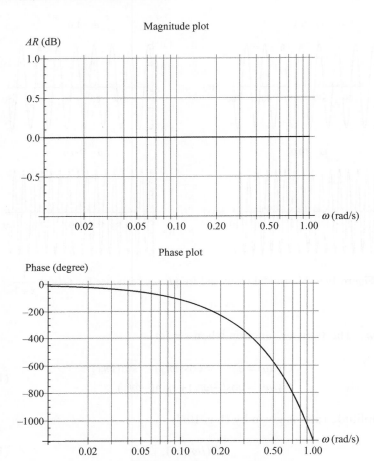

Figure 10.7. Bode plot for e^{-20s}.

information directly from the Bode plot. In Chapter 11, this tool will be implemented to study the stability of closed-loop systems.

Example 10.6 Draw the Nyquist diagram for

$$G(s) = \frac{1}{s+1}.$$

Solution It can be shown that

$$G(i\omega) = \frac{1}{1+\omega^2} - i\frac{\omega}{1+\omega^2}. \tag{10.40}$$

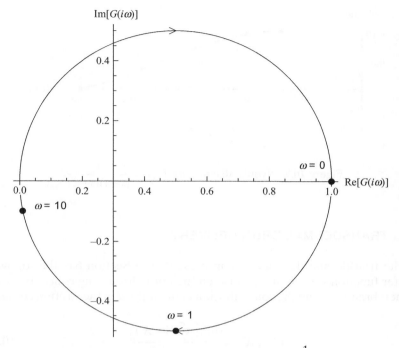

Figure 10.8. Nyquist diagram for $G(s) = \dfrac{1}{s+1}$.

Figure 10.8 shows the Nyquist plot for $G(0i) = 1$, $G(i) = \left(\frac{1}{2}\right) - i\left(\frac{1}{2}\right)$, and $G(10i) = \left(\frac{1}{101}\right) - i(101)$.

Example 10.7 Draw the Nyquist diagram for

$$G(s) = \frac{1}{(s+1)(s-2)}.$$

Solution The transfer function is

$$G(i\omega) = -\frac{2 + \omega^2}{4 + 5\omega^2 + \omega^4} + i \frac{\omega}{4 + 5\omega^2 + \omega^4}. \tag{10.41}$$

The Nyquist plot is shown in Figure 10.9.

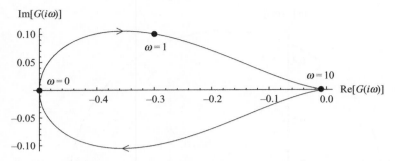

Figure 10.9. Nyquist diagram for $G(s) = \dfrac{1}{(s+1)(s-2)}$.

10.4 TRANSDERMAL DRUG DELIVERY

For the transdermal delivery system presented in Section 6.6, the following transfer function was obtained between the normalized drug concentration in the membrane (U) and the concentration at the surface of the stratum corneum:

$$G(s, X) = \frac{\sinh\left[\sqrt{s}\,(1-X)\right]}{\sinh\left(\sqrt{s}\right)}. \tag{10.42}$$

It would be difficult to analyze the dynamic behavior of $G(i\omega, 0.5)$ (e.g., stability) based on the poles. In such situations, graphical methods are viable alternatives. The Bode and Nyquist diagrams for the transfer function can be calculated by first replacing the variable s with $i\omega$:

$$G(i\omega, X) = \frac{\sinh\left[\sqrt{i\omega}\,(1-X)\right]}{\sinh\left(\sqrt{i\omega}\right)}. \tag{10.43}$$

When $X = 0.5$ is considered, the transfer function is

$$G(i\omega, 0.5) = \frac{\sinh\left(0.5\sqrt{i\omega}\right)}{\sinh\left(\sqrt{i\omega}\right)}. \tag{10.44}$$

10.4.1 Method 1

Computational software tools, such as *Mathematica*® (Wolfram Research, Inc.), can be used to plot the frequency response. The norm and argument of $G(i\omega, 0.5)$ are calculated using the commands *Abs* and *Arg*, respectively. Figure 10.10 shows the Bode diagram. Similarly, the Nyquist plot is obtained

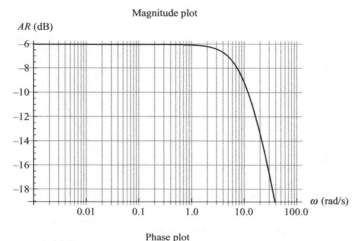

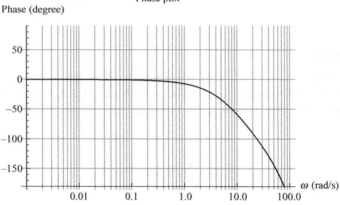

Figure 10.10. Bode plot for $G(s, 0.5) = \dfrac{\sinh(0.5\sqrt{s})}{\sinh(\sqrt{s})}$.

by drawing the curve described by the parametric equation $\{\mathrm{Re}[G(i\omega, 0.5)],$ $\mathrm{Im}[G(i\omega, 0.5)]\}$ as ω varies from 0 to 500 (Fig 10.11). It should be noted that $G(i\omega, 0.5)$ is first converted to a complex number before implementing both procedures.

Because of the presence of transcendental expressions in $G(i\omega, 0.5)$, rational functions are frequently written to generate approximate and simpler forms. For example, in expressions containing a delay term, the Padé formula can be applied to construct rational forms that are readily analyzed with frequency response tools available in MATLAB® (MathWorks, Inc.) and *Mathematica* (Wolfram Research, Inc.). A second method is outlined below to illustrate how these estimation techniques are used to compute magnitude and phase values.

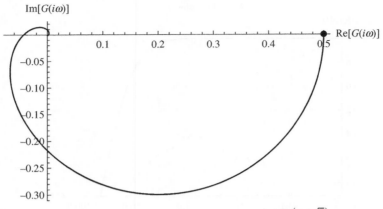

Figure 10.11. Nyquist plot for $G(s, 0.5) = \dfrac{\sinh\left(0.5\sqrt{s}\right)}{\sinh\left(\sqrt{s}\right)}$.

10.4.2 Method 2

The reduced-order methodologies, outlined in Section 8.2, are only applicable to higher-order rational functions with or without a time-delay term. In the case of $G(s, 0.5)$, one option is to minimize the squared magnitude of a complex error [2]:

$$\wp = \sum_{j=1}^{N} |\varepsilon(i\omega_j)|^2 \tag{10.45}$$

$$\varepsilon(i\omega_j) = G(i\omega_j, 0.5) - \tilde{G}(i\omega_j, 0.5), \tag{10.46}$$

where $\tilde{G}(s, 0.5)$ is the estimated transfer function and N is the number of data points. The values of the parameters are obtained after minimizing Equation (10.45). For example, if a second-order system of the form

$$\tilde{G}(s, 0.5) = \frac{K_p}{\tau^2 s^2 + 2\varsigma\tau s + 1} \tag{10.47}$$

is assumed, the optimization will yield values of K_p, ς, and τ. A more conventional approach is to use the generalized version of the Padé approximation, which converts a function into a quotient of two polynomials. The ratio is called a *Padé approximant* and is obtained from the coefficients of a Taylor series expansion. The details of the procedure are beyond the scope of this book. However, methods available in computational software can be used to derive Padé approximants. The function *PadeApproximant[expr,{x,x₀,{m,n}}]* is used in *Mathematica*. It represents the Padé approximant to the expression *expr* about the point x_0, where the numerator and denominator are of the orders m and n, respectively. For $m = 0$, $n = 2$, and $s_0 = 0$, we have

$$\tilde{G}(s, 0.5) = \frac{0.5}{0.00260s^2 + 0.125s + 1}. \tag{10.48}$$

The Bode plots for $G(s, 0.5)$ and $\tilde{G}(s, 0.5)$ are shown in Figure 10.12, while the Nyquist diagrams are plotted in Figure 10.13.

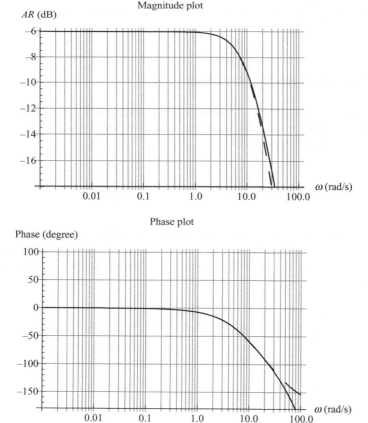

Figure 10.12. Bode plot for $G(s, 0.5) = \sinh\left(0.5\sqrt{s}\right)/\sinh\left(\sqrt{s}\right)$ (solid line) and the approximated transfer function $\tilde{G}(s, 0.5) = 0.5/(0.00260s^2 + 0.125s + 1)$ (dashed lines).

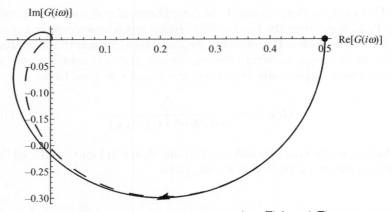

Figure 10.13. Nyquist diagram for $G(s, 0.5) = \sinh\left(0.5\sqrt{s}\right)/\sinh\left(\sqrt{s}\right)$ (solid line) and the approximated transfer function $\tilde{G}(s, 0.5) = 0.5/(0.00260s^2 + 0.125s + 1)$ (dashed lines).

10.5 COMPARTMENTAL MODELS

For anesthetic drugs, such as propofol, the relationship between the drug plasma concentration and the administered dose can be represented by the following transfer function:

$$G(s) = \frac{A_1(s)}{D} = \frac{(s + k_{21})(s + k_{31})}{(s + \lambda_1)(s + \lambda_2)(s + \lambda_3)}, \tag{10.49}$$

where A_1 is the amount of drug (i.e., mass) in compartment 1 and D is the administered dose (Fig. 10.14). The parameters λ_1, λ_2, and λ_3 are functions of the rate constants. The transfer function is derived from the mass balance equation [3]

$$\frac{dA_1}{dt} = D\delta(t) + k_{21}A_2 - k_{12}A_1 + k_{31}A_3 - k_{13}A_1 - k_{el}A_1$$

$$\frac{dA_2}{dt} = k_{12}A_1 - k_{21}A_2 \tag{10.50}$$

$$\frac{dA_3}{dt} = k_{13}A_1 - k_{31}A_3.$$

with the following rate constants (minute^{-1}): $k_{12} = 0.105, k_{21} = 0.064, k_{13} = 0.022,$ $k_{31} = 0.0034,$ and $k_{el} = 0.0827$ [3]. The poles are $\lambda_1 = 0.2468, \lambda_2 = 0.02769,$ and $\lambda_3 = 0.002633.$

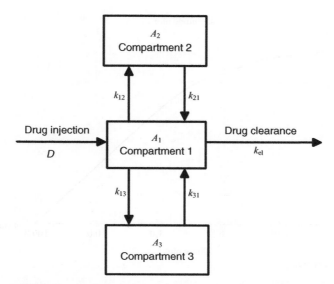

Figure 10.14. Three-compartment model as a function of the mass of drug.

To develop the frequency response for the model, we replace s by $i\omega$:

$$G(i\omega) = \frac{(i\omega + k_{21})(i\omega + k_{31})}{(i\omega + \lambda_1)(i\omega + \lambda_2)(i\omega + \lambda_3)} \quad (10.51)$$

and obtain the following expression:

$$G(i\omega) = a + ib, \quad (10.52)$$

where

$$a = \frac{\left\{ \begin{array}{l} \omega^2 \left(k_{31} (\lambda_2 \lambda_3 + \lambda_1 (\lambda_2 + \lambda_3) - \omega^2) + (\lambda_2 + \lambda_3)\omega^2 + \lambda_1 (\omega^2 - \lambda_2 \lambda_3) \right) \\ + k_{21} (\omega^2 (\lambda_2 \lambda_3 + \lambda_1 (\lambda_2 + \lambda_3) - \omega^2) - k_{31} ((\lambda_2 + \lambda_3)\omega^2 + \lambda_1 (\omega^2 - \lambda_2 \lambda_3))) \end{array} \right\}}{(\lambda_1^2 + \omega^2)(\lambda_2^2 + \omega^2)(\lambda_3^2 + \omega^2)}$$

$$(10.53)$$

and

$$b = -\frac{\left\{ \begin{array}{l} \omega \left(k_{31} ((\lambda_2 + \lambda_3)\omega^2 + \lambda_1 (\omega^2 - \lambda_2 \lambda_3)) + k_{21} (k_{31} (\lambda_2 \lambda_3 + \lambda_1 (\lambda_2 + \lambda_3) - \omega^2) \right. \\ \left. + (\lambda_2 + \lambda_3)\omega^2 + \lambda_1 (\omega^2 - \lambda_2 \lambda_3)) + \omega^2 (-\lambda_2 \lambda_3 - \lambda_1 (\lambda_2 + \lambda_3) + \omega^2)) \end{array} \right\}}{(\lambda_1^2 + \omega^2)(\lambda_2^2 + \omega^2)(\lambda_3^2 + \omega^2)}.$$

$$(10.54)$$

The Bode and Nyquist diagrams are shown in Figures 10.15 and 10.16.

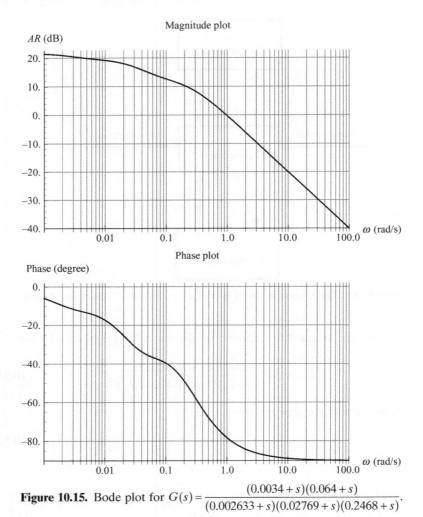

Figure 10.15. Bode plot for $G(s) = \dfrac{(0.0034 + s)(0.064 + s)}{(0.002633 + s)(0.02769 + s)(0.2468 + s)}$.

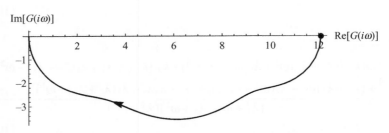

Figure 10.16. Nyquist diagram for $G(s) = \dfrac{(0.0034 + s)(0.064 + s)}{(0.002633 + s)(0.02769 + s)(0.2468 + s)}$.

10.6 SUMMARY

Techniques for conducting a frequency response analysis were proposed in this chapter. The transfer function is written as a complex number, z. The modulus and angle made by z with the positive real axis are the amplitude ratio and argument of the transfer function, respectively. Bode and Nyquist diagrams can be generated for both rational and transcendental functions. In the latter case, a mathematical software may be necessary to generate the plots. Illustrations of controlled-release systems and multicompartment models were featured.

PROBLEMS

10.1. Draw the Nyquist diagram for $G(s)=5/(s + 3)$.

10.2. Draw the Nyquist diagram for $G(s)=1/[(s + 5)(s - 2)]$.

10.3. Draw the Nyquist diagram for $G(s)=3/[(s + 2)(s - 4)]$.

10.4. Construct the Bode plot for $G(s)=1/[(s + 5)(s - 2)]$.

10.5. Construct the Bode plot for $G(s)=3(s + 3)/(s^2 - 2)$.

10.6. Construct the Bode plot for $G(s) = 1/\left[s\left(1+\tfrac{1}{2}s\right)\left(1+\tfrac{1}{3}s\right)\right]$.

10.7. The transfer function for a three-compartment model is given by (Section 10.5)

$$G(i\omega) = \frac{(i\omega+k_{21})(i\omega+k_{31})}{(i\omega+\lambda_1)(i\omega+\lambda_2)(i\omega+\lambda_3)}.$$

With $G(i\omega) = a + ib$, derive the following expressions for a and b:

$$a = \frac{\left\{\begin{array}{l}\omega^2(k_{31}(\lambda_2\lambda_3 + \lambda_1(\lambda_2 + \lambda_3) - \omega^2) + (\lambda_2 + \lambda_3)\omega^2 + \lambda_1(\omega^2 - \lambda_2\lambda_3)) \\ + k_{21}(\omega^2(\lambda_2\lambda_3 + \lambda_1(\lambda_2 + \lambda_3) - \omega^2) - k_{31}((\lambda_2 + \lambda_3)\omega^2 + \lambda_1(\omega^2 - \lambda_2\lambda_3)))\end{array}\right\}}{(\lambda_1^2 + \omega^2)(\lambda_2^2 + \omega^2)(\lambda_3^2 + \omega^2)}$$

and

$$b = -\frac{\left\{\begin{array}{l}\omega(k_{31}((\lambda_2 + \lambda_3)\omega^2 + \lambda_1(\omega^2 - \lambda_2\lambda_3)) + k_{21}(k_{31}(\lambda_2\lambda_3 + \lambda_1(\lambda_2 + \lambda_3) - \omega^2) \\ + (\lambda_2 + \lambda_3)\omega^2 + \lambda_1(\omega^2 - \lambda_2\lambda_3)) + \omega^2(-\lambda_2\lambda_3 - \lambda_1(\lambda_2 + \lambda_3) + \omega^2))\end{array}\right\}}{(\lambda_1^2 + \omega^2)(\lambda_2^2 + \omega^2)(\lambda_3^2 + \omega^2)}.$$

10.8. Given the transfer function in Problem 10.7,

$$G(i\omega) = \frac{(i\omega + k_{21})(i\omega + k_{31})}{(i\omega + \lambda_1)(i\omega + \lambda_2)(i\omega + \lambda_3)},$$

derive expressions for the modulus ($|G(i\omega)|$) and argument ($\angle[G(i\omega)]$) of $G(i\omega)$.

10.9. Show that the first-order Padé approximation (i.e., $m = n = 1$) of $G(s) = e^{-t_d s}$,

$$\tilde{G}(s) \approx \frac{1 - \dfrac{st_d}{2}}{1 + \dfrac{st_d}{2}},$$

can be derived from the formula

$$\tilde{G}(s) \approx \frac{G(0)(sG''(0) - 2G'(0)) - 2sG'(0)^2}{sG''(0) - 2G'(0)}.$$

10.10. The changes in drug concentrations in a three-compartment model are given by (Section 10.5)

$$\frac{dA_1}{dt} = D\delta(t) + k_{21}A_2 - k_{12}A_1 + k_{31}A_3 - k_{13}A_1 - k_{el}A_1$$

$$\frac{dA_2}{dt} = k_{12}A_1 - k_{21}A_2$$

$$\frac{dA_3}{dt} = k_{13}A_1 - k_{31}A_3.$$

(a) Evaluate the transfer function relating A_1, the amount of drug in compartment 1, to D, the administered dose for the following rate constants (minute^{-1}): $k_{12} = 0.105, k_{21} = 0.064, k_{13} = 0.032, k_{31} = 0.0045$, and $k_{el} = 0.071$.

(b) Draw the Bode plot of the transfer function.

(c) Construct the Nyquist diagram of the transfer function.

REFERENCES

1. Marlin TE. *Process Control: Designing Processes and Control Systems for Dynamic Performance*. Boston: McGraw-Hill, 2000.

2. Simon L, Goyal A. Dynamics and control of percutaneous drug absorption in the presence of epidermal turnover. *Journal of Pharmaceutical Sciences* 2009; 98:187–204.

3. Upton RN. Calculating the hybrid (macro) rate constants of a three-compartment mamillary pharmacokinetic model from known micro-rate constants. *Journal of Pharmacological and Toxicological Methods* 2004; 49:65–68.

CHAPTER 11

STABILITY ANALYSIS OF FEEDBACK SYSTEMS

We learned in Chapter 4 how to test the stability of linear systems. The models were written in state-space forms and the eigenvalues λ_i of a matrix were calculated. The stability of steady-state points was assessed based on the sign of the real part of λ_i, $\text{Re}(\lambda_i)$. Tools are also available to determine stability directly from transfer functions. These methods can be applied to examine the roots of the *characteristic equation*: $1 + G_{OL}(s) = 0$.

11.1 ROUTH–HURWITZ STABILITY CRITERION

The inverse Laplace transform of the output response to a unit impulse function contains exponential terms of the form $e^{-s_i t}$, where s_i is a pole of the closed-loop system. According to the *bounded-input bounded-output* stability criterion discussed in Chapter 4, a bounded response is obtained when all the poles of the closed-loop system lie in the left-half plane (LHP). Simply put, *a feedback system is stable if all the poles of the system transfer function have negative real parts*. This necessary and sufficient condition should not come as a surprise because the eigenvalues correspond to the poles in the s-plane. The system is not stable if at least one of the poles is in the right-half plane. As before, marginal stability is achieved when $\text{Re}(s_i) \leq 0$ for all i. Note that only simple roots are permitted on the imaginary axis [1].

Control of Biological and Drug-Delivery Systems for Chemical, Biomedical, and Pharmaceutical Engineering, First Edition. Laurent Simon.
© 2013 John Wiley & Sons, Inc. Published 2013 by John Wiley & Sons, Inc.

If the closed-loop transfer function is represented by $Q(s)/P(s)$, the characteristic equation is $P(s) = 1 + G_{OL}(s) = 0$. The system is stable when all the roots of $P(s) = 0$ have negative real parts. Thus, one method of testing the stability is to calculate the roots. The *Hurwitz* method makes it possible to determine if all poles are negative without solving the equation. Although this technique does not provide the actual values of the poles, it is a useful tool in stability analysis.

Consider

$$P(s) = s^n + a_1 s^{n-1} + \cdots + a_{n-1} s + a_n. \tag{11.1}$$

The characteristic equation is written as

$$s^n + a_1 s^{n-1} + \cdots + a_{n-1} s + a_n = 0. \tag{11.2}$$

Before applying the *Hurwitz stability criterion*, all the $n + 1$ terms are included in Equation (11.2) and the coefficient of the leading term (i.e., s^n) is 1. According to this criterion, all the poles s have negative real parts iff $\Delta_1 > 0, \Delta_2 > 0, \ldots,$ $\Delta_n > 0$. In addition, Δ_m is a Hurwitz determinant associated with $P(s)$:

$$\Delta_m = \det(H_m) \tag{11.3}$$

and

$$H_m = \begin{bmatrix} a_1 & 1 & 0 & 0 & 0 & 0 & \cdots & 0 \\ a_3 & a_2 & a_1 & 1 & 0 & 0 & \cdots & 0 \\ a_5 & a_4 & a_3 & a_2 & a_1 & 1 & \cdots & 0 \\ \vdots & \vdots & \vdots & \vdots & \vdots & \vdots & \ddots & \vdots \\ 0 & 0 & 0 & 0 & 0 & 0 & \cdots & a_m \end{bmatrix}, \tag{11.4}$$

where a_i is replaced by 0 for $i > m$ [2]; H_m is called the *Hurwitz matrix* of the polynomial $P(s)$. Although a detailed discussion of the Hurwitz matrix is beyond the scope of this book, some examples are given below to help assess the stability of a system and to show how a_i is calculated. Note that n such Hurwitz matrices should be calculated for a polynomial of order n.

For example, when $n = 1$,

$$\Delta_1 = \det[a_1] = a_1 > 0. \tag{11.5}$$

For $n = 2$,

$$\Delta_1 = a_1 > 0,$$

$$\Delta_2 = \det\left(\begin{bmatrix} a_1 & 1 \\ 0 & a_2 \end{bmatrix}\right) = \begin{vmatrix} a_1 & 1 \\ 0 & a_2 \end{vmatrix} = a_1 a_2 > 0. \tag{11.6}$$

For $n = 3$,

$$\Delta_1 = a_1 > 0,$$

$$\Delta_2 = \begin{vmatrix} a_1 & 1 \\ a_3 & a_2 \end{vmatrix} = a_1 a_2 - a_3 > 0,$$

$$\Delta_3 = \det\left(\begin{bmatrix} a_1 & 1 & 0 \\ a_3 & a_2 & a_1 \\ 0 & 0 & a_3 \end{bmatrix}\right) = \begin{vmatrix} a_1 & 1 & 0 \\ a_3 & a_2 & a_1 \\ 0 & 0 & a_3 \end{vmatrix} = a_1 \begin{vmatrix} a_2 & a_1 \\ 0 & a_3 \end{vmatrix} - a_3 \begin{vmatrix} 1 & 0 \\ 0 & a_3 \end{vmatrix} \qquad (11.7)$$

$$= a_1 a_2 a_3 - a_3^2 = a_3 (a_1 a_2 - a_3) > 0.$$

When $n = 4$,

$$\Delta_1 = a_1 > 0,$$

$$\Delta_2 = \begin{vmatrix} a_1 & 1 \\ a_3 & a_2 \end{vmatrix} > 0,$$

$$\Delta_3 = \begin{vmatrix} a_1 & 1 & 0 \\ a_3 & a_2 & a_1 \\ 0 & a_4 & a_3 \end{vmatrix} > 0 \qquad (11.8)$$

$$\Delta_4 = \det\left(\begin{bmatrix} a_1 & 1 & 0 & 0 \\ a_3 & a_2 & a_1 & 1 \\ 0 & a_4 & a_3 & a_2 \\ 0 & 0 & 0 & a_4 \end{bmatrix}\right) = \begin{vmatrix} a_1 & 1 & 0 & 0 \\ a_3 & a_2 & a_1 & 1 \\ 0 & a_4 & a_3 & a_2 \\ 0 & 0 & 0 & a_4 \end{vmatrix} = a_1 a_2 a_3 a_4 - a_3^2 a_4 - a_1^2 a_4^2$$

$$= a_4 \left(a_1 a_2 a_3 - a_3^2 - a_1^2 a_4\right) = a_4 \left(a_3 (a_1 a_2 - a_3) - a_1^2 a_4\right) > 0.$$

Remarks

- The Hurwitz stability criterion involves the computation of Hurwitz determinants.
- It is necessary (but not sufficient) that all of the polynomial coefficients $a_1, a_2, \ldots, a_n$ be positive in order for the equation $P(s) = 0$ to have only roots with negative real parts.
- The method is time-consuming for higher-order systems and does not predict the number of poles with positive real parts.
- The prediction of marginal stability is difficult.

Example 11.1 Study the stability of a fourth-order system with the following characteristic equation:

$$1 + G_{\text{OL}}(s) = s^4 + 2s^3 + s^2 + s + 5 = 0. \qquad (11.9)$$

Solution   This is the case of $n = 4$. Thus, $a_1 = 2, a_2 = 1, a_3 = 1, a_4 = 5$:

$$\Delta_1 = 2 > 0,$$

$$\Delta_2 = \begin{vmatrix} 2 & 1 \\ 1 & 1 \end{vmatrix} = 1 > 0,$$

$$\Delta_3 = \begin{vmatrix} 2 & 1 & 0 \\ 1 & 1 & 2 \\ 0 & 5 & 1 \end{vmatrix} = -19 < 0.$$

The system is unstable and there is no need to compute Δ_4.

The *Routh–Hurwitz stability criterion* is usually applied instead of the Hurwitz stability criterion. The first step is to generate a *Routh array* from the coefficients of the characteristic equation, $a_0 s^n + a_1 s^{n-1} + \cdots + a_{n-1} s + a_n = 0$:

$$
\begin{array}{c|ccccc}
s^n & a_0 & a_2 & a_4 & a_6 & \cdots \\
s^{n-1} & a_1 & a_3 & a_5 & a_7 & \cdots \\
s^{n-2} & b_1 & b_2 & b_3 & \cdots & \cdots \\
s^{n-3} & c_1 & c_2 & \cdots & \cdots & \cdots \\
\vdots & \vdots & \vdots & \vdots & \vdots & \vdots \\
s^0 & \cdots & \cdots & \cdots & \cdots & \cdots
\end{array}
\tag{11.10}
$$

where

$$b_1 = \frac{a_1 a_2 - a_0 a_3}{a_1}, \; b_2 = \frac{a_1 a_4 - a_0 a_5}{a_1}, \; b_3 = \frac{a_1 a_6 - a_0 a_7}{a_1},$$

$$c_1 = \frac{b_1 a_3 - a_1 b_2}{b_1}, \; c_2 = \frac{b_1 a_5 - a_1 b_3}{b_1}.$$

The procedure is repeated to compute the remaining coefficients until all additional entries in the table become zero. According to the Routh–Hurwitz stability criterion, *the number of roots, with positive real parts, of the characteristic equation $a_0 s^n + a_1 s^{n-1} + \cdots + a_{n-1} s + a_n = 0$ is equal to the number of sign changes observed in the first column of the array.* A feedback system is stable if there are no sign changes in the first column of the array.

Remarks   It is customary to make a_0 positive by multiplying the characteristic equation by -1, if necessary. The following observations are made, assuming that a_0 is positive:

- Just as in the Hurwitz test, all the coefficients $a_1, a_2, \ldots, a_n$ must be positive. However, this is only a necessary condition. The *Routh table* should be formed before assessing the system stability.
- If any coefficient is negative, the system is unstable and there is no need to form the Routh array.
- If any element in the first column is negative, the system is unstable.
- The number of sign changes in the first column of the array is equal to the number of roots with positive real parts (i.e., unstable pole).

Example 11.2 Study the stability of the problem given in Example 11.1 by using the Routh–Hurwitz criterion:

$$s^4 + 2s^3 + s^2 + s + 5 = 0. \tag{11.11}$$

Solution In this case, $a_0 = 1$, $a_1 = 2$, $a_2 = 1$, $a_3 = 1$, $a_4 = 5$. Because the coefficients are positive, we build the array:

$$
\begin{array}{c|ccc}
s^4 & a_0 & a_2 & a_4 \\
s^3 & a_1 & a_3 & 0 \\
s^2 & b_1 & b_2 & 0 \\
s & c_1 & 0 & 0 \\
s^0 & d_1 & 0 & 0
\end{array}
$$

The elements are

$$b_1 = \frac{a_1 a_2 - a_0 a_3}{a_1} = \frac{1}{2} \quad \text{and} \quad b_2 = \frac{a_1 a_4 - a_0 a_5}{a_1} = 5 \quad (\text{Note: } a_5 = 0),$$

$$c_1 = \frac{b_1 a_3 - a_1 b_2}{b_1} = -19 \quad \text{and} \quad d_1 = \frac{c_1 b_2 - b_1 c_2}{c_1} = 5.$$

As a result, the array is

$$
\begin{array}{c|ccc}
s^4 & 1 & 1 & 5 \\
s^3 & 2 & 1 & 0 \\
s^2 & \dfrac{1}{2} & 5 & 0 \\
s & -19 & 0 & 0 \\
s^0 & 5 & 0 & 0
\end{array}
$$

The system is unstable because of -19 in the first column. Furthermore, there are two sign changes (i.e., $+\frac{1}{2}$ to -19 and -19 to 5). As a result, we expect two roots with positive real parts. The roots can be obtained using *Mathematica*® (Wolfram Research, Inc.): $s_1 = 0.5950 - 1.085i$, $s_2 = 0.5950 + 1.085i$, $s_3 = -1.595 - 0.8490i$, and $s_3 = -1.595 + 0.8490i$.

11.2 ROOT LOCUS ANALYSIS

The Routh–Hurwitz stability criterion is based on the signs of the real parts of the roots of the characteristic equation: $1 + G_{OL}(s) = 0$. An alternative and straightforward method, when feasible, is to calculate all of the roots, which helps to reach similar conclusions. The *root-locus* approach is a graphical method that is applied to determine stability when the equation contains a parameter (*usually positive*) such as a proportional gain. The path followed by the roots (i.e., *root locus*) of $1 + G_{OL}(s) = 0$ is shown in a complex plane.

Example 11.3   Consider the following open-loop transfer function:

$$G_{OL}(s) = \frac{K_c(s+1)}{s(s+3)}. \qquad (11.12)$$

Construct the root-locus diagram.

Solution   The characteristic equation is $sK_c + K_c + s^2 + 3s = 0$ and the roots (or closed-loop poles) are

$$s_1 = \frac{1}{2}\left(-K_c - \sqrt{K_c^2 + 2K_c + 9} - 3\right) \quad \text{and} \quad s_2 = \frac{1}{2}\left(-K_c + \sqrt{K_c^2 + 2K_c + 9} - 3\right).$$

The following table shows a range of $[s_1 \ s_2]$ when K_c varies from 0 to 100 in increments of 10. MATLAB® (MathWorks, Inc.) is used to construct the root locus with the command "*rlocus*" from the control toolbox (Fig. 11.1):

$$\begin{bmatrix} -3. & 0. \\ -12.2 & -0.821 \\ -22.1 & -0.905 \\ -32.1 & -0.936 \\ -42. & -0.951 \\ -52. & -0.961 \\ -62. & -0.967 \\ -72. & -0.972 \\ -82. & -0.975 \\ -92. & -0.978 \\ -102. & -0.98 \end{bmatrix}.$$

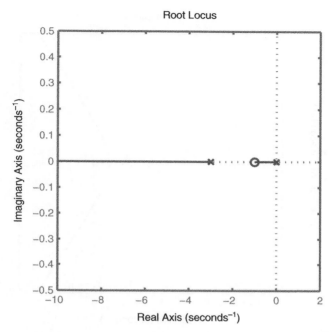

Figure 11.1. *Root locus* of the open-loop transfer function $G_{OL}(s) = \dfrac{K_c(s+1)}{s(s+3)}$.

Example 11.4 Consider the following open-loop transfer function:

$$G_{OL}(s) = \frac{K_c}{s(s+3)(s+2)}. \qquad (11.13)$$

Construct the root-locus diagram.

Solution The root locus is shown in Figure 11.2. The characteristic equation is $s^3 + 5s^2 + 6s + k = 0$. For values of K_c greater than 30.0, the closed-loop system is unstable (i.e., positive real part).

11.3 BODE STABILITY CRITERION

A closed-loop system is unstable if the frequency response of the open-loop transfer function $G_{OL}(s)$ has an amplitude ratio larger than 1 at the *phase crossover frequency* (also called *critical frequency*) ω_c, defined as the

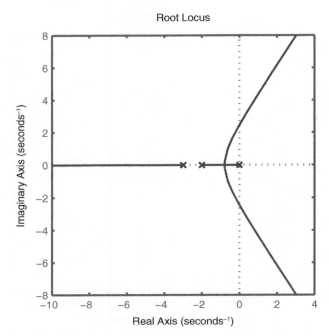

Figure 11.2. Root locus of the open-loop transfer function $G_{OL}(s) = \dfrac{K_c}{s(s+3)(s+2)}$.

frequency where the phase angle is $-180°$: $\angle[G_{OL}(j\omega)] = -180°$. Otherwise, the closed-loop system is stable. This criterion only applies to open-loop stable systems. In addition, the phase angles and amplitude ratio plots of $G_{OL}(s)$ should be monotonic [3]. The parameter ω_c should not be confused with the gain frequency ω_g, the value where the amplitude ratio is 1.

An experiment is described below to help understand the Bode stability criterion:

1. The loop of a closed-loop system is initially open between the comparator and the measuring device (Fig. 11.3). A sinusoidal wave is introduced as the reference input $r(s) = A\omega/(s^2 + \omega^2)$. Because the plant is linear and stable, a frequency response $y_{m,ss}$ with constant amplitude is obtained after a long time, as shown in Chapter 10.
   - The open-loop transfer function, $G_{OL}(s){=}G_p(s)G_f(s)G_c(s)G_m(s)$, relates $\bar{y}_m(s)$ to $\bar{r}(s)$: $\bar{y}_m(s) = G_{OL}(s)\bar{r}(s)$.
   - The frequency response is $y_{m,ss}(t) = A|G_{OL}(j\omega)|\sin(\omega t + \phi)$, where ϕ is the argument of $G_{OL}(j\omega)$: $\angle[G_{OL}(j\omega)]$.
   - The amplitude ratio is $|G_{OL}(j\omega)|$.

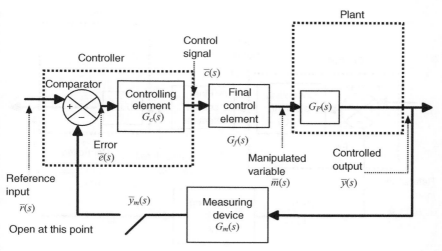

Figure 11.3. Closed-loop feedback system opened between the comparator and the measuring device.

- The frequency ω is selected so that the phase delay $\phi = -180°$. Assume $A = 1$ for simplicity. As a result, $y_{m,ss}(t) = |G_{OL}(j\omega)| \sin(\omega_c t - 180°) = -|G_{OL}(j\omega)| \sin(\omega_c t)$.

2. The system is closed and the reference input is set at 0: $r = 0$ (Fig. 11.4). Now the error is $e(t) = |G_{OL}(j\omega_c)| \sin(\omega_c t)$ because of the negative feedback.

   - If $|G_{OL}(j\omega_c)| > 1$, the magnitude of the error is amplified as the sine wave continues to travel around the loop. In this case, the system is unstable because the amplitude of the signal becomes infinitely large.
   - If $|G_{OL}(j\omega_c)| < 1$, the magnitude of the error signal decreases as the sine wave passes around the loop. In this case, the system is stable because the amplitude of the signal becomes vanishingly small.
   - If $|G_{OL}(j\omega_c)| = 1$, the system oscillates continuously with constant amplitude.

The following steps should be followed when applying the Bode stability criterion:

1. Derive the open-loop transfer function $G_{OL}(s)$.
2. Derive an expression for the phase angle ϕ.
3. Find the critical frequency ω_c from the phase angle equation.
4. Derive the expression for the amplitude ratio of $G_{OL}(s)$.
5. Test if $|G_{OL}(j\omega_c)| > 1$.

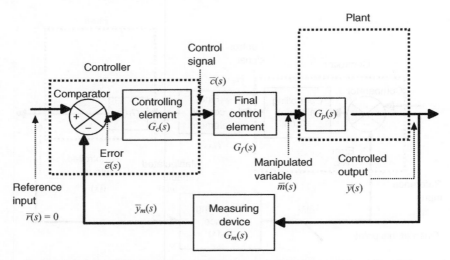

Figure 11.4. Closed-loop feedback system with the set point set at 0.

Two parameters are defined in the context of the Bode stability criterion: the *gain margin GM*:

$$GM = \frac{1}{|G_{OL}(j\omega_c)|} \qquad (11.14)$$

and the *phase margin PM*:

$$PM = 180° + \phi_1, \qquad (11.15)$$

where ϕ_1 is the phase angle at which $|G_{OL}(j\omega_1)| = 1$ with ω_1 being the corresponding frequency. A system is stable if $GM > 1$. The gain margin is an indicator of how far away the system is from instability. With a high GM, a process control designer has considerable flexibility in tuning the controller. The gain margin is generally chosen greater than 1.7 and can reach 3.0 if the modeling error is large (e.g., uncertain parameters) [4].

Similarly, the phase margin represents *the additional lag required for the system to become unstable.* For a well-tuned controller, phase margins should range from 30° to 45° [5].

Example 11.5 A process is defined by

$$G_P(s) = \frac{3}{4s+1}. \qquad (11.16)$$

Study the stability of the closed-loop system when $G_m(s) = 1$, $G_c(s) = K_c = 10$, $G_f(s) = 1$.

Solution

1. The open-loop transfer function is

$$G_{OL}(s) = G_P(s)G_f(s)G_c(s)G_m(s) = \frac{30}{4s+1}.$$

2. The phase angle is $\phi = \tan^{-1}(-\omega\tau_p) = \tan^{-1}(-4\omega)$.
3. Because the phase angle of a first-order system only varies from 0 to $-90°$ (see Section 10.2), there is no phase crossover frequency. As a result, the system is stable.

Example 11.6 A process is defined by

$$G_P(s) = \frac{3e^{-0.4s}}{4s+1}. \tag{11.17}$$

Study the stability of the closed-loop system when $G_m(s) = 1$, $G_c(s) = K_c = 10$, $G_f(s) = 1$.

Solution

1. The open-loop transfer function is $G_{OL}(s) = G_p(s)G_f(s)G_c(s)G_m(s) = 30e^{-0.4s}/4s + 1$.
2. The phase angle is $\phi = \tan^{-1}(-\omega\tau_p) - t_d\omega = \tan^{-1}(-4\omega) - 0.4\omega$.
3. To find the critical frequency, we have $-\pi = \tan^{-1}(-4\omega_c) - 0.4\omega_c$. The solution to this equation is $\omega_c = 4.08$.
4. The amplitude ratio of $G_{OL}(s)$ is
$$|G_{OL}(j\omega)| = \left(30/\sqrt{16\omega^2 + 1}\right) \times 1 = 30/\sqrt{16\omega^2 + 1}.$$
5. $|G_{OL}(j\omega_c)| = 30/\sqrt{16\omega_c^2 + 1} = 1.83$.

Thus, the system is unstable. The gain margin is $GM = \frac{1}{1.83} = 0.545$. In decibels, $GM = 20 \log_{10}(0.545) = -5.27$ dB.

The phase margin is $PM = 180° + \phi_1$. The frequency ω_1 is obtained by solving the equation $30/\sqrt{16\omega^2 + 1} = 1$. The two values are -7.496 and 7.496. Since $\omega_1 \geq 0$, the solution is $\omega_1 = 7.496$. Consequently, $\phi_1 = \tan^{-1}(-4\omega_1) - 0.4\omega_1 = -4.536$. The phase angle, expressed in degrees, is $\phi_1 = -259.88°$. Then, $PM = 180° - 259.88° = -79.88°$. The Control toolbox in MATLAB is used to generate Figure 11.5.

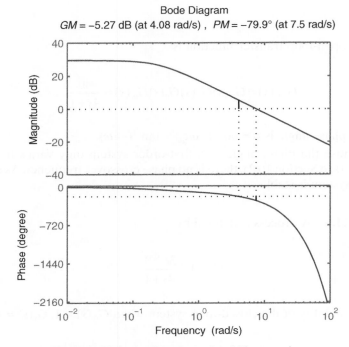

Figure 11.5. Bode diagram and stability margins.

Remarks

- Compared to the Routh–Hurwitz stability criterion, the Bode approach does not involve the analysis of roots.
- The method is feasible for transfer functions containing transcendental expressions.
- With the gain margin, it is possible to assess how far away the system is from instability. This information may be used as a measure of robustness.
- A numerical solution procedure may be necessary to estimate the critical frequency.

11.4 NYQUIST STABILITY CRITERION

It was mentioned that, in cases where the phase angles and amplitude ratio plots of $G_{OL}(j\omega)$ were not monotonic with respect to ω, the Bode stability criterion could not be applied. However, the Nyquist diagram can be used to analyze those frequency data. The method is also relevant for open-loop unstable systems. We have shown, in Section 10.3, how a Nyquist diagram can be generated by varying the frequency from 0 to ∞. Also, recall that a closed-loop system is stable if the roots of the characteristic equation $1 + G_{OL}(s) = 0$ are in the LHP. The *Nyquist stability criterion* projects the characteristics of the roots on the diagram to help establish a stability test that is based on the Nyquist plot.

The open-loop transfer function can be written as a rational function: $G_{OL}(s) = N(s)/D(s)$. We notice that the zeros of $1 + G_{OL}(s)$ are the poles of the closed-loop transfer function. The equation $1 + G_{OL}(s) = 0$ becomes $1 + N(s)/D(s) = 0$ or $(D(s) + N(s))/D(s) = 0$. In terms of zeros and poles, the form $(D(s) + N(s))/D(s)$ can be adopted:

$$1 + G_{OL}(s) = \frac{D(s) + N(s)}{D(s)} = \frac{\prod_{i=1}^{n}(s - z_i)}{\prod_{j=1}^{n}(s - p_j)}, \tag{11.18}$$

where n is the order of the numerator and denominator. The zeros and poles are represented by z_i and p_j, respectively. *Let P be the number of poles of $1 + G_{OL}(s)$ in the right-half plane.* Note that $1 + G_{OL}(s)$ and $G_{OL}(s)$ have the same poles, which can be found by solving $D(s) = 0$. *The number of zeros of $1 + G_{OL}(s)$ in the right-half plane (or the number of unstable poles of the closed-loop transfer function) is represented by Z.* The Nyquist stability criterion states that the number of encirclements (N) of the point $(-1, 0)$ made by the Nyquist plot in the clockwise direction is

$$N = Z - P. \tag{11.19}$$

The number Z can easily be determined from the Nyquist plot (i.e., N) and knowledge of P. According to the Nyquist stability test, *given an open-loop transfer function, the closed-loop system is unstable if there is any net encirclement of the point $(-1, 0)$.* Note that the transfer function $G_{OL}(s) = -1 + 0i$ has an amplitude ratio of 1 and a phase angle of $-180°$.

Example 11.7   Consider the open-loop transfer function $G_{OL}(s) = K_c/(s + 1)$. Study the stability of the closed-loop system for $K_c = 1$ using the Nyquist stability criterion.

Solution   The Nyquist plot of this transfer function was drawn in Example 10.6. Since $P = 0$ and $Z = 0$ (Fig. 10.8), the feedback closed-loop system is stable.

Example 11.8   Consider the open-loop transfer function

$$G_{OL}(s) = \frac{K_c}{(s + 1)(3s + 1)(5s + 1)}.$$

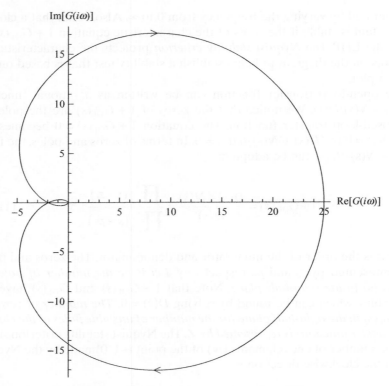

Figure 11.6. Nyquist plot of $G_{\text{OL}}(s) = \dfrac{25}{(s+1)(3s+1)(5s+1)}$.

Study the stability of the closed-loop system for $K_c = 25$ using the Nyquist–Bode stability criterion.

Solution   The Nyquist diagram, described in Figure 11.6, shows that the closed-loop system is unstable with two encirclements around -1 (i.e., $N = 2$). In addition, the number of unstable closed-loop poles is 2 since the number of poles in the RHP of the open-loop system is 0. The closed-loop response to a unit step change in the set point $\overline{r}(s) = 1/s$ is

$$\overline{y}(s) = \frac{G_{\text{OL}}(s)}{1+G_{\text{OL}}(s)} \frac{1}{s}.$$

Figure 11.7, depicting the time profile of $y(t)$, shows that the closed-loop system is indeed unstable. The poles of the closed-loop transfer function, $p_1 = -1.75$, $p_2 = 0.11 + 0.99i$, and $p_3 = 0.11 - 0.99i$, are in agreement with the above results: $Z = 2$.

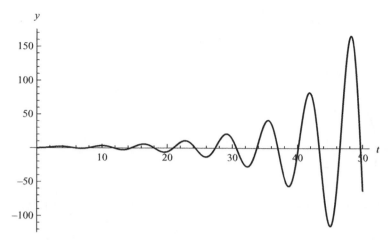

Figure 11.7. Closed-loop response to a step change in the set point.

Example 11.9   Repeat Example 11.8 with $K_c = 5$.

Solution   Following the procedure above, we have $N = 0$, $P = 0$, and $Z = 0$ (Fig. 11.8). The closed-loop system is stable.

The poles of the closed-loop transfer function are $p_1 = -1.31$, $p_2 = -0.112 + 0.541i$, and $p_3 = -0.112 - 0.541i$. Figure 11.9 shows the closed-loop response to a unit step change in the set point. An offset remains as a result of the proportional controller. Note that the gain cannot be increased indefinitely to reduce the offset because of stability issues associated with large K_c values.

11.5 CHEYNE–STOKES RESPIRATION

Under normal oxygen concentration (*normoxic condition*), which is about 20–21% in the atmosphere, breathing is almost entirely controlled by the partial pressure of carbon dioxide in the arterial blood P_{aCO2} [6]. *Periodic breathing* (BP) is a cyclic breathing pattern described by smooth increases and decreases in ventilation with cycle lengths ranging from ~25 to 100 seconds (0.01–0.04 Hz) [7]. Such patterns may occur during sleep, when engaging in mental activity, or when climbing to a higher altitude to make up for variations in serum partial pressures of oxygen and carbon dioxide. The oscillations are between states of hyperpnea (e.g., rise in the rate or the depth of breathing) and hypopnea (e.g., very shallow breathing or an abnormally low respiratory rate). *Cheyne–Stokes respiration*, an excessive form of BP, is routinely observed in patients with congestive heart failure. This condition is characterized by a cyclic modulation between hyperpnea and apnea, which is a

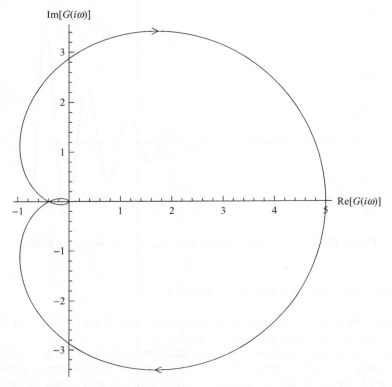

Figure 11.8. Nyquist plot of $G_{OL}(s) = \dfrac{5}{(s+1)(3s+1)(5s+1)}$.

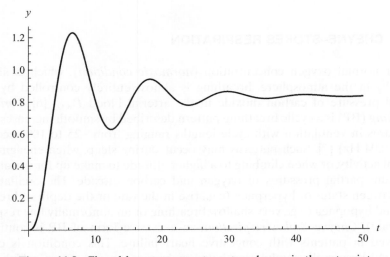

Figure 11.9. Closed-loop response to a step change in the set point.

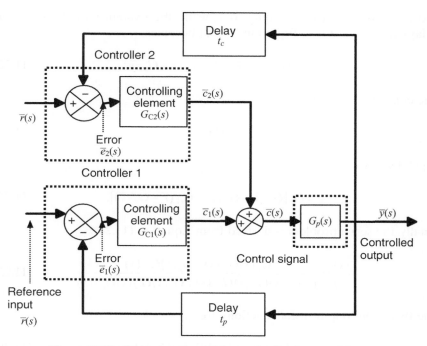

Figure 11.10. Control system with two feedback control loops.

temporary cessation of respiration. In Khoo [6], a Nyquist stability analysis is conducted to test the assumption that BP is the consequence of instability in the feedback control system that governs ventilation and arterial blood gases. Comparisons are drawn between normal subjects and patients with congestive heart failure.

Before performing the analysis, it is important to derive equations for closed-loop systems when *two feedback loops are involved* (Fig. 11.10). A review of the *set point tracking*, or *servo* problems, may be necessary before proceeding with the rest of this section. The net control is described by the sum of the two controllers' outputs:

$$\bar{c}(s) = \bar{c}_1(s) + \bar{c}_2(s) \tag{11.20}$$

or

$$\bar{c}(s) = G_{c1}(s)\bar{e}_1(s) + G_{c2}(s)\bar{e}_2(s), \tag{11.21}$$

where the controller transfer functions are given by $G_{c1}(s)$ and $G_{c2}(s)$. For simplicity, we assume that the transfer functions of the final control element and measuring device are equal to 1, while the dynamics of the transport delays are G_{m1} and G_{m2}. Equation (11.21) becomes

$$\bar{c} = G_{c1}(\bar{r} - G_{m1}\bar{y}) + G_{c2}(\bar{r} - G_{m2}\bar{y}).$$ (11.22)

The controlled output is

$$\bar{y} = G_p\bar{c},$$ (11.23)

which leads to

$$\bar{y} = G_p[G_{c1}(\bar{r} - G_{m1}\bar{y}) + G_{c2}(\bar{r} - G_{m2}\bar{y})].$$ (11.24)

Finally, the servo problem is derived from Equation (11.24):

$$\frac{\bar{y}(s)}{\bar{r}(s)} = \frac{G_p(s)G_{c1}(s) + G_p(s)G_{c2}(s)}{1 + G_p(s)G_{c1}(s)G_{m1}(s) + G_p(s)G_{c2}(s)G_{m2}(s)}.$$ (11.25)

The two open-loop transfer functions are

$$H_{OL1}(s) = G_p(s)G_{c1}(s)G_{m1}(s)$$ (11.26)

and

$$H_{OL2}(s) = G_p(s)G_{c2}(s)G_{m2}(s).$$ (11.27)

Consequently, the overall transfer function is given by

$$H_{OL}(s) = H_{OL1}(s) + H_{OL2}(s).$$ (11.28)

The Nyquist plot can be applied to $H_{OL}(s)$ when assessing the stability of the closed-loop system.

In the analysis conducted by Khoo, the two feedback loops are the *central chemoreflex* and the *peripheral chemoreflex* [6]. The dynamics of the *peripheral controller* (i.e., controller 1) is

$$G_{c1}(s) = G_{V_pP_p}(s) = \frac{\bar{V}_p(s)}{\bar{P}_{pCO_2}(s)} = \frac{K_p}{\tau_p s + 1},$$ (11.29)

where $\bar{V}_p(s)$ is the *peripheral* response (i.e., $\bar{c}_1(s) = \bar{V}_p(s)$) and $\bar{P}_{pCO_2}(s)$ is the input to the peripheral controller. Similarly, the transfer function of the *central controller* is

$$G_{c2}(s) = G_{V_c P_c}(s) = \frac{\bar{V}_c(s)}{\bar{P}_{cCO_2}(s)} = \frac{K_c}{\tau_c s + 1},$$ (11.30)

with $\bar{V}_c(s)$ representing the *central* response (i.e., $\bar{c}_2(s) = \bar{V}_c(s)$) and $\bar{P}_{cCO_2}(s)$ is the input to the central controller. The Laplace transform of the total ventilation $\bar{V}_E(s)$ is related to $\bar{P}_{aCO_2}(s)$ via the following relationship:

$$G_p = G_L(s) = \frac{\bar{P}_{aCO_2}(s)}{\bar{V}_E(s)} = \frac{K_{pL}}{\tau_L s + 1},$$ (11.31)

where K_{pL} and τ_L depend on the steady-state levels of total ventilation and carbon dioxide partial pressure and other parameters. The two delay functions are $G_{m1} = e^{-l_p s}$ and $G_{m2} = e^{-l_c s}$, which give

$$\bar{P}_{pCO_2}(s) = \bar{P}_{aCO_2}(s)e^{-l_p s}$$ (11.32)

and

$$\bar{P}_{cCO_2}(s) = \bar{P}_{aCO_2}(s)e^{-l_c s}.$$ (11.33)

The two open-loop transfer functions become

$$H_{OLp}(s) = G_L(s)G_{V_p P_p}(s)e^{-l_p s} = \frac{K_{pL}K_p e^{-l_p s}}{(\tau_L s + 1)(\tau_p s + 1)}$$ (11.34)

and

$$H_{OLc}(s) = G_L(s)G_{V_c P_c}(s)e^{-l_c s} = \frac{K_{pL}K_c e^{-l_c s}}{(\tau_L s + 1)(\tau_c s + 1)}.$$ (11.35)

The Nyquist stability criterion can be applied to the open-loop transfer function:

$$H_{OL}(s) = \frac{K_{pL}K_p e^{-l_p s}}{(\tau_L s + 1)(\tau_p s + 1)} + \frac{K_{pL}K_c e^{-l_c s}}{(\tau_L s + 1)(\tau_c s + 1)}.$$ (11.36)

Parameters, representative of typical normal subjects and patients with congestive heart failure, are given in Khoo [6].

11.6   REGULATION OF BIOLOGICAL PATHWAYS

Buzi [8] considered a simplified and linearized two-state representation of autocatalytic networks (see glycolysis in Sections 3.3 and 4.6) involving ATP (y) and a lumped intermediate metabolite (x) to denote the states [8]. The model,

$$\frac{dx}{dt} = -k_x x + \frac{1}{b} r_y \left(q - \hat{h} \right) y$$

$$\frac{dy}{dt} = (b+a)k_x x + \left(-\frac{a}{b} r_y \left(q - \hat{h} \right) - k_y \right) y, \tag{11.37}$$

can be viewed as a closed-loop process regulated by the enzyme phosphofructokinase with the feedback gain given by $\hat{h}$. The parameter q denotes the strength of autocatalysis; a is the number of ATP molecules consumed by the pathway, while b is the number produced; k_x and k_y are reaction rate constants; r_y is the ATP consumption rate evaluated at $y = 1$. It is of interest to determine the values of the gain that makes the closed-loop system stable.

After making the substitution $u = -\hat{h}y$, we have

$$\frac{dx}{dt} = -k_x x + \frac{1}{b} r_y qy + \frac{1}{b} r_y u$$

$$\frac{dy}{dt} = (b+a)k_x x + \left(-\frac{a}{b} r_y q - k_y \right) y - \frac{a}{b} r_y u, \tag{11.38}$$

where u is the input.

Applying Laplace transforms to Equation (11.38) results in

$$\bar{x}(s) = \frac{\left(k_y r_Y + s r_Y \right) \bar{u}}{-aqsr_Y + bqk_x r_Y - bsk_x - bsk_y - bk_x k_y - bs^2} \tag{11.39}$$

and

$$\bar{y}(s) = \frac{\left(bk_y r_Y - asr_Y \right) \bar{u}(s)}{-aqsr_Y + bqk_x r_Y - bsk_x - bsk_y - bk_x k_y - bs^2}. \tag{11.40}$$

The transfer function relating $\bar{y}(s)$ to $\bar{u}(s)$ is

$$G(s) = \frac{-\dfrac{asr_y}{b} + k_x r_y}{s^2 + \left(\dfrac{aqr_y}{b} + k_x + k_y \right) s - qk_x r_y + k_x k_y}, \tag{11.41}$$

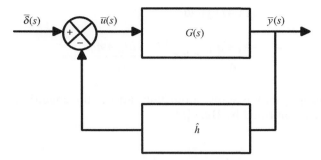

Figure 11.11. Closed-loop system subjected to a disturbance $\bar{\delta}(s)$.

and the closed-loop transfer function relating $\bar{y}(s)$ to a disturbance $\bar{\delta}(s)$ in ATP consumption is (Fig. 11.11)

$$G_\delta(s) = \frac{G(s)}{1 + \hat{h}G(s)}. \tag{11.42}$$

The characteristic equation is

$$1 + \hat{h}G(s) = \frac{(h-q)r_Y(bk_x - as) + b(k_x + s)(k_y + s)}{qr_Y(as - bk_x) + b(k_x + s)(k_y + s)} = 0 \tag{11.43}$$

or

$$s^2 + sk_x + sk_y + k_xk_y - \frac{ahsr_y}{b} + \frac{aqsr_y}{b} + hk_xr_y - qk_xr_y = 0, \tag{11.44}$$

which can be written as

$$s^2 + \left(k_x + k_y - \frac{a}{b}hr_y + \frac{a}{b}qr_y \right)s + k_xk_y + hk_xr_y - qk_xr_y = 0. \tag{11.45}$$

From the Hurwitz stability criterion

$$k_x + k_y - \frac{a}{b}hr_y + \frac{a}{b}qr_y > 0 \tag{11.46}$$

and

$$k_xk_y + hk_xr_y - qk_xr_y > 0. \tag{11.47}$$

Equations (11.46) and (11.47) yield

$$\frac{-k_y + qr_Y}{r_Y} < h < \frac{bk_x + bk_y + aqr_Y}{ar_Y}. \tag{11.48}$$

The closed-loop system is stable if the gain satisfies the condition set in Equation (11.48) as confirmed by Buzi [8].

11.7 PUPILLARY LIGHT REFLEX

The concept of pupillary light reflex, as an example of physiological control, was discussed in Chapter 1. The block diagram of a linearized approximation of the control system is given in Figure 11.12. Laplace transforms of deviation variables are included in the figure. The variable $\bar{A}(s)$ is the output (i.e., change in area) of the controller; A_{ref} is the reference value of the area; A_c is the controlled area. The controlled light flux is obtained by multiplying A_c by I_{AV}, an average intensity value [9]. The open-loop transfer function is

$$G_{OL}(s) = G_c(s)I_{AV} \tag{11.49}$$

and, according to findings reported in Sherman and Stark [9], $G_{OL}(s)$ is expressed by

$$G_{OL}(s) = \frac{Ke^{-0.18s}}{(1+0.1s)^3}, \tag{11.50}$$

where $K = 0.16$.

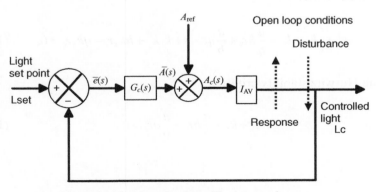

Figure 11.12. Linearized pupillary control system.

The stability of pupillary reflex can be analyzed using the Bode stability criterion. A procedure outlined in Example 11.6 is shown below:

1. The open-loop transfer function is

$$G_{OL}(s) = \frac{0.16e^{-0.18s}}{(1+0.1s)^3}.$$

2. The phase angle is $\phi = n\tan^{-1}(-\omega\tau_p) - t_d\omega = 3\tan^{-1}(-0.1\omega) - 0.18\omega$.
3. To estimate the critical frequency, we set $-\pi = 3\tan^{-1}(-0.1\omega_c) - 0.18\omega_c$. The solution to this equation is $\omega_c = 7.13$.
4. The amplitude ratio of $G_{OL}(s)$ is

$$|G_{OL}(j\omega)| = 0.16\left(\frac{1}{\sqrt{0.01\omega^2+1}}\right)^3 \cdot 1 = \frac{0.16}{(0.01\omega^2+1)^{3/2}}.$$

5. $|G_{OL}(j\omega_c)| = 0.16/(0.01\omega_c^2+1)^{3/2} = 0.086.$

Therefore, the system is stable.

11.8 SUMMARY

Several analytical tools were given to assess the stability of closed-loop systems. While the Routh–Hurwitz stability criterion and the root locus approach are based on the poles of a characteristic polynomial equation, the Bode and Nyquist stability conditions make use of the frequency response. Gain and phase margins, indicating how far away the closed-loop system is from instability, are also defined. Several examples are provided to show the implementation of these tools in the control of biological processes.

PROBLEMS

11.1. Study the stability of a closed-loop feedback system with the following characteristic equation using the Hurwitz stability criterion:

$$1 + G_{OL}(s) = s^4 + s^3 + 2s^2 + s + 3 = 0.$$

11.2. Study the stability of a closed-loop feedback system with the following characteristic equation using the Routh–Hurwitz stability criterion:

$$1 + G_{OL}(s) = s^3 + 3s^2 + 5s + 11 = 0.$$

11.3. Consider the following open-loop transfer function:

$$G_{\mathrm{OL}}(s) = \frac{K_c}{s(s+4)(s+5)}.$$

Construct the root-locus diagram.

11.4. Consider the following open-loop transfer function:

$$G_{\mathrm{OL}}(s) = \frac{K_c(s+2)}{s(s+7)}.$$

Construct the root-locus diagram.

11.5. A process is defined by

$$G_P(s) = \frac{4}{5s+1}.$$

Study the stability of the closed-loop system when $G_m(s) = 1$, $G_c(s) = K_c = 9$, and $G_f(s) = 1$ using the Bode stability criterion.

11.6. A process is defined by

$$G_P(s) = \frac{6e^{-0.5s}}{4s+1}.$$

Study the stability of the closed-loop system when $G_m(s) = 1$, $G_c(s) = K_c = 14$, and $G_f(s) = 1$ using the Bode stability criterion.

11.7. Consider the open-loop transfer function $G_{\mathrm{OL}}(s) = 2K_c/(s+3)$. Study the stability of the closed-loop system for $K_c = 5$ using the Nyquist stability criterion.

11.8. Derive the stability result obtained in Section 11.6 using the Routh–Hurwitz stability criterion:

$$\frac{-k_y + qr_Y}{r_Y} < \hat{h} < \frac{bk_x + bk_y + aqr_Y}{ar_Y}.$$

11.9. The stability of the pupillary light reflex system was shown in Section 11.7 using the Bode stability criterion. Demonstrate the system stability using the Nyquist stability criterion.

11.10. The stability of the pupillary light reflex system was verified in Section 11.7 for $K = 0.16$. Test the stability of the system for $K = 0.5$ using the Bode stability criterion.

REFERENCES

1. Levine WS, ed. *Control System Fundamentals*. Boca Raton, FL: CRC Press, 2000.
2. Allen LJS. *An Introduction to Mathematical Biology*. Upper Saddle River, NJ: Pearson/Prentice Hall, 2007.
3. Marlin TE. *Process Control: Designing Processes and Control Systems for Dynamic Performance*. Boston: McGraw-Hill, 2000.
4. Stephanopoulos G. *Chemical Process Control: an Introduction to Theory and Practice*. Englewood Cliffs, NJ: Prentice Hall, 1984.
5. Seborg DE, Edgar TF, Mellichamp DA. *Process Dynamics and Control*. Hoboken, NJ: Wiley, 2004.
6. Khoo MCK. *Physiological Control Systems: Analysis, Simulation, and Estimation*. New York: IEEE Press, 2000.
7. Pinna GD, Maestri R, Mortara A, La Rovere MT, Fanfulla F, Sleight P. Periodic breathing in heart failure patients: testing the hypothesis of instability of the chemoreflex loop. *Journal of Applied Physiology* 2000; 89:2147–2157.
8. Buzi G. Control theoretic analysis of autocatalytic networks in biology with applications to glycolysis. Dissertation (PhD), California Institute of Technology, 2010.
9. Sherman PM, Stark L. A servoanalytic study of consensual pupil reflex to light. *Journal of Neurophysiology* 1957; 20:17–26.

CHAPTER 12

DESIGN OF FEEDBACK CONTROLLERS

In previous chapters, we have studied how to derive mathematical models and to analyze process dynamics. If the controller parameters are available, techniques, learned in Chapter 11, can be used to determine the stability of the feedback closed-loop system. One important aspect of controlling a system is the ability to determine the best *tuning parameters*. In the case of a proportional–integral–derivative (PID) controller, for instance, the gain K_c, integral time τ_I, and derivative time τ_D have to be tuned to ensure stability and to yield the desired performance. A small subset of tuning guidelines, available in the literature and in the industry, is outlined below.

12.1 TUNING METHODS FOR FEEDBACK CONTROLLERS

Two of the most widely used tuning procedures are the *Cohen–Coon* and *Ziegler–Nichols* methods.

12.1.1 The Cohen–Coon Method

The Cohen–Coon method assumes that the controller is detached from the closed-loop system in such a way that a step change is made in the controller signal $\bar{c}(s)$ (before the actuator) and the measured variable recorded: $\bar{y}_m(s)$.

Control of Biological and Drug-Delivery Systems for Chemical, Biomedical, and Pharmaceutical Engineering, First Edition. Laurent Simon.
© 2013 John Wiley & Sons, Inc. Published 2013 by John Wiley & Sons, Inc.

TABLE 12.1 Cohen–Coon Tuning Rules

Controller	K_c	τ_I	τ_D
P	$\dfrac{1}{K_p}\dfrac{\tau}{\theta}\left(1+\dfrac{\theta}{3\tau}\right)$	–	–
PI	$\dfrac{1}{K_p}\dfrac{\tau}{\theta}\left(\dfrac{9}{10}+\dfrac{\theta}{12\tau}\right)$	$\theta\left(\dfrac{30+3\theta/\tau}{9+20\theta/\tau}\right)$	–
PID	$\dfrac{1}{K_p}\dfrac{\tau}{\theta}\left(\dfrac{4}{3}+\dfrac{\theta}{4\tau}\right)$	$\theta\left(\dfrac{32+6\theta/\tau}{13+8\theta/\tau}\right)$	$\theta\left(\dfrac{4}{11+2\theta/\tau}\right)$

This response is called a *process reaction curve* and can be represented by a *first-order plus dead-time* (FOPDT) model:

$$G_{\text{RC}} = \frac{\bar{y}_m(s)}{\bar{c}(s)} = \frac{K_p e^{-\theta s}}{\tau s + 1}. \tag{12.1}$$

Note that the process needs to be stable to use this method. Cohen–Coon settings are derived for the closed-loop responses to have one-fourth decay ratio, minimum offset, and minimum integral squared error (Table 12.1). The integral of the squared error is defined as $\int_0^\infty [y_{\text{sp}} - y(t)]^2\, dt$, where y_{sp} is the set point.

The parameters of G_{RC} can be evaluated by regressing the time-domain response equation to experimental data generated after a step point change A in the controller output. However, graphical methods are preferred over the regression approach. The tangent line technique is described in Figure 12.1. After generating the step response $y(t)$, a tangent line y_{tan} is drawn at the inflexion point. The gain K_p is readily available by measuring the steady-state y_{ss} and applying the following relation: $y_{\text{ss}} = K_p A$. To estimate θ, the intersection between y_{tan} and the time axis is noted. The time constant τ is calculated from the intersection of y_{tan} with the horizontal line $y = y_{\text{ss}}$ (i.e., $\theta + \tau$). Another way to determine τ is to observe that $\theta + \tau$ is the time when $y(t)$ is at 63% of y_{ss}. In this approach, the initial time corresponds to when the step input is applied.

Example 12.1 In Example 8.2, we show that a transfer function

$$G(s) = \frac{4}{(0.1s+1)(0.2s+1)(1.0s+1)} \tag{12.2}$$

can be approximated as

$$G(s) = \frac{4e^{-0.3s}}{1.0s+1} \tag{12.3}$$

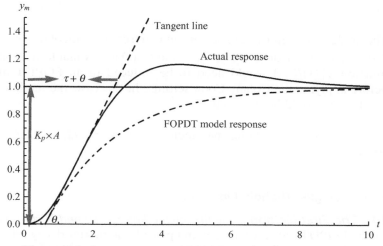

Figure 12.1. Step responses of FOPDT model and actual system.

using a model reduction method. Let us see if the tangent line approach leads to a similar FOPDT model. We assume that

$$\frac{\bar{y}_m(s)}{\bar{c}(s)} = G_f(s)G_p(s)G_m(s) = \frac{4}{(0.1s+1)(0.2s+1)(1.0s+1)}. \qquad (12.4)$$

Solution   The time response to a step change in the input $c(t)$ is first generated. In this case, we choose $\bar{c}(s) = 1/s$ and obtain the following expression for $y_m(t)$:

$$y_m(t) = 4(1 - 0.11e^{-10t} + 0.5e^{-5t} - 1.39e^{-t}). \qquad (12.5)$$

Although $y_m(t)$ is given here, data points would be available in practice. The maximum slope corresponds to the slope at the inflection point. After taking the second time derivative of $y_m(t)$, setting it equal to zero and solving numerically for t, we have $t_{in} = 0.533$. Note that $y_m(t_{in}) = 0.878$ and the slope is $dy_m/dt|_{t=t_{in}} = 2.586$. The equation of the tangent line is $y_{tan}(t) = 2.586t - 0.501$ and the intersection with the t-axis is $\theta = 0.194$. Also, the intersection with the horizontal line $y_m = \lim_{t \to \infty}(y_m(t)) = 4$ is $\tau + \theta = 1.741$, which yields a time constant of $\tau = 1.547$. Finally, the transfer function for the process reaction is

$$G_{RC} = \frac{4e^{-\theta s}}{1.547s + 1}. \qquad (12.6)$$

Remarks

1. The model parameter would be determined from the plot of the response.
2. If the transfer functions $G_f(s)$, $G_p(s)$, and $G_m(s)$ are available, the transfer function $y_m(s)/\bar{c}(s)$ would need to be reduced to an FOPDT model before applying the Cohen–Coon settings. In this case, the method outlined above could be followed.

The Cohen–Coon PID tuning parameters are $K_c = 2.72$, $\tau_I = 0.45$, and $\tau_D = 0.069$.

12.1.2 The Ziegler–Nichols Methods

12.1.2.1 The First Method The *first method of Ziegler and Nichols* is also based on an FOPDT model of a plant and provides a quarter decay ratio. The ideal form of the controller is used:

$$G_c(s) = K_c\left(1 + \frac{1}{\tau_I s} + \tau_D s\right), \tag{12.7}$$

and the suggested tuning parameters are listed in Table 12.2 [1]:

12.1.2.2 The Second Method The *Ziegler–Nichols closed-loop technique* (or *ultimate cycle method*) is applied directly to the frequency response [1]. After switching to a proportional (P) controller, a set point, or load change, is made. The goal is to vary the controller gain K_c, incrementally, until a sustained oscillation is achieved. This controller gain is known as the *ultimate gain K_u*, which is equal to the reciprocal of the magnitude of G_{OL}:

$$K_u = \frac{1}{|G_{OL}(j\omega_c)|} \quad \text{or} \quad K_u|G_{OL}(j\omega_c)| = 1$$

TABLE 12.2 Ziegler–Nichols Tuning Rules Based on the First Method

Controller	K_c	τ_I	τ_D
P	$\dfrac{1}{K_p}\dfrac{\tau}{\theta}$	∞	0
PI	$\dfrac{0.9}{K_p}\dfrac{\tau}{\theta}$	$\dfrac{\theta}{0.3}$	0
PID	$\dfrac{1.2}{K_p}\dfrac{\tau}{\theta}$	2θ	0.5θ

TABLE 12.3   Ziegler–Nichols Tuning Rules Based on the Second Method

Controller	K_c	τ_I	τ_D
P	$0.5K_u$	∞	0
PI	$0.45K_u$	$P_u/1.2$	0
PID	$0.6K_u$	$P_u/2$	$P_u/1.8$

Note: The ideal form of the PID controller is used.

where G_{OL} is the open-loop transfer function $G_{OL}(s) = G_P(s)G_f(s)G_m(s)$ defined for $K_c = 1$. The oscillation period is

$$P_u = \frac{2\pi}{\omega_c}. \tag{12.8}$$

In practice, if the transfer functions are known, the Bode stability criterion can be used to calculate K_u directly because ω_c is the phase crossover frequency or the frequency that induces continuous oscillation. The tuning parameters are listed in Table 12.3.

Example 12.2   Consider the following transfer functions in a closed-loop system:

$$G_f = 1, G_m = \frac{1}{5s+1} \quad \text{and} \quad G_p = \frac{1}{(3s+1)(s+1)}.$$

Derive P, proportional–integral (PI), and PID controller tuning settings using the Ziegler–Nichols second method.

Solution   The open-loop transfer function is obtained:

$$G_{OL}(s) = \frac{1}{(5s+1)(3s+1)(s+1)}. \tag{12.9}$$

The phase angle is $\phi = \tan^{-1}(-5\omega) + \tan^{-1}(-3\omega) + \tan^{-1}(-\omega)$. Solving the equation $-\pi = \tan^{-1}(-5\omega_c) + \tan^{-1}(-3\omega_c) + \tan^{-1}(-\omega_c)$ yields $\omega_c = 0.77$. The amplitude ratio at this point is

$$|G_{OL}(j\omega_c)| = \frac{1}{\left(\sqrt{25\omega_c^2 + 1}\right)\left(\sqrt{9\omega_c^2 + 1}\right)\left(\sqrt{\omega_c^2 + 1}\right)} = 0.078.$$

As a result, $K_u = 12.8$ after using the formula $K_u|G_{OL}(j\omega)| = 1$ and $P_u = 8.11$. Finally, the tuning parameters are as follows:

P controller: $K_c = 6.4$.
PI controller: $K_c = 5.76$, $\tau_I = 6.76$.
PID controller: $K_c = 7.68$, $\tau_I = 4.05$, and $\tau_D = 1.01$.

12.1.3   Internal Model Control (IMC) Tuning

The design of an IMC is based on a model of the plant (Fig. 12.2). Assuming $G_m = G_f = 1$, an open-loop response in the set point $\bar{r}(s)$ gives

$$\bar{y}(s) = G_c(s)G_p(s)\bar{r}(s). \tag{12.10}$$

In this case, perfect control is achieved by setting the controller transfer function equal to the inverse of the plant transfer function $G_c(s) = G_p^{-1}(s)$. No feedback loop is necessary. The IMC algorithm implements this idea by using an approximate model of the plant $\tilde{G}_p(s)$. Both the original and an estimated plant receives an input signal $\bar{m}(s)$. The error between $\bar{y}(s)$ and the estimated response $\bar{y}^*(s)$ is

$$\bar{e}_y(s) = \bar{y}(s) - \bar{y}^*(s) = \bar{d}(s) + G_p(s)\bar{m}(s) - \tilde{G}_p(s)\bar{m}(s) \tag{12.11}$$

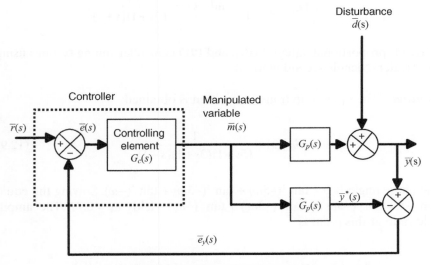

Figure 12.2. The IMC controller.

or

$$\bar{e}_y(s) = \bar{d}(s) + \left[G_p(s) - \tilde{G}_p(s)\right]\bar{m}(s). \tag{12.12}$$

The manipulated variable is

$$\bar{m}(s) = G_c(s)\left[\bar{r}(s) - \bar{e}_y(s)\right], \tag{12.13}$$

which becomes

$$\bar{m}(s) = G_c(s)\left\{\bar{r}(s) - \left[\bar{d}(s) + \left[G_p(s) - \tilde{G}_p(s)\right]\right]\bar{m}(s)\right\}$$
$$\bar{m}(s) = G_c(s)\bar{r}(s) - G_c(s)\bar{d}(s) - G_c(s)G_p(s)\bar{m}(s) + G_c(s)\tilde{G}_p(s)\bar{m}(s)$$
$$\bar{m}(s) = \frac{G_c(s)\bar{r}(s) - G_c(s)\bar{d}(s)}{1 + G_c(s)G_p(s) - G_c(s)\tilde{G}_p(s)} \tag{12.14}$$
$$\bar{m}(s) = \frac{\left[\bar{r}(s) - \bar{d}(s)\right]G_c(s)}{1 + \left[G_p(s) - \tilde{G}_p(s)\right]G_c(s)}$$

Since $\bar{y}(s) = \bar{d}(s) + G_p(s)\bar{m}(s)$, the following equation holds:

$$\bar{y}(s) = \bar{d}(s) + \frac{\left[\bar{r}(s) - \bar{d}(s)\right]G_c(s)G_p(s)}{1 + \left[G_p(s) - \tilde{G}_p(s)\right]G_c(s)} \tag{12.15}$$

or

$$\bar{y}(s) = \frac{\bar{d}(s) + \left\{1 + \left[G_p(s) - \tilde{G}_p(s)\right]G_c(s)\right\} + \left[\bar{r}(s) - \bar{d}(s)\right]G_c(s)G_p(s)}{1 + \left[G_p(s) - \tilde{G}_p(s)\right]G_c(s)}$$
$$\bar{y}(s) = \frac{\bar{d}(s) - \bar{d}(s)\tilde{G}_p G_c(s) + \bar{r}(s)G_c(s)G_p(s)}{1 + \left[G_p(s) - \tilde{G}_p(s)\right]G_c(s)}, \tag{12.16}$$

which yields

$$\bar{y}(s) = \frac{\left[1 - \tilde{G}_p G_c(s)\right]\bar{d}(s) + \bar{r}(s)G_c(s)G_p(s)}{1 + \left[G_p(s) - \tilde{G}_p(s)\right]G_c(s)}$$
$$\bar{y}(s) = \frac{G_c(s)G_p(s)}{1 + \left[G_p(s) - \tilde{G}_p(s)\right]G_c(s)}\bar{r}(s) + \frac{\left[1 - \tilde{G}_p G_c(s)\right]}{1 + \left[G_p(s) - \tilde{G}_p(s)\right]G_c(s)}\bar{d}(s). \tag{12.17}$$

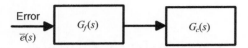

Figure 12.3. The low-pass filter.

Note that if $G_p(s) = \tilde{G}_p(s)$ and $G_c(s) = \tilde{G}_p^{-1}(s)$, we have $\bar{y}(s) = \bar{r}(s)$. A low-pass filter, $G_f(s)$, is selected in order to have a proper controller [2]:

$$G_f(s) = \frac{1}{(1+\tau_f s)^n} \tag{12.18}$$

where τ_f is a tuning parameter and n is the order of the filter transfer function. This filter diminishes high-frequency fluctuations. Additional information on the influence of $G_f(s)$ on the stability of the closed-loop response, the behavior of the manipulated variable, and the performance of the controller variable can be found in Marlin [3]. In Figure 12.2, the filter block is placed between the error $\bar{e}(s)$ and the "controlling element" block (Fig. 12.3).

The tuning parameter τ_f determines the speed of the closed-loop response. The IMC controller is

$$G_{\text{cont}}(s) = G_c(s)G_f(s). \tag{12.19}$$

In Equation (12.17), G_c is replaced by $G_{\text{cont}}(s)$:

$$\bar{y}(s) = \frac{G_{\text{cont}}(s)G_p(s)}{1+\left[G_p(s)-\tilde{G}_p(s)\right]G_{\text{cont}}(s)}\bar{r}(s) + \frac{\left[1-\tilde{G}_p G_{\text{cont}}(s)\right]}{1+\left[G_p(s)-\tilde{G}_p(s)\right]G_{\text{cont}}(s)}\bar{d}(s). \tag{12.20}$$

When dealing with a perfect model (i.e., $G_p(s) = \tilde{G}_p(s)$), the response becomes

$$\bar{y}(s) = G_{\text{cont}}(s)\tilde{G}_p(s)\bar{r}(s) + \left[1-\tilde{G}_p G_{\text{cont}}(s)\right]\bar{d}(s). \tag{12.21}$$

To implement the IMC controller,

- $\tilde{G}_p(s)$ is written as the product of noninvertible $\tilde{G}_p^-(s)$ (e.g., positive zeros and dead time) and invertible components $\tilde{G}_p^+(s)$: $\tilde{G}_p(s) = \tilde{G}_p^-(s)\tilde{G}_p^+(s)$.
- The controller takes the form of $G_c(s) = 1/\tilde{G}_p^+(s)$, which yields

$$G_{\text{cont}}(s) = \frac{G_f(s)}{\tilde{G}_p^+(s)}. \tag{12.22}$$

Example 12.3 A process is given by $G_p(s) = e^{-4s}/(1+10s)$. Design an internal model controller.

Solution Assuming $G_p(s) = \tilde{G}_p(s)$, we have

$$\bar{y}(s) = G_{\text{cont}}(s)\tilde{G}_p(s)\bar{r}(s) + \left[1 - \tilde{G}_p(s)G_{\text{cont}}(s)\right]\bar{d}(s). \quad (12.23)$$

The function $\tilde{G}_p(s)$ can be written as $\tilde{G}_p(s) = \tilde{G}_p^-(s)\tilde{G}_p^+(s)$, where $\tilde{G}_p^-(s) = e^{-4s}$ and $\tilde{G}_p^+(s) = 1/(1+10s)$. As a result, $G_{\text{cont}}(s) = (1+10s)G_f(s)$. The low-pass filter is taken as

$$G_f(s) = \frac{1}{(1+\tau_f s)^n}.$$

If we choose $n = 1$ and $\tau_f = \frac{10}{2} = 5$ (response is twice as fast as the open-loop dynamics), we have

$$G_f(s) = \frac{1}{1+5s}. \quad (12.24)$$

Therefore,

$$\bar{y}(s) = \frac{1+10s}{1+5s}\frac{e^{-4s}}{1+10s}\bar{r}(s) + \left[1 - \frac{e^{-4s}}{1+10s}\frac{1+10s}{1+5s}\right]\bar{d}(s) \quad (12.25)$$

or

$$\bar{y}(s) = \frac{e^{-4s}}{1+5s}\bar{r}(s) + \left(1 - \frac{e^{-4s}}{1+5s}\right)\bar{d}(s). \quad (12.26)$$

For the servo problem, if $\bar{r}(s) = 1/s$, the closed-loop response becomes

$$\bar{y}(s) = \frac{4^{-4s}}{1+5s}\frac{1}{s}$$

and its ultimate value is evaluated by the final value theorem $\lim_{s \to 0}[s\bar{y}(s)]$ $= 1$. The steady-state offset is zero for set-point step changes. Similarly, for a unit step change in the load, the steady-state offset is zero.

Example 12.4 Derive the PI tuning parameters for an FOPDT model using the IMC method.

Solution   The process is $\tilde{G}_p(s) = K_p e^{-\theta s} / (\tau s + 1)$. Using the Taylor series expansion, we have $e^{-\theta s} = 1 - s\theta$. As a result, $\tilde{G}_p(s)$ becomes

$$\tilde{G}_p(s) = \frac{K_p(1-s\theta)}{\tau s + 1}. \tag{12.27}$$

In this case,

$$\tilde{G}_p^-(s) = 1 - s\theta \tag{12.28}$$

and

$$\tilde{G}_p^+(s) = \frac{K_p}{\tau s + 1}. \tag{12.29}$$

Using $G_f(s) = 1/(1 + \tau_f s)$, the IMC controller becomes

$$G_{\text{cont}}(s) = \frac{G_f(s)}{\tilde{G}_p^+(s)} = \frac{1}{1 + \tau_f s} \frac{\tau s + 1}{K_p}. \tag{12.30}$$

If G_{1c} represents the transfer function of the PI controller, the block diagram of a feedback control equivalent to Figure 12.2 is shown in Figure 12.4. The next step is to find the expression for G_{1c} so that the two block diagrams are

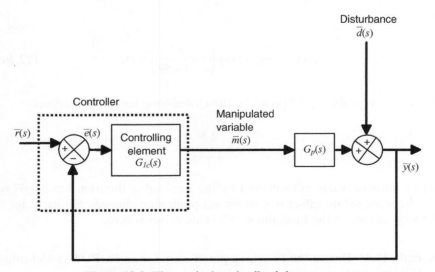

Figure 12.4. The equivalent feedback loop system.

equivalent. Mathematically, we simply compare the closed-loop transfer functions of the block diagrams. A servo problem is selected in this example. As a result, the following equation holds:

$$G_{SP} = \frac{G_p G_{1c}}{1 + G_p G_{1c}} = \frac{G_{cont} G_p}{1 + (G_p - \tilde{G}_p) G_{cont}}. \qquad (12.31)$$

Finally,

$$G_{1c} = \frac{G_{cont}}{1 - \tilde{G}_p G_{cont}}. \qquad (12.32)$$

After using the equations for $\tilde{G}_p(s)$ and $G_{cont}(s)$, we obtain

$$G_{1c} = \frac{\tau s + 1}{K_p} \frac{1}{s(\tau_f + \theta)}, \qquad (12.33)$$

which yields, after factorization,

$$G_{1c} = \frac{s\tau + 1}{K_p s(\tau_f + \theta)} = \frac{1}{K_p(\tau_f + \theta)} \left(\frac{s\tau + 1}{s} \right) \qquad (12.34)$$

or

$$G_{1c} = \frac{\tau}{K_p(\tau_f + \theta)} \left(1 + \frac{1}{s\tau} \right). \qquad (12.35)$$

Consequently, the tuning parameters for the PI controller are

$$K_c = \frac{\tau}{K_p(\tau_f + \theta)} \qquad (12.36)$$

and

$$\tau_I = \tau. \qquad (12.37)$$

12.2   REGULATION OF GLYCEMIA

Figure 1.8 describes a feedback control mechanism for blood glucose by insulin injection in a diabetic person. In Percival et al. [4], a regimen was developed to ensure that insulin is delivered at levels sufficient to sustain a normal plasma glucose concentration. The following model was derived:

$$Y(s) = \frac{K_M e^{-\tau_{D,M}s}}{s(\tau_M s + 1)} U_M(s) + \frac{K_I e^{-\tau_{D,I}s}}{s(\tau_I s + 1)} U_I(s), \tag{12.38}$$

where U and Y represent the input (meal: U_M or insulin U_I) and output (glucose) variables, respectively. In this equation, K is the process gain, τ is the process time constant, and τ_D is the process time delay. The subscripts M and I are used for meal and insulin, respectively; U_I is the manipulated variable and U_M is the load.

Note: $K_M = 4.2$ mg/dL/g CHO, $K_I = -0.075$ mU/min/mg/dL, $\tau_{D,M} = 12$ minutes, $\tau_{D,I} = 27$ minutes, $\tau_M = 53$ minutes, and $\tau_I = 146$ minutes.

A closed-loop system can be designed to control the level of blood glucose for the servo problem using the Ziegler–Nichols closed-loop technique. We assume that $G_f = G_m = 1$. Since the plant transfer function is

$$G_p(s) = \frac{-0.075 e^{-27s}}{s(146s + 1)}, \tag{12.39}$$

we have

$$G_{OL}(s) = \frac{-0.075 e^{-27s}}{s(146s + 1)}. \tag{12.40}$$

The phase angle is

$$\phi = -\frac{\pi}{2} + \tan^{-1}(-146\omega) - 27\omega.$$

Thus,

$$-\pi = -\frac{\pi}{2} + \tan^{-1}(-146\omega_c) - 27\omega_c,$$

which gives $\omega_c = 0.015$. The amplitude ratio at the phase crossover frequency is

$$|G_{OL}(j\omega_c)| = \frac{-0.075}{\omega_c \left(\sqrt{(146)^2 \omega_c^2 + 1} \right)} = -1.97$$

or $K_u = -0.508$ after using the formula $K_u |G_{OL}|(jw_c)| = 1$; $P_u = 406.61$. The PID settings from Table 12.2 are $K_c = -0.305$, $\tau_I = 203.30$, and $\tau_D = 225.89$. To test the controller, let us analyze the closed-loop response to a step change of size 15 mg/dL (i.e., $U_1(s) = 15/s$). The closed-loop transfer function is

$$G_{SP} = \frac{G_p G_c}{1 + G_p G_c} \tag{12.41}$$

or

$$G_{SP}(s) = \frac{5.17s^2 + 0.023s + 0.00011}{146s^3 e^{27s} + s^2(e^{27s} + 5.17) + 0.023s + 0.00011} \tag{12.42}$$

after making appropriate substitutions. A Padé approximation of e^{27s} leads to

$$G_{SP}(s) = \frac{-139.55s^3 + 9.72s^2 + 0.043s + 0.000225}{3942s^4 + 179.45s^3 + 11.72s^2 + 0.043s + 0.000225}. \tag{12.43}$$

We take the inverse Laplace transform of $(15/s)G_{sp}$ and obtain

$$y(t) = 125 \begin{bmatrix} 0.168e^{-0.00178t}\sin(0.0041t) - 1.1396e^{-0.0210t}\sin(0.0486t) \\ -0.0878e^{-0.00178t}\cos(0.0041t) - 0.912e^{-0.0210t}\cos(0.0486t) + 1 \end{bmatrix}.$$
$$\tag{12.44}$$

The $y(t)$ profile in Figure 12.5 shows that the controller is able to track the set-point change.

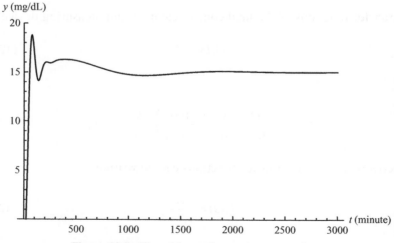

Figure 12.5. Closed-loop glucose concentration.

12.3 DISSOLVED OXYGEN CONCENTRATION

In Chapter 9, the oxygen mass balance in the liquid phase of a cell-free cultivation medium resulted in

$$\frac{dC_{O_2,L}}{dt} = k_L a_B \left(\frac{P}{H} y_{O_2,B} - C_{O_2,L} \right) + k_L a_H \left(\frac{P}{H} y_{O_2,H} - C_{O_2,L} \right). \quad (12.45)$$

Note:

$C_{O_2,L}$: oxygen concentration (output variable) in the liquid phase (mol/L)
$k_L a_B$: mass transfer coefficient in the bubble phase (second^{-1})
P: total pressure (bar)
H: Henry's constant (bar L/mol)
$y_{O_2,B}$: oxygen mole fraction in the bubble phase (manipulated variable)
$y_{O_2,H}$: oxygen mole fraction in the head phase (disturbance)
$k_L a_H$: mass transfer coefficient in the headspace.

For the servo problem (i.e., $\bar{y}_{O_2,H}(s) = 0$), we have

$$\bar{C}_{O_2,L}(s) = G_p(s) \bar{y}_{O_2,B}(s), \quad (12.46)$$

where

$$G_p(s) = \frac{P}{H} \frac{k_L a_B}{(s + k_L a_B + k_L a_H)}. \quad (12.47)$$

The transfer functions of the final control element and measuring device are

$$G_f(s) = 1 \quad (12.48)$$

and

$$G_m(s) = \frac{\bar{D}_{O_2}(s)}{\bar{C}_{O_2,L}(s)} = \left(\frac{100}{\bar{C}_{O_2,sat}} \right) \frac{k_m}{\tau_m s + 1} e^{-t_d s}, \quad (12.49)$$

respectively. The above transfer functions can be written as

$$G_p(s) = \frac{K_1}{\tau_1 s + 1} \quad (12.50)$$

and

$$G_m(s) = \frac{K_2 e^{-t_d s}}{\tau_2 s + 1}.$$

Since the measured output variable is D_{O2}, the process transfer function is

$$G(s) = \frac{K e^{-t_d s}}{(\tau_1 s + 1)(\tau_2 s + 1)}, \tag{12.51}$$

where $K = K_1 K_2$. An IMC is designed to derive PID tuning parameters. A first-order Taylor series expansion gives

$$e^{-t_d s} = 1 - t_d s \tag{12.52}$$

and

$$G(s) = \frac{K(1 - t_d s)}{(\tau_1 s + 1)(\tau_2 s + 1)}. \tag{12.53}$$

In this case,

$$\tilde{G}_p^-(s) = 1 - t_d s \tag{12.54}$$

and

$$\tilde{G}_p^+(s) = \frac{K}{(\tau_1 s + 1)(\tau_2 s + 1)}. \tag{12.55}$$

Based on $G_f(s) = 1/1 + \tau_f s$, the IMC controller takes the form

$$G_{cont}(s) = \frac{G_f(s)}{\tilde{G}_p^+(s)} = \frac{1}{1 + \tau_f s} \frac{(\tau_1 s + 1)(\tau_2 s + 1)}{K}. \tag{12.56}$$

Similar to Example 12.4, G_{1c} is determined from Equation (12.32):

$$G_{1c} = \frac{G_{cont}}{1 - G_p G_{cont}}$$

$$G_{1c}(s) = \frac{(s\tau_1 + 1)(s\tau_2 + 1)}{Ks(t_d + \tau_f)}. \tag{12.57}$$

Equation (12.57) is the ideal PID controller:

$$G_{1c}(s) = K_c\left(1 + \frac{1}{\tau_I s} + \tau_D s\right),$$ (12.58)

with

$$K_c = \frac{1}{K}\frac{\tau_1 + \tau_2}{\tau_f + t_d}; \quad \tau_I = \tau_1 + \tau_2; \quad \tau_D = \frac{\tau_1 \tau_2}{\tau_1 + \tau_2}.$$ (12.59)

12.4   CONTROL OF BIOMASS IN A CHEMOSTAT

In Chapter 2, component mass balance equations were derived for a chemostat. Such equations are often used when designing control schemes to improve the performance of these bioreactors. Consider the following expression for the specific growth rate(/hour):

$$\mu = \frac{\mu_{max} C_s}{K_M + C_s + K_1 C_s^2},$$ (12.60)

where C_s is the limiting substrate concentration (g/L), $\mu_{max} = 0.53$ hour^{-1} is the maximum specific growth rate, $K_M = 0.12$ g/L is the substrate saturation constant, and K_1 is the substrate inhibition constant [5]. The transfer function, relating the dilution rate to the biomass, is

$$G_p(s) = \frac{-0.6758e^{-1s}}{0.4417s + 1}$$ (12.61)

after linearizing the state equations and using the method of least squares to derive the transfer function. Table 12.2 can be used immediately to obtain the Ziegler–Nichols tuning rules of the PI controller based on the first method:

$$G_c(s) = -0.588\left(\frac{0.3}{s} + 1\right).$$ (12.62)

The following closed-loop transfer function $G_{SP}(s)$ is obtained:

$$G_{SP}(s) = -\frac{0.398(s - 2.)(s + 0.3)}{0.44s^3 + 1.49s^2 + 2.68s + 0.24}.$$ (12.63)

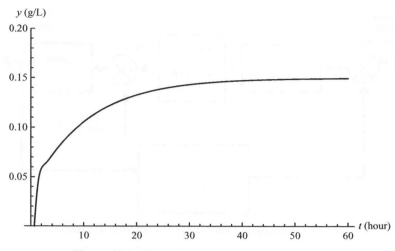

Figure 12.6. Closed-loop biomass concentration.

A step change in the set point of size 0.15 g/L yields the following response (Fig. 12.6):

$$y(t) = -0.11e^{-0.094t} - 0.12e^{-1.64t}\sin(1.75t) - 0.036e^{-1.64t}\cos(1.75t) + 0.15. \quad (12.64)$$

12.5 CONTROLLED INFUSION OF VASOACTIVE DRUGS

In Chapter 1, feedback regulation of blood pressure using the *vasomotor center* (VMC) was provided as an illustration of biological control. To simplify the analysis, the VMC in the brain is the only controller. Resistance and capacitance vessels play a key role in the process. Figure 1.6 shows that the baroreceptor sensor monitors the *mean arterial pressure* (MAP). The MAP, in addition to heart rate and pupil size, is one measure of the depth of anesthesia (DOA) and is the most credible guide for providing inhaled anesthetics to the patient [6]. Several methods of controlling the DOA, which is correlated to the dose provided, have been proposed in the literature to help keep the MAP within a specific value. Automated feedback control of the DOA would enable the anesthetist to focus on other charges such as controlling fluid balance [6].

To explain the change in the MAP of a patient under the influence of *sodium nitroprusside* (SNP), Sheppard used the following transfer function [7]:

$$G_p(s) = \frac{\overline{P}(s)}{\overline{u}(s)} = \frac{Ke^{-T_i s}\left(1 + \alpha e^{-T_c s}\right)}{(1 + \tau s)}, \quad (12.65)$$

where P is the MAP (mmHg) change due to SNP infusion; u denotes the drug infusion rate in deviation form (mL/h); K stands for the sensitivity to the drug

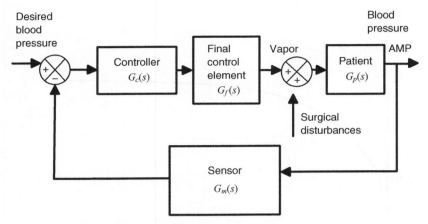

Figure 12.7. Feedback control of the mean arterial pressure.

(−0.72 mmHg/mL/h); α is the recirculation constant (0.4); T_i is the initial transport lag (30 seconds); T_c is the recirculation time (45 seconds); and τ is the time constant of the response (50 seconds) [7]. For this illustration, assume that the measuring device (noise free) and final control element transfer function are $G_m(s) = 1$ and $G_f(s) = 1$, respectively (Fig. 12.7). After substituting the parameter values and using Padé approximation, $G_p(s)$ becomes

$$G_p(s) = \frac{291.6s^2 + 10.8s - 2.0}{33750s^3 + 4425s^2 + 175s + 2}. \tag{12.66}$$

The FOPDT model is developed:

$$\tilde{G}_p(s) = -\frac{1.0e^{-27.7s}}{97.2s + 1}. \tag{12.67}$$

The controller tuning settings can be calculated using Tables 12.1 and 12.2. A computer-aided control system design (CACSD) project, similar to Jones and Tham [8], can be considered using tools available in MATLAB® and Simulink® (MathWorks, Inc.).

12.6   BONE REGENERATION

The influence of stress stimulation on fracture strength was explained in Chapter 8 by a first-order equation:

$$T\frac{dc(t)}{dt} + c(t) = Kr(t), \tag{12.68}$$

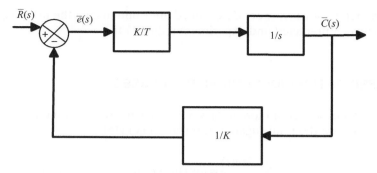

Figure 12.8. Feedback control representation of a first-order system.

where *r* is the mechanical stress stimulus and *c* is the fracture strength during bone healing [9]. In this equation, *T* and *K* represent the time constant and gain, respectively. Considering that a first-order process is adequately represented by a closed-loop feedback control system, experiments can be conducted to estimate the values of *T* and *K*. In this case, the role of the process control engineer is to identify the parameter settings and not to tune the controller.

The block diagram of the first-order system is shown in Figure 12.8 [9]. The transfer function developed in Chapter 8 can be defined as the ratio $\bar{C}(s) / \bar{R}(s)$:

$$\frac{\bar{C}(s)}{\bar{R}(s)} = \frac{K}{Ts+1}. \tag{12.69}$$

After conducting experiments with 70 adult New Zealand rabbits, Wang et al. concluded that the model is appropriate to explain the influence of stress stimulation on fracture strength during healing [9]. In this case, regression is more suitable for this type of analysis than the methods outlined above. The feedback conceptualization is in line with research conducted by Frost, who promoted the idea of a mechanical feedback system that regulates bone mass [10]. According to the proposal, bone mass is adjusted ensuring that peak mechanical strains in the bone are maintained within a satisfactory range. For studies focusing on the electrical response of bone to a mechanical deformation *in vivo*, the work of Levy is a good starting point [11].

Bone is also involved in calcium homeostasis in mammals. The set points for calcium concentration are 0.085–0.105 g/L [12] in humans and 0.08–0.1 g/L in dairy cows [13]. Instead of calculating the best controller settings by the methods outlined in the previous sections, research in this area focuses on the elucidation of the underlying homeostatic mechanism that governs the process. An investigation, using calving cows, showed the existence of a PI controller, the parameters of which were determined after minimizing the mean squared error between predicted and experimental data [14].

12.7 FED-BATCH BIOCHEMICAL PROCESSES

Section 9.5 introduced a bioreactor where the substrate feeding rate (u) was used to regulate the total sugar concentration (y) [15]:

$$\frac{dy(t)}{dt} = ay(t) + bu(t).$$

The transfer functions were

$$G_p(s) = \frac{K_p}{\tau_p s + 1}, \quad G_m(s) = 1, \quad \text{and} \quad G_f(s) = 1.$$

An IMC-based PI controller can be designed for this first-order process. After assuming a perfect model: $G_p(s) = \tilde{G}_p(s)$ and using

$$G_f(s) = \frac{1}{1 + \tau_f s},$$

the IMC controller is written as

$$G_{\text{cont}}(s) = \frac{G_f(s)}{\tilde{G}_p^+(s)} = \frac{1}{1 + \tau_f s} \frac{\tau_p s + 1}{K_p}. \tag{12.70}$$

Following Example 12.4, we have

$$G_{\text{PI}} = \frac{G_{\text{cont}}}{1 - \tilde{G}_p G_{\text{cont}}}, \tag{12.71}$$

which leads to

$$G_{\text{PI}} = \frac{\tau_p s + 1}{K_p} \frac{1}{s \tau_f} \tag{12.72}$$

or

$$G_{\text{PI}} = \frac{\tau_p}{K_p \tau_f} \left(1 + \frac{1}{s \tau_p} \right). \tag{12.73}$$

The PI tuning parameters are

$$K_c = \frac{\tau_p}{K_p \tau_f} \tag{12.74}$$

and

$$\tau_I = \tau_p. \tag{12.75}$$

12.8 SUMMARY

Tuning methods for feedback controllers were offered. Most of these techniques are based either on the frequency response or a process reaction curve, which is a FOPDT representation of the original process. The IMC approach can also be adopted and can lead to a PID-type controller in some cases. In biological systems, the role of the process control engineer is to identify the type of controllers that can explain the underlying mechanism. Nonlinear regression techniques can be implemented to help in the analysis.

PROBLEMS

12.1. A process reaction curve is given by

$$G_{RC} = \frac{0.5e^{-0.7s}}{(3.5s+1)(0.3s+1)}.$$

Find the PID controller settings using the Cohen–Coon method.

12.2. The following process reaction curve is obtained:

$$G_{RC} = \frac{1.5}{(5s+1)(3s+1)(s+1)}.$$

Find the PI controller parameters using the Cohen–Coon tuning rules.

12.3. The transfer functions of the measuring device, the final control element, and the process in a feedback control system are given by

$$G_f = 1, G_m = \frac{2}{3s+1}, \quad \text{and} \quad G_p = \frac{2}{(4s+1)(s+1)}, \text{respectively.}$$

Derive P, PI, and PID controller parameters using the Ziegler–Nichols second method.

12.4. A process is given by

$$G_p(s) = \frac{e^{-3s}}{1 + 20s}.$$

Assume that $G_f = G_m = 1$ for the final control element and measuring device. Derive the PI tuning parameters using the IMC method.

Note: the closed-loop response should be twice as fast as the open-loop response.

12.5. The following model was derived for the regulation of glycemia (Section 12.2):

$$Y(s) = \frac{K_M e^{-\tau_{D,M}s}}{s(\tau_M s + 1)} U_M(s) + \frac{K_I e^{-\tau_{D,I}s}}{s(\tau_I s + 1)} U_I(s).$$

(a) Derive a PI controller to regulate the blood glucose level for the servo problem using the Ziegler–Nichols closed-loop technique. Assume that $G_f = G_m = 1$. The load is U_M.

(b) Plot the closed-loop response to a step change of size 15 mg/dL in U_I (i.e., $U_I(s) = 15/s$)

Note: The model parameters are given in Section 12.2.

12.6. The following model was written for the control of glycemia (Section 12.2):

$$Y(s) = \frac{K_M e^{-\tau_{D,M}s}}{s(\tau_M s + 1)} U_M(s) + \frac{K_I e^{-\tau_{D,I}s}}{s(\tau_I s + 1)} U_I(s).$$

(a) Derive a PID controller for disturbance compensation (regulator problem) using the Ziegler–Nichols closed-loop technique. Assume that $G_f = G_m = 1$. The disturbance is U_M.

(b) Plot the closed-loop response to a step change of size 0.2 g CHO in U_M (i.e. $U_M(s) = 0.2/s$).

Note: The model parameters are listed in Section 12.2.

12.7. In Section 12.3, the following IMC-based PID settings were obtained:

$$K_c = \frac{1}{K} \frac{\tau_1 + \tau_2}{\tau_f + \tau_d}; \quad \tau_I = \tau_1 + \tau_2; \quad \tau_D = \frac{\tau_1 \tau_2}{\tau_1 + \tau_2}.$$

Write these PID tuning parameters in terms of $k_L a_B$, $k_L a_H$, P, H, $C_{O_2,\text{sat}}$, k_m and τ_m.

12.8. In Section 12.4 (see [5] for details of the model), the following closed-loop transfer function was obtained for the control of biomass in a chemostat:

$$G_{SP}(s) = -\frac{0.398(s-2.)(s+0.3)}{0.44s^3 + 1.49s^2 + 2.68s + 0.24}$$

after using the Ziegler–Nichols tuning rules. Derive G_{sp} using the Cohen–Coon tuning rules for a PI controller.

Note: The transfer functions are given in Section 12.4.

12.9. The following FOPDT model was developed for the controlled infusion of vasoactive drugs (Section 12.5):

$$\tilde{G}_p = -\frac{1.0e^{27.7s}}{97.2s+1}.$$

Derive the P, PI and PID tuning parameters using the Ziegler–Nichols first method. Assume that $G_f = G_m = 1$.

12.10. Redo Problem 12.9 using the Ziegler–Nichols second method.

REFERENCES

1. Ziegler J, Nichols N. Optimum settings for automatic controllers. *Transactions of the American Society of Mechanical Engineers* 1942; 64:759–768.
2. Rivera DE, Skogestad S, Morari M. Internal model control 4: PID controller design. *Industrial and Engineering Chemistry Process Design and Development* 1986; 25:252–265.
3. Marlin TE. *Process Control: Designing Processes and Control Systems for Dynamic Performance*. Boston: McGraw-Hill, 2000.
4. Percival M, Dassau E, Zisser H, Jovanovic L, Doyle FJ III. Practical approach to design and implementation of a control algorithm in an artificial pancreatic beta cell. *Industrial & Engineering Chemistry Research* 2009; 48:6059–6067.
5. Giriraj Kumar SM, Jain R, Anantharaman N, Dharmalingam V, Sheriffa Begum KMM. Genetic algorithm based PID controller tuning for a model bioreactor. *Indian Chemical Engineering*. 2008; 50:214–226.
6. Meier R, Nieuwland J, Zbinden AM, Hacisalihzade SS. Fuzzy logic control of blood pressure during anesthesia. *IEEE Control Systems Magazine*. 1992; 12:12–17.
7. Sheppard LC. Computer control of the infusion of vasoactive drugs. *Annal of Biomedical Engineering*. 1980; 8:431–434.
8. Jones RW, Tham MT. An undergraduate CACSD project: the control of mean arterial blood pressure during surgery. *International Journal of Engineering Education* 2005; 21:1043–1049.

9. Wang X, Zhang X, Li Z, Yu X. A first order system model of fracture healing. *Journal of Zhejiang University Science B* 2005; 6:926–930.

10. Frost HM. Bone "mass" and the "mechanostat": a proposal. *Anatomical Record* 1987; 219:1–9.

11. Levy DD. A pulsed electrical stimulation technique for inducing bone growth. *The Annals of the New York Academy of Sciences* 1974; 238:478–490.

12. Griffin JE, Ojeda SR. *Textbook of Endocrine Physiology*. Oxford: Oxford University Press, 1996.

13. Goff JP, Horst RL, Jardon PW, Borelli C, Wedam J. Field trials of an oral calcium propionate paste as an aid to prevent milk fever in periparturient dairy cows. *Journal of Dairy Science.* 1996; 79:378–383.

14. Khammash M, El-Samad H. Systems biology: from physiology to systems biology: from physiology to gene regulation. *IEEE Control Systems Magazine.* 2004; 24: 62–76.

15. Tsao J, Chuang H, Wu W. On-line state estimation and control in glutamic acid production. *Bioprocess Engineering* 1991; 7:35–39.

CHAPTER 13

FEEDBACK CONTROL OF DEAD-TIME SYSTEMS

Example 11.5 illustrates a case where no crossover frequency is observed for the open-loop transfer function

$$G_{OL}(s) = G_p(s)G_f(s)G_c(s)G_m(s) = \frac{30}{4s+1},$$

resulting in a stable closed-loop system. However, the introduction of a dead time,

$$G_{OL}(s) = \frac{30e^{-0.4s}}{4s+1},$$

as shown in Example 11.6, causes the closed-loop system to become unstable with gain margin $GM = 0.545$ and phase margin $PM = -79.88°$. As the dead time increases, the PM decreases, which may lead to instability (Fig. 13.1). Even when the closed-loop system is stable, it takes the process variable some time to respond to the controller output or to a disturbance. While a high GM allows for more flexibility in tuning the controller, a reduced GM restricts the ability of the control engineer to regulate the process efficiently. The *Smith predictor*, or *dead-time compensator*, is a method developed to help mitigate the effects of a pure time delay on closed-loop performance.

Control of Biological and Drug-Delivery Systems for Chemical, Biomedical, and Pharmaceutical Engineering, First Edition. Laurent Simon.
© 2013 John Wiley & Sons, Inc. Published 2013 by John Wiley & Sons, Inc.

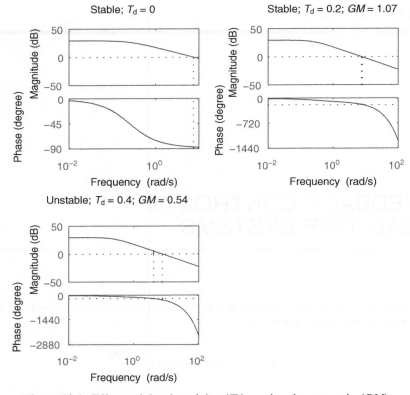

Figure 13.1. Effects of the time delay (T_d) on the phase margin (PM).

13.1 SMITH PREDICTOR-BASED METHODS

Similar to the internal model control (IMC) approach, the Smith predictor is derived directly from the plant model [1]. Figure 13.2 represents a feedback loop with a long dead time. The open-loop response to a change in the reference signal is

$$\bar{y}(s) = G_c(s)G_p(s)e^{-t_d s}\bar{r}(s). \tag{13.1}$$

The goal of the Smith predictor is to remove the delay from the feedback loop so that the controller is no longer acting on a delayed response. Figure 13.3 represents such an ideal system. In this case, $\bar{y}^*(s)$ is the open-loop delay-free response:

$$\bar{y}^*(s) = G_c(s)G_p(s)\bar{r}(s). \tag{13.2}$$

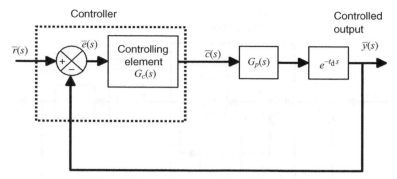

Figure 13.2. Feedback loop with a dead time.

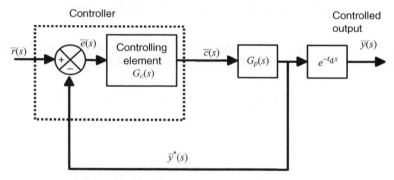

Figure 13.3. Feedback loop without a dead time.

The output of the controlling element is fed back to the system (Fig. 13.4). Note that $\bar{y}^*(s)$ is the sum of $G_c(s)(1-e^{-t_d s})G_p(s)\bar{r}(s)$ and $\bar{y}(s)$. The transfer function of the compensator block is

$$G_{dc}(s)=\tilde{G}_p(s)\left(1-e^{-\tilde{t}_d s}\right), \tag{13.3}$$

where $\tilde{G}_p(s)$ and $\tilde{t}_d$ approximate $G_p(s)$ and t_d, respectively. The remainder of this chapter assumes a perfect model: $\tilde{G}_p(s)=G_p(s)$ and $\tilde{t}_d=t_d$.

To show how the compensator $G_{dc}(s)$ works to eliminate the impact of dead time, the input to the controlling element is first calculated:

$$\bar{e}_c(s)=[\bar{r}(s)-\bar{y}(s)]-\bar{y}_{dc}(s)$$

or

$$\bar{e}_c(s)=[\bar{r}(s)-\bar{y}(s)]-G_p(s)\left(1-e^{-t_d s}\right)\bar{c}(s),$$

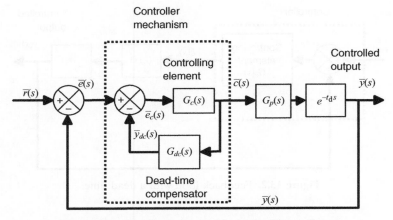

Figure 13.4. Feedback loop with the Smith predictor.

where $\bar{c}(s)$ is the output of the controller mechanism. Since $\bar{y}(s)$ (*the current response*) is represented by $G_p(s)e^{-t_d s}\bar{c}(s)$ in Figure 13.3, we have

$$\bar{e}_c(s) = \left[\bar{r}(s) - G_p(s)e^{-t_d s}\bar{c}(s)\right] - G_p(s)\left(1 - e^{-t_d s}\right)\bar{c}(s)$$

or

$$\bar{e}_c(s) = \bar{r}(s) - G_p(s)\bar{c}(s),$$

where $G_p(s)\bar{c}(s)$ represents the *predicted response* (the name Smith predictor). Therefore, the dead time is effectively removed by introducing the function $G_{dc}(s) = G_p(s)\left(1 - e^{-t_d s}\right)$ around the controller.

To derive the closed-loop transfer function, the output of the controller mechanism is expressed as a function of the input $\bar{e}(s)$:

$$\bar{c}(s) = G_{cm}(s)\bar{e}_c(s), \tag{13.4}$$

where $G_{cm}(s)$ is defined as

$$G_{cm}(s) = \frac{G_c(s)}{1 + G_c(s)G_{dc}(s)} \tag{13.5}$$

or

$$G_{cm}(s) = \frac{G_c(s)}{1 + G_c(s)G_p(s)\left(1 - e^{-t_d s}\right)}. \tag{13.6}$$

Now that we have a single transfer function to represent the controller mechanism block, the following expression is obtained:

$$\frac{\bar{y}(s)}{\bar{r}(s)} = \frac{G_{cm}(s)G_p(s)e^{-t_d s}}{1+G_{cm}(s)G_p(s)e^{-t_d s}}. \tag{13.7}$$

After further simplification, substituting the expression for $G_{cm}(s)$ into Equation (13.7) gives

$$\frac{\bar{y}(s)}{\bar{r}(s)} = \frac{G_c(s)G_p(s)e^{-t_d s}}{1+G_c(s)G_p(s)}. \tag{13.8}$$

When compared to the closed-loop transfer function without the compensator,

$$G_{\text{SP-comp}} = \frac{G_c(s)G_p(s)e^{-t_d s}}{1+G_c(s)G_p(s)e^{-t_d s}}, \tag{13.9}$$

the Smith predictor eliminates the time delay in the feedback loop. For the regulator problem, the following relationship is derived:

$$\frac{\bar{y}(s)}{\bar{d}(s)} = \frac{G_p(s)e^{-t_d s}}{1+G_c(s)G_p(s)} + \frac{G_p(s)G_c(s)e^{-t_d s}}{1+G_c(s)G_p(s)}G_p(s)(1-e^{-t_d s}), \tag{13.10}$$

where $\bar{d}(s)$ is a disturbance (Fig. 13.5). This scheme performs very well when the model accurately represents the plant dynamics. One of the advantages of

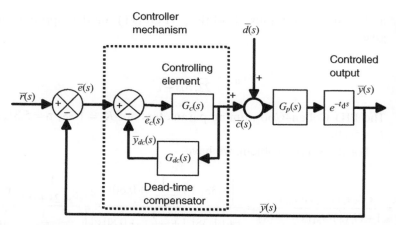

Figure 13.5. Feedback loop with the Smith predictor when a disturbance is added.

the approach is that larger controller gains can be selected without running the risk of destabilizing the system. However, this strategy is sensitive to parameter uncertainties.

Example 13.1   Design a Smith predictor for the following time-delayed process:

$$G_{p1}(s) = \frac{1}{s^2 + 2s + 0.1}; \quad G_{p2}(s) = e^{-2s}. \tag{13.11}$$

Solution

(a) First, we will investigate the application of a conventional proportional–integral–derivative (PID) controller. If we assume that $G_f = G_m(s) = 1$, we have

$$\frac{\overline{y}_m(s)}{\overline{c}(s)} = G_f(s)G_p(s)G_m(s) = \frac{\overline{y}(s)}{\overline{c}(s)} = G_p(s) = \frac{e^{-2s}}{s^2 + 2s + 0.1},$$

with

$$G_p(s) = G_{p1}(s)G_{p2}(s) = \frac{e^{-2s}}{s^2 + 2s + 0.1}.$$

A Padé approximation of e^{-2s} leads to

$$G_p \approx \frac{1-s}{s^3 + 3s^2 + 2.1s + 0.1}$$

and a first-order plus dead-time (FOPDT) model approximation gives

$$G_{RC1} \approx \frac{9.97e^{-2.24s}}{22.70s + 1}.$$

The PID Cohen–Coon tuning parameters are $K_c = 1.38$, $\tau_I = 5.31$, and $\tau_D = 0.80$.

For the servo problem, we have

$$\frac{\overline{y}(s)}{\overline{r}(s)} = \frac{G_c(s)G_{p1}(s)G_{p2}(s)}{1 + G_c(s)G_{p1}(s)G_{p2}(s)} = \frac{\left[1.38\left(1 + \dfrac{1}{5.31s} + 0.80s\right)\right]\left(\dfrac{1}{s^2 + 2s + 0.1}\right)e^{-2s}}{1 + \left[1.38\left(1 + \dfrac{1}{5.31s} + 0.80s\right)\right]\left(\dfrac{1}{s^2 + 2s + 0.1}\right)e^{-2s}}.$$

If $\bar{r}(s) = 1/s$, the closed-loop response is

$$y(t) = \left[1 - 0.00048e^{-1.03t_1} + 0.31e^{-0.33t_1} + 1.03e^{-0.27t_1} \sin(0.83t_1)\right. \\ \left. - 1.31e^{-0.27t_1} \cos(0.83t_1)\right]\psi(t_1),$$

(13.12)

where $t_1 = t - 2$ and ψ represents the unit step function. Note that the Padé formula is only applied in the denominator.

(b) We now implement the Smith predictor. An FOPDT approximation of

$$G_{p1}(s) = \frac{1}{s^2 + 2s + 0.1}$$

yields

$$G_{RC2} = \frac{10.0e^{-0.41s}}{21.56s + 1}.$$

The Cohen–Coon tuning settings are $K_c = 7.02$, $\tau_I = 1.00$, and $\tau_D = 0.15$. When $\bar{d}(s) = 0$,

$$\frac{\bar{y}(s)}{\bar{r}(s)} = \frac{G_c(s)G_{p1}(s)G_{p2}(s)}{1 + G_c(s)G_{p1}(s)} = \frac{\left[7.02\left(1 + \dfrac{1}{1.00s} + 0.15s\right)\right]\left(\dfrac{1}{s^2 + 2s + 0.1}\right)}{1 + \left[7.02\left(1 + \dfrac{1}{1.00s} + 0.15s\right)\right]\left(\dfrac{1}{s^2 + 2s + 0.1}\right)}e^{-2s},$$

and after a unit step change in the set point, we have

$$\bar{y}(s) = \frac{G_c(s)G_{p1}(s)G_{p2}(s)}{1 + G_c(s)G_{p1}(s)} = \frac{\left[7.02\left(1 + \dfrac{1}{1.00s} + 0.15s\right)\right]\left(\dfrac{1}{s^2 + 2s + 0.1}\right)}{1 + \left[7.02\left(1 + \dfrac{1}{1.00s} + 0.15s\right)\right]\left(\dfrac{1}{s^2 + 2s + 0.1}\right)}\frac{e^{-2s}}{s}.$$

The inverse Laplace transform is

$$y(t) = \left[1 + 0.15e^{-1.46t_1} + 0.17e^{-0.80t_1} \sin(2.04t_1) - 1.15e^{-0.80t_1} \cos(2.04t_1)\right]\psi(t_1),$$

(13.13)

with $t_1 = t - 2$. The control performances of both schemes are shown in Figure 13.6.

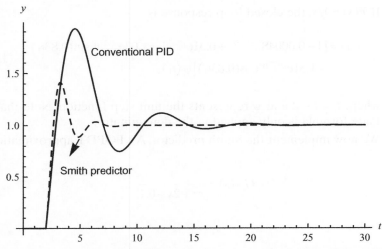

Figure 13.6. Comparison of the Smith predictor and a conventional PID controller.

13.2   CONTROL OF BIOMASS

In Section 12.4, the transfer function, representing the relationship between the dilution rate and the biomass concentration, was given by

$$G_p(s) = \frac{-0.6758e^{-s}}{0.4417s + 1}. \tag{13.14}$$

The function $G_p(s)$ is written as $G_p(s) = G_{p1}(s)G_{p2}(s)$ with

$$G_{p1}(s) = \frac{-0.6758}{0.4417s + 1}$$

and $G_{p2}(s) = e^{-s}$. Application of the Smith predictor produces

$$\frac{\overline{y}^*(s)}{\overline{c}(s)} = \frac{-0.6758}{0.4417s + 1}.$$

The Cohen–Coon or Ziegler–Nichols first method is not appropriate because the process has no time delay. The IMC-based proportional–integral (PI) controller for a first-order process was derived in Section 12.7:

$$K_c = \frac{\tau_p}{K_p\tau_f} = -2.9595 \quad \text{and} \quad \tau_I = \tau_p = 0.4417$$

when $\tau_f = \frac{0.4417}{2} = 0.2208$ (response is twice as fast as the open-loop response). For the servo problem $\bar{d}(s) = 0$,

$$\frac{\bar{y}(s)}{\bar{r}(s)} = \frac{G_c(s)G_{p1}(s)G_{p2}(s)}{1+G_c(s)G_{p1}(s)} = \frac{\left[-2.9595\left(1+\dfrac{1}{0.4417s}\right)\right]\left(\dfrac{-0.6758}{0.4417s+1}\right)}{1+\left[-2.9595\left(1+\dfrac{1}{0.4417s}\right)\right]\left(\dfrac{-0.6758}{0.4417s+1}\right)}e^{-s},$$

and with a step change in the set point of size 0.15, we have

$$\bar{y}(s) = \frac{G_c(s)G_{p1}(s)G_{p2}(s)}{1+G_c(s)G_{p1}(s)} = \frac{\left[-2.9595\left(1+\dfrac{1}{0.4417s}\right)\right]\left(\dfrac{-0.6758}{0.4417s+1}\right)}{1+\left[-2.9595\left(1+\dfrac{1}{0.4417s}\right)\right]\left(\dfrac{-0.6758}{0.4417s+1}\right)}\frac{0.15e^{-s}}{s}.$$

The inverse Laplace transform is

$$y(t) = \left(0.15 - 0.15e^{-4.53t_1}\right)\psi(t_1), \tag{13.15}$$

with $t_1 = t - 1$.

For comparison, the control scheme for the original system with no dead-time compensator is developed. The IMC-based PI controller for a first-order process with a time delay is

$$K_c = \frac{\tau_p}{K_p(\tau_f + \theta)} = -0.5354$$

and $\tau_I = \tau_p = 0.4417$ (see Example 12.4). The closed-loop Laplace transform response is

$$\bar{y}(s) = \frac{G_c(s)G_{p1}(s)G_{p2}(s)}{1+G_c(s)G_{p1}(s)G_{p2}(s)}\frac{0.15}{s},$$

which yields

$$y(t) = [0.15 + 0.03e^{-0.59t_1}\sin(1.14t_1) - 0.15e^{-0.59t_1}\cos(1.14t_1)]\psi(t_1), \tag{13.16}$$

with $t_1 = t - 1$. The performances of the two strategies are shown in Figure 13.7.

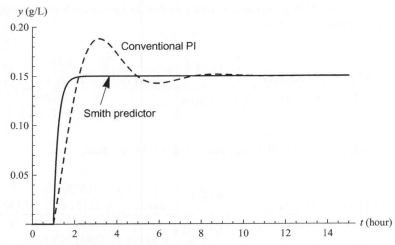

Figure 13.7. Comparison of the Smith predictor and a conventional PI controller.

13.3 *ZYMOMONAS MOBILIS* FERMENTATION FOR ETHANOL PRODUCTION

Karim and Traugh investigated the fermentation of *Z. mobilis* for ethanol production [2]. The goal of the study was to develop software tools for data acquisition, analysis, processing, and automation of fermentation systems. The computer program would also contain algorithms that would make it easy to control biomass or ethanol concentration in a bioreactor. When the time lag of an enzyme analyzer was considered, the following transfer function was obtained to represent the relationship between ethanol concentration (control variable: C [g/L]) and a feed-flow rate (manipulated variable: F [L/h]):

$$\frac{C(s)}{F(s)} = \frac{-2.65\exp(-0.48s)}{(2.4s+1)(0.88s+1)}.$$

A Smith predictor can be used to maintain the ethanol concentration at 24.0 g/L.

This problem is similar to the one described in Example 13.1. In this case,

$$G_{p1}(s) = \frac{-2.65}{(2.4s+1)(0.88s+1)}; \quad G_{p2}(s) = e^{-0.48s}$$

and

$$\frac{\overline{y}^{*}(s)}{\overline{c}(s)} = \frac{-2.65}{(2.4s+1)(0.88s+1)}.$$

The derived FOPDT model is

$$G_{RC} = \frac{-2.64e^{-0.38s}}{4.28s+1}.$$

The Cohen–Coon PID tuning parameters are $K_c = -5.73$, $\tau_I = 0.91$, and $\tau_D = 0.14$. When the servo problem is considered, we have

$$\frac{\overline{y}(s)}{\overline{r}(s)} = \frac{\left[-5.73\left(1+\dfrac{1}{0.91s}+0.14s\right)\right]\left(\dfrac{-2.65}{(2.4s+1)(0.88s+1)}\right)}{1+\left[-5.73\left(1+\dfrac{1}{0.91s}+0.14s\right)\right]\left(\dfrac{-2.65}{(2.4s+1)(0.88s+1)}\right)}e^{-0.48s}$$

or

$$\overline{y}(s) = \frac{49.98s^2 + 364.27s + 400.51}{2.11s^4 + 5.36s^3 + 16.18s^2 + 16.69s}e^{-0.48s}$$

for $\overline{r}(s) = 24/s$. The concentration is (Fig. 13.8)

$$y(t) = \left[24 - 0.59e^{-1.31t_1} + 3.55e^{-0.62t_1}\sin(2.38t_1) - 23.41e^{-0.62t_1}\cos(2.38t_1)\right]\psi(t_1), \tag{13.17}$$

with $t_1 = t - 0.48$.

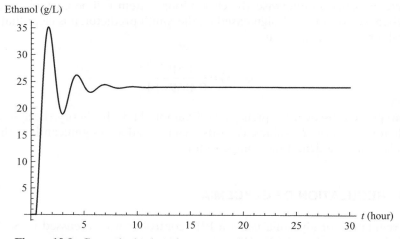

Figure 13.8. Control of ethanol concentration using a Smith predictor.

13.4 FED-BATCH CULTIVATION OF *ACINETOBACTER CALCOACETICUS* RAG-1

Choi et al. developed a feedback control mechanism to regulate ethanol concentration in the fed-batch cultivation of *A. calcoaceticus* RAG-1 [3]. A transfer function was obtained that linked the ethanol concentration in the bioreactor (y) to the manipulated variable (u: feeding rate of ethanol):

$$G_p(s) = \frac{Y(s)}{U(s)} = \frac{1.5197e^{-0.1667s}}{0.4659s + 1}.$$

Let us examine the effects of the Smith predictor on the Ziegler–Nichols tuning rule for a proportional (P) controller based on the second method. Assuming that $G_f = G_m = 1$, we have

$$G_{OL}(s) = \frac{1.5197e^{-0.1667s}}{0.4659s + 1}.$$

The phase angle is $\phi = \tan^{-1}(-0.4659\omega) - 0.1667\omega$. Therefore, $-\pi = \tan^{-1}(-0.4659\omega_c) - 0.1667\omega_c$, which yields $\omega_c = 10.62$ and

$$|G_{OL}(j\omega_c)| = \frac{1.5197}{\left(\sqrt{(0.4659)^2 \omega_c^2 + 1}\right)} = 0.30.$$

As a result, $K_u = \frac{1}{0.301} = 3.32$ and $P_u = 2\pi/\omega_c = 0.59$. The proportional gain is $K_c = 0.5K_u = 1.66$. Even if the gain is increased in order to decrease the offset in the ethanol concentration, K_c has to remain less than 3.32. (*Question: Why is there an offset?*) Otherwise, the closed-loop system will be unstable.

When we consider a design based on the Smith predictor, it is clear that the open-loop transfer function

$$G_{OL}(s) = \frac{1.5197}{0.4659s + 1}$$

has no phase crossover frequency (see Example 11.5). In theory, the system is stable for very large K_c values. The offset can be reduced significantly with the introduction of a dead-time compensator.

13.5 REGULATION OF GLYCEMIA

The regulation of glycemia using a PID controller was discussed in Section 12.2. The transfer function is $G_p(s) = G_{p1}(s)G_{p2}(s)$ with

$$G_{p1}(s) = \frac{-0.075}{s(146s+1)}; \quad G_{p2}(s) = e^{-27s}.$$

Application of the Smith predictor yields

$$\frac{\bar{y}^*(s)}{\bar{c}(s)} = \frac{-0.075}{s(146s+1)}.$$

The Ziegler–Nichols closed-loop technique shows that there is no phase cross-over frequency (*why?*). Let us change the gain from $K_c = -0.305$ to $K_c = -10.0$ and keep the other settings from Section 12.2: $\tau_I = 203.30$ and $\tau_D = 225.89$. A step change of size 15 mg/dL gives

$$\bar{y}(s) = \frac{2541.26s^2 + 11.25s + 0.055}{146s^4 + 170.42s^3 + 0.75s^2 + 0.0037s} e^{-27s}$$

and

$$y(t) = \left[15 - 14.97e^{-1.16t_1} + 0.085e^{-0.0022t_1} \sin(0.0041t_1) \right. \\ \left. - 0.032e^{-0.0022t_1} \cos(0.0041t_1)\right] \psi(t_1). \tag{13.18}$$

with $t_1 = t - 27$ (Fig. 13.9).

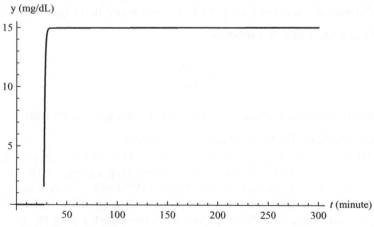

Figure 13.9. Control of glucose concentration using a dead-time compensator.

13.6 SUMMARY

A Smith predictor was used to control processes exhibiting a transportation lag. This method is based on a model of the system and involves a prediction step. Larger gains, which would otherwise lead to instability, are feasible with the use of the dead-time compensator. Applications in bioreactor and blood glucose regulations are outlined.

PROBLEMS

13.1. Consider a mixing process with two inlet mass flow rates into a tank (F_1 and F_2). The mass fractions of a component A in the two streams are x_1 and x_2.

(a) By performing total and component mass balances around the tank, show that the process dynamics can be written as

$$\bar{x}(s) = \frac{K_1}{\tau s+1}\bar{x}_1(s) + \frac{K_2}{\tau s+1}\bar{x}_2(s) + \frac{K_3}{\tau s+1}\bar{F}_1(s) + \frac{K_4}{\tau s+1}\bar{F}_2(s),$$

where x is the mixture composition.

(b) Using constant values of x_2, F_1, and x_1, and allowing for a time delay, we have

$$\bar{x}(s) = \frac{K_4 e^{-\theta_4 s}}{\tau s+1}\bar{F}_2(s).$$

The objective is to control x by manipulating F_2. Draw a block diagram showing the feedback loop with the Smith predictor (see Fig. 13.4).

13.2. Given the following process:

$$\bar{x}(s) = \frac{e^{-s}}{25s+1}\bar{m}(s),$$

where $\bar{m}$ is the Laplace transform of the manipulated variable:

(a) Design a PID controller for the system.

(b) Compare the performance of the PID controller in part (a) with that of a Smith predictor. Use a unit step change in the set point and $G_f = G_m(s) = 1$. Note: Apply IMC tuning rules when designing a PI controller for the process described by $1/(25s + 1)$ and make sure that the response is six times faster than the open-loop dynamics.

13.3. A system is given by

$$G_p(s) = \frac{5}{4s+1} e^{-3s}.$$

(a) Design a PID controller for the system.

(b) Compare the performance of the PI controller in part (a) with that of the Smith predictor. Use a unit step change in the set point and $G_f = G_m(s) = 1$.

Note: Apply IMC tuning rules when designing a PI controller for the process described by $5/(4s + 1)$ and make sure that the response is four times faster than the open-loop dynamics.

13.4. Consider the following process:

$$\bar{x}(s) = \frac{0.05}{(6s+1)^3} \bar{m}(s),$$

where $\bar{m}$ is the Laplace transform of the manipulated variable.

(a) Derive an FOPDT approximation of the transfer function $0.05/(6s + 1)^3$.

(b) Design a Smith predictor for the FOPDT model obtained in part (a).

Note: Apply IMC tuning rules when designing a PI controller. The response should be three times faster than the open-loop dynamics.

13.5. A process is described by the following transfer function:

$$\frac{\bar{x}(s)}{\bar{m}(s)} = \frac{5}{0.2s^2 + 0.9s + 1} e^{-0.2s}.$$

(a) Design a PID controller for the system using Cohen–Coon tuning rules.

(b) Compare the performance of the PID controller in part (a) with that of the Smith predictor. Use a unit step change in the set point and $G_f = G_m(s) = 1$.

Note: Apply Cohen–Coon tuning rules when designing the PID controller for the process described by $5/(0.2s^2 + 0.9s + 1)$.

13.6. Redo the *Z. mobilis* fermentation example in Section 13.3:

$$\frac{C(s)}{F(s)} = \frac{-2.65\exp(-0.48s)}{(2.4s+1)(0.88s+1)}.$$

Design a Smith predictor for the system using the Ziegler–Nichols first method to tune the PID controller. Plot the response after a step change of size 24 g/L in the set point.

13.7. Redo Problem 13.6 using IMC methods instead of the Ziegler–Nichols technique to tune the PID controller. Choose $\tau_f = 0.4$ for the low-pass filter.

13.8. In the fed-batch cultivation of *A. calcoaceticus* RAG-1 (Section 13.4), the following transfer function relating the ethanol concentration in the bioreactor (y [g/L]) to the feeding rate of ethanol (u [L/h]) was obtained:

$$G_p(s) = \frac{Y(s)}{U(s)} = \frac{1.5197e^{-0.1667s}}{0.4659s + 1}.$$

With a Smith predictor, the closed-loop system remains stable for a very large gain $K_c(\text{L}^2/\text{g/h})$.

(a) Plot the closed-loop response to a step change of size 1.0 g/L in the set point of glucose. Use a proportional controller gain, $K_c = 2.0$, without a Smith predictor.

(b) Sketch the closed-loop response to a step change of size 1.0 g/L in the set point of glucose. Use a proportional controller gain, $K_c = 5.0$, with a Smith predictor.

(c) Plot the closed-loop response to a step change of size 1.0 g/L in the set point of glucose. Use a proportional controller gain, $K_c = 5.0$, without a Smith predictor.

13.9. Using the process in Problem 13.8,

$$G_p(s) = \frac{Y(s)}{U(s)} = \frac{1.5197e^{-0.1667s}}{0.4659s + 1},$$

design a Smith predictor for the system. Apply a step change of size 1.0 g/L in the set point. Plot the result.

Note: Use IMC methods to tune a PI controller for $1.5197/(0.4659s + 1)$. Choose $\tau_f = 0.2$ for the low-pass filter.

13.10. Given the process in Section 13.5,

$$G_p(s) = G_{p1}(s)G_{p2}(s),$$

with

$$G_{p1}(s) = \frac{-0.075}{s(146s + 1)}; \quad G_{p2}(s) = e^{-27s},$$

design a Smith predictor for the system using a P controller with $K_c = -10$. Apply a step change of size 15.0 mg/dL in the set point. Plot the result.

REFERENCES

1. Smith OM. Closer control of loops with dead time. *Chemical Engineering Progress* 1957; 53:217–219.
2. Karim M, Traugh G. Data acquisition and control of a continuous fermentation unit. *Journal of Industrial Microbiology* 1987; 2:305–317.
3. Choi J-W, Choi H-G, Lee K-S, Lee W-H. Control of ethanol concentration in a fed-batch cultivation of *Acinetobacter calcoaceticus* RAG-1 using a feedback-assisted iterative learning algorithm. *Journal of Biotechnology* 1996; 49:29–43.

CHAPTER 14

CASCADE AND FEEDFORWARD CONTROL STRATEGIES

So far, we have covered single-input and single-output (SISO) systems. There are also processes that require the use of several manipulated variables or where the control objective consists of maintaining various outputs near set-point values. Although this textbook does not deal with multi-input, multi-output (MIMO) plants, this chapter offers two examples of *multiloop* control configurations: cascade and feedforward controls.

14.1 CASCADE CONTROL

A controller may be designed to suppress the effects of disturbances or to track set-point changes. However, the examples presented up to now assumed that the manipulated input variable was unaffected by disturbances. A cascade design provides a way to reduce the effects of perturbations in the manipulated input on closed-loop performance.

To illustrate the applications of cascade control, the stirred-tank heater model is reintroduced. We now assume that the manipulated variable, steam flow rate F, is subject to fluctuations (Fig. 14.1) due to variations in the steam header pressure. Holding F_i constant, the other input disturbance is represented by T_i. The controlled variable is the temperature of the liquid in the tank. A cascade control configuration is shown in Figure 14.2. The measured flow from the transmitter FT is compared to the output (F_{SP}) of the temperature

Control of Biological and Drug-Delivery Systems for Chemical, Biomedical, and Pharmaceutical Engineering, First Edition. Laurent Simon.
© 2013 John Wiley & Sons, Inc. Published 2013 by John Wiley & Sons, Inc.

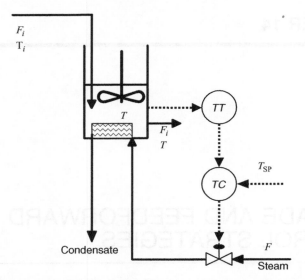

Figure 14.1. A temperature control system with a conventional SISO feedback loop.

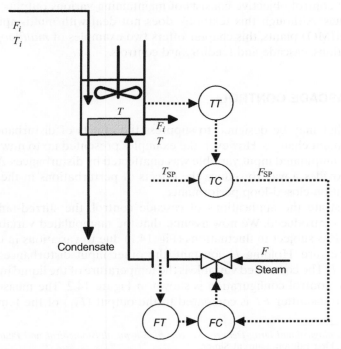

Figure 14.2. A temperature control system with a cascade control scheme.

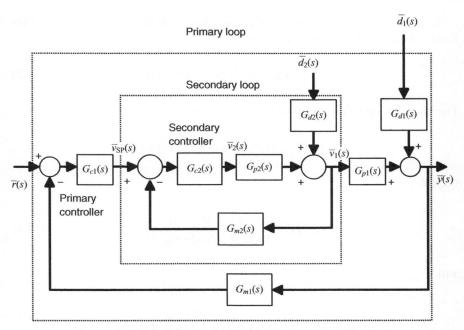

Figure 14.3. Block diagram of a cascade control architecture.

controller (*TC*). The flow controller (*FC*) then sends a command signal to the valve, which is adjusted in order to keep *T* as close as possible to T_{SP}.

A block diagram of a generalized cascade control is described in Figure 14.3. Considering the primary (open) loop, the output variable *y* is affected by changes in d_1 and v_1. From the figure, the following equation can be written:

$$\bar{y}(s) = G_{d1}(s)\bar{d}_1(s) + G_{p1}(s)\bar{v}_1(s). \tag{14.1}$$

In comparison to Figure 14.2, $d_1 = T_i$, $v_1 = F$; d_2 is a disturbance that affects *F*. The transfer function $G_{m2}(s)$ describes the dynamics of the transmitter *FT*, which measures the input v_1, while *TT* is represented by $G_{m1}(s)$. The *master* or *primary* controller *TC* (transfer function $G_{c1}(s)$) adjusts the set point of the *slave* or *secondary* controller *FC* (transfer function $G_{c2}(s)$). This inner loop responds quickly to fluctuations (d_2) to make sure that the impact on the manipulated variable, and eventually the whole process, is reduced. The valve dynamics is represented by $G_{p2}(s)$. A conventional feedback mechanism would try to attenuate the influence of the disturbance on the controlled output. However, the controller may be too slow to take corrective action. In the cascade control architecture, the flow rate is measured so that any deviation from the flow requirement, because of changes in upstream pressure, is detected early.

To derive the closed-loop transfers, we notice, in the secondary loop, that $\bar{v}_1(s)$ is related to $\bar{v}_2(s)$ and $\bar{d}_2(s)$ by

$$\bar{v}_1(s) = G_{d2}(s)\bar{d}_2(s) + G_{p2}(s)\bar{v}_2(s). \tag{14.2}$$

The closed-loop relationship in the inner loop is

$$\bar{v}_1(s) = G_{\text{SPsec}}(s)\bar{v}_{\text{1SP}}(s) + G_{\text{loadsec}}(s)\bar{d}_2(s), \tag{14.3}$$

where

$$G_{\text{SPsec}}(s) = \frac{G_{c2}(s)G_{p2}(s)}{1 + G_{c2}(s)G_{p2}(s)G_{m2}(s)} \tag{14.4}$$

and

$$G_{\text{loadsec}}(s) = \frac{G_{d2}(s)}{1 + G_{c2}(s)G_{p2}(s)G_{m2}(s)}. \tag{14.5}$$

The closed-loop transfer functions can be obtained by manipulating the elements of the block diagram:

$$\bar{y}(s) = G_{\text{SP}}(s)\bar{r}(s) + G_{\text{load1}}(s)\bar{d}_1(s) + G_{\text{load2}}(s)\bar{d}_2(s), \tag{14.6}$$

where

$$G_{\text{SP}}(s) = \frac{G_{p1}(s)G_{c1}(s)G_{\text{SPsec}}(s)}{1 + G_{p1}(s)G_{c1}(s)G_{\text{SPsec}}(s)G_{m1}(s)}, \tag{14.7}$$

$$G_{\text{load1}}(s) = \frac{G_{d1}(s)}{1 + G_{p1}(s)G_{c1}(s)G_{\text{SPsec}}(s)G_{m1}(s)}, \tag{14.8}$$

and

$$G_{\text{load2}}(s) = \frac{G_{p1}(s)G_{\text{loadsec}}(s)}{1 + G_{p1}(s)G_{c1}(s)G_{\text{SPsec}}(s)G_{m1}(s)}. \tag{14.9}$$

The last three transfer functions can also be derived by drawing a block diagram (Fig. 14.4) equivalent to Figure 14.3. When Figure 14.4 is compared

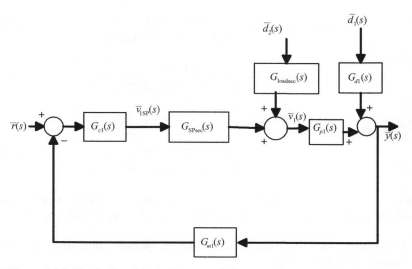

Figure 14.4. Equivalent block diagram of the cascade control configuration.

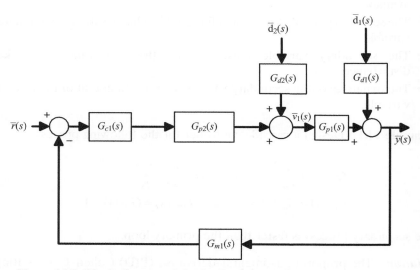

Figure 14.5. Block diagram of the feedback loop with no cascade control.

to a standard feedback loop (Fig. 14.5), the main differences between the two architectures become evident.

Based on Figure 14.5, the controlled response is

$$\bar{y}(s) = G_{\text{SP-C}}(s)\bar{r}(s) + G_{\text{load1-C}}(s)\bar{d}_1(s) + G_{\text{load2-C}}(s)\bar{d}_2(s), \qquad (14.10)$$

with

$$G_{\text{SP-C}}(s) = \frac{G_{p1}(s)G_{c1}(s)G_{p2}(s)}{1 + G_{p1}(s)G_{c1}(s)G_{p2}(s)G_{m1}(s)}, \qquad (14.11)$$

$$G_{\text{load1-C}}(s) = \frac{G_{d1}(s)}{1 + G_{p1}(s)G_{c1}(s)G_{p2}(s)G_{m1}(s)}, \qquad (14.12)$$

and

$$G_{\text{load2-C}}(s) = \frac{G_{p1}(s)G_{d2}(s)}{1 + G_{p1}(s)G_{c1}(s)G_{p2}(s)G_{m1}(s)}, \qquad (14.13)$$

where $-C$ is used to indicate the absence of a cascade controller. Marlin established the following criteria for designing a cascade control scheme [1]:

- Cascade control is a good option when the performance of a single-loop control is not adequate.
- Cascade control is appropriate when a secondary measurement is available.
- The secondary variable must be influenced by changes in the manipulated variable.
- The secondary variable must indicate the appearance of a key disturbance.
- The dynamics of the secondary variable is faster than that of the primary variable.

Example 14.1 Design a cascade controller for the following process:

$$G_{p1}(s) = \frac{e^{-0.5s}}{1 + 0.9s}, \quad G_{p2}(s) = \frac{e^{-0.05s}}{1 + 0.09s},$$
$$G_{d1}(s) = G_{d2}(s) = 1, \quad \text{and} \quad G_{m1}(s) = G_{m2}(s) = 1.$$

The secondary process is faster than the primary loop.

Solution The proportional–integral–derivative (PID) Cohen–Coon settings are $K_{c1} = 2.6500$, $\tau_{I1} = 1.0127$, and $\tau_{D1} = 0.1651$ for $G_{p1}(s)$. The tuning parameters for $G_{p2}(s)$ are $K_{c2} = 2.6500$, $\tau_{I2} = 0.1013$, and $\tau_{D1} = 0.0165$. If we assume that $\overline{r}(s) = \overline{d}_1(s) = 0$, the following transfer function is obtained:

$$G_{\text{load2}}(s) = \frac{G_{d2}(s)G_{p1}(s)}{1 + G_{c2}(s)G_{p2}(s)[G_{c1}(s)G_{m1}(s)G_{p1}(s) + G_{m2}(s)]}.$$

For a unit step change in d_2, the closed-loop responses are shown in Figure 14.6. The cascade control outperforms the single-loop control. In the

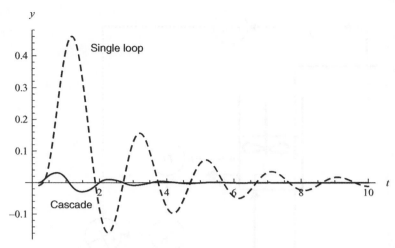

Figure 14.6. Cascade versus single-loop control for a unit step change in d_2.

simulation, a second-order Padé approximation was selected. A moving average filter (`MovingAverage` in *Mathematica*®) was also applied to process the data further.

14.2 FEEDFORWARD CONTROL

The goal of the *feedforward control* is to prevent a measured disturbance from influencing the output. This scheme assumes the existence of a model equation that connects the disturbance to the controlled variable. Contrary to a feedback mechanism, adequate action is taken before allowing the perturbation to disrupt the operation of the plant. The inlet stream temperature T_i (Fig. 14.1) is now a measured disturbance. Figure 14.7 shows a feedforward–feedback control system. The outputs of the feedforward and feedback controllers (*FFC* and *TC*, respectively) are combined and a signal is sent to a valve, which adjusts the steam flow rate. For the procedure to be effective against random variations in T_i, the *FFC* algorithm needs to relate T_i to T via a model equation. The block diagram of the system is shown in Figure 14.8.

The closed-loop transfer functions in Figure 14.8 can be derived by inspecting Figure 14.5 and applying Equation (14.10). Following a procedure similar to the one outlined in Section 14.1, the response is (Figure 14.8)

$$\overline{y}(s) = G_{\text{SP}}(s)\,\overline{r}(s) + G_{\text{load1}}(s)\,\overline{d}(s) + G_{\text{load2}}(s)\,\overline{d}(s), \tag{14.14}$$

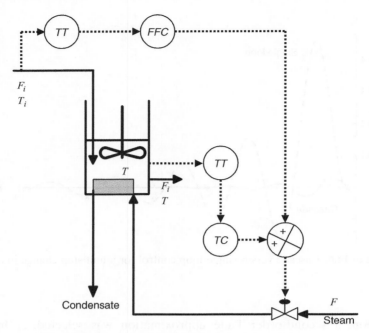

Figure 14.7. The feedforward–feedback temperature control.

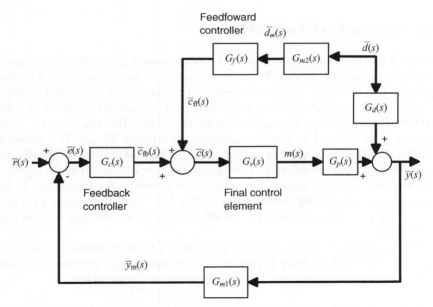

Figure 14.8. Block diagram of the feedforward–feedback temperature control.

where

$$G_{SP}(s) = \frac{G_p(s)G_c(s)G_v(s)}{1 + G_p(s)G_c(s)G_v(s)G_{m1}(s)}, \qquad (14.15)$$

$$G_{\text{load1}}(s) = \frac{G_d(s)}{1 + G_p(s)G_c(s)G_v(s)G_{m1}(s)}, \qquad (14.16)$$

and

$$G_{\text{load1}}(s) = \frac{G_{m2}(s)G_f(s)G_v(s)G_p(s)}{1 + G_p(s)G_c(s)G_v(s)G_{m1}(s)}. \qquad (14.17)$$

As a result,

$$\bar{y}(s) = G_{SP}(s)\bar{r}(s) + G_{\text{load}}(s)\bar{d}(s), \qquad (14.18)$$

where

$$G_{\text{load}}(s) = G_{\text{load1}}(s) + G_{\text{load2}}(s) \qquad (14.19)$$

or

$$G_{\text{load}}(s) = \frac{G_d(s) + G_{m2}(s)G_f(s)G_v(s)G_p(s)}{1 + G_p(s)G_c(s)G_v(s)G_{m1}(s)}. \qquad (14.20)$$

The feedforward control strategy is traditionally used for disturbance rejection. In order to keep the controlled variable at the reference point (i.e., $\bar{r}(s) = 0$), the following condition holds:

$$G_{\text{load}}(s) = \frac{\bar{y}(s)}{\bar{d}(s)} = \frac{G_d(s) + G_{m2}(s)G_f(s)G_v(s)G_p(s)}{1 + G_p(s)G_c(s)G_v(s)G_{m1}(s)} = 0, \qquad (14.21)$$

which gives

$$G_d(s) + G_{m2}(s)G_f(s)G_v(s)G_p(s) = 0. \qquad (14.22)$$

As a result, the feedforward controller design is

$$G_f(s) = -\frac{G_d(s)}{G_{m2}(s)G_v(s)G_p(s)}. \qquad (14.23)$$

Note that Equation (14.23) assumes a perfect model of the plant, which is generally not satisfied in practice. To account for this, a feedback controller, with the configuration described in Figure 14.8, is often added. Application of Equation (14.23) may lead to a *physically unrealizable controller*, that is, the occurrence of a positive time delay or a situation where the degree of the numerator is higher than the degree of the denominator. For more implementation issues, consult Marlin [1] and Ogunnaike [2].

Example 14.2   Design the feedforward–feedback control structure described in Figure 14.8.

$$G_p(s) = \frac{1}{s+1}, \quad G_{m1}(s) = 1, \quad G_{m2}(s) = 1, \quad G_v(s) = 1, \quad G_d(s) = \frac{1}{0.25s+1},$$

and

$$G_c(s) = K_c\left(1 + \frac{1}{\tau_1 s}\right).$$

Solution   From Section 12.7, the tuning parameters of the feedback internal model controller are

$$K_c = \frac{\tau_p}{K_p \tau_f} \quad \text{and} \quad \tau_1 = \tau_p.$$

If we choose $\tau_f = \frac{1}{2}$ so that the closed-loop response is twice as fast as the open-loop response, the controller settings are $K_c = 2$ and $\tau_1 = 1$.

Application of Equation (14.23) leads to the following lead–lag controller:

$$G_f = -\frac{s+1}{0.25s+1}. \tag{14.24}$$

Based on this scheme, the effects of the disturbance are suppressed immediately before it has a chance to affect the operation of the process. For comparison, we derive $G_{\text{load}}(s)$ when the feedforward controller is removed:

$$G_{\text{load}}(s) = \frac{G_d(s)}{1 + G_p(s)G_c(s)G_v(s)G_{m1}(s)}$$

$$G_{\text{load}}(s) = \frac{s}{0.25s^2 + 1.5s + 2}. \tag{14.25}$$

A unit step change in the disturbance gives

$$y(t) = 2\left(e^{-2t} - e^{-4t}\right) \tag{14.26}$$

and, in particular, $\lim_{t\to\infty} y(t) = 0$. In the feedforward–feedback architecture, the corrective action is instantaneous. The process variable never deviates from its original value of zero. Without the feedforward mechanism, perturbation of $y(t)$ is inevitable, although ultimately, the feedback action manages to bring the system back to zero (e.g., $\lim_{t\to\infty} y(t) = 0$).

14.3 INSULIN INFUSION

The idea of an "artificial organ" for type 1 diabetes has received increased attention and would consist of an automated insulin delivery system (portable or implantable) composed of a glucose sensor, an insulin infusion pump, and a control law that calculates the appropriate insulin dosage. Marchetti et al. proposed a feedforward–feedback glucose control strategy that considered the inexactness of the individual's estimation of the carbohydrate content (CHO) of meals [3]. This section deals with the feedforward controller design. Additional information on the feedback controller can be found in Marchetti et al. [3].

The glucose open-loop response, $\bar{y}(s)$, expressed as a function of the gut absorption rate of the CHO content $\bar{d}(s)$, a measured disturbance, and $\bar{u}(s)$, the insulin infusion rate, is

$$\bar{y}(s) = G_d \bar{d}(s) + G_p \bar{u}(s), \tag{14.27}$$

with

$$G_d = \frac{K_d}{\tau_d s + 1} \tag{14.28}$$

and

$$G_p(s) = \frac{K e^{-\theta s}}{(\tau_1 s + 1)(\tau_2 s + 1)}. \tag{14.29}$$

When $G_{m2}(s) = G_v(s) = 1$, the ideal feedforward controller is

$$G_f(s) = -\frac{G_d(s)}{G_p(s)} = -\frac{K_d}{K} \frac{(\tau_1 s + 1)(\tau_2 s + 1) e^{+\theta s}}{\tau_d s + 1}, \tag{14.30}$$

which is physically unrealizable. The transfer function $G_f(s)$ can be approximated by using a lead–lag unit [3]:

$$G_f(s) = -K_f \frac{\tau_3 s + 1}{\tau_4 s + 1}, \tag{14.31}$$

where the following relations $K_f = K_d/K$, $\tau_3 = \tau_1 + \tau_2 + \theta$, and $\tau_4 = \tau_d$ can be used as first estimates of the tuning parameters. More accurate results, which consider the effects of meal sizes, are given in Marchetti et al. [3].

The equation relating $\bar{d}(s)$ to $\bar{y}(s)$ without the feedback controller is (i.e., using only the feedforward loop of Fig. 14.8)

$$\bar{y}(s) = [G_d(s) + G_{m2}(s)G_f(s)G_v(s)G_p(s)]\bar{d}(s), \qquad (14.32)$$

which simplifies to

$$\bar{y}(s) = [G_d(s) + G_f(s)G_p(s)]\bar{d}(s), \qquad (14.33)$$

with $G_{m2}(s) = G_v(s) = 1$. As a result, the response is

$$\frac{\bar{y}(s)}{\bar{d}(s)} = \frac{e^{-\theta s}[K_d e^{\theta s}(\tau_1 s + 1)(\tau_2 s + 1)(\tau_4 s + 1) - KK_f(\tau_3 s + 1)(\tau_d s + 1)]}{(\tau_1 s + 1)(\tau_2 s + 1)(\tau_4 s + 1)(\tau_d s + 1)}. \qquad (14.34)$$

The following parameters are used by Marchetti et al. [3]:

$K = -24$ mg min/mU dL
$K_d = 375$ mg min/mmol
$\tau_1 = 366$ minutes
$\tau_2 = 151$ minutes
$\theta = 25$ minutes
$\tau_d = 313$ minutes.

The response to a step change of size 1.00 mmol/min in the gut glucose absorption rate (i.e., $\bar{d}(s) = 1/s$) is

$$\bar{y}(s) = \frac{1,125s(460,550s + 45,669)}{(25s + 2)(151s + 1)(313s + 1)(366s + 1)} \qquad (14.35)$$

or

$$y(t) = 1,125\left(\frac{441,250e^{-2t/25}}{117,699,239} - \frac{6,435,469e^{-t/151}}{9,647,910} + \frac{13,833,847e^{-t/313}}{5,160,186}\right.$$
$$\left. - \frac{1,6254,304e^{-t/366}}{8,056,265}\right). \qquad (14.36)$$

in the time domain. In spite of the lead–lag approximation, the controller was able to suppress the effects of the disturbance (i.e., $\lim_{t \to \infty} y(t) = 0$). Figure 14.9 shows the glucose concentration (not in deviation form) for the step change.

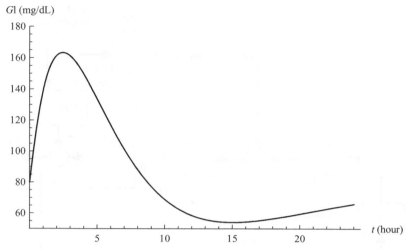

Figure 14.9. Glucose concentration profile after activating the feedforward controller.

14.4 A GAZE CONTROL SYSTEM

Inspired by the *vestibulo-ocular reflex* (VOR) found in humans, Viollet and Franceschini constructed a high-speed gaze control system for robotic platforms [4]. The VOR is credited for aiding in the stability of images on the retina even in the presence of rotational and translational disturbances [5]. Because of an elaborate biological control system, both eyeballs rotate to the left when the head moves right. The net result is that the gaze is stabilized during activities such as walking and running.

A block diagram of the miniature oculomotor system built by Viollet and Franceschini [4] is shown in Figure 14.10. The gaze θ_{gaze}, expressed in an angular unit, is to be maintained on a target θ_t. A rate gyro $G_{m2}(s)$, installed on printed circuit boards, measures the angular speed of the head Ω_{head}, which originates from the inertial disturbance (i.e., head rotation θ_{head}). The output of the rate gyro is transmitted to a feedforward controller, $G_f(s)$. A measuring device, the optical sensor for the control of autonomous robots (OSCAR), converts the displacement input ($\theta_t - \theta_{gaze}$) into an angular speed before emitting a voltage signal that is nearly proportional to this velocity. The output of the OSCAR visual system first goes into a zero-setting limiter (ZSL) before entering the feedback controller $G_c(s)$. The difference between the feedback (c_{fb}) and feedforward controllers (c_{ff}) enters the plant (i.e., the eye). Transfer functions $G_v(s)$ and $G_d(s)$, included for comparison with Figure 14.7, can be omitted. $G_{m1}(s)$ and the block ZSL are placed after the comparator in Figure 14.8.

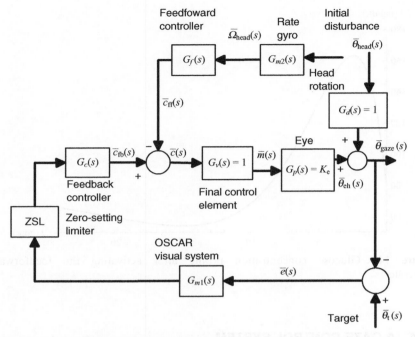

Figure 14.10. Block diagram of the feedforward–feedback gaze control system (adapted from Viollet and Franceschini [4, fig. 5]).

The following models were used:

$$G_{m2}(s) = K_g \frac{\tau_1 s - 1}{(\tau_2 s + 1)(\tau_3 s + 1)} \qquad (14.37)$$

$$G_c(s) = \frac{K_v}{s} \qquad (14.38)$$

$$G_{m1}(s) = K_0 e^{-\tau s} \qquad (14.39)$$

$$G_p(s) = K_e \qquad (14.40)$$

with the parameters defined as [4]:

$K_g = 1.1$ seconds^{-1}
$\tau_1 = 2.8$ seconds
$\tau_2 = 2.1 \times 10^{-3}$ seconds
$\tau_3 = 1.9 \times 10^{-3}$ seconds
$K_v = 300$ steps/V

τ = 0.05 second

K_0 = 0.05 V per degree

K_e = 0.15 degrees per step.

One step of a micro stepper motor (not to be confused with a step change in an input variable), a component of an "eye-in-head" apparatus, causes the eye to rotate by 0.15°. To derive the feedforward controller, Equation (14.23) is used after considering the negative sign in the feedforward control loop:

$$-G_f(s) = -\frac{G_d(s)}{G_{m2}(s)G_v(s)G_p(s)} \tag{14.41}$$

or

$$G_f = \frac{1}{K_g K_e}\frac{(\tau_2 s+1)(\tau_3 s+1)}{(\tau_1 s-1)}. \tag{14.42}$$

Viollet and Franceschini [4] approximated G_f using a physically realizable rational function:

$$G_f = K_{ff}\frac{(as+1)(-cs+1)}{(bs+1)(ds+1)}, \tag{14.43}$$

where

K_{ff} = 7.1 steps/degree/second

a = 7×10^{-3} seconds

b = 5 seconds

c = 5.6×10^{-4} seconds

d = 1.2×10^{-3} seconds.

The transfer function G_{load} is

$$G_{\text{load}}(s) = \frac{1 - G_{m2}(s)G_f(s)G_v(s)G_p(s)}{1 + G_p(s)G_c(s)G_{m1}(s)}$$

or

$$G_{\text{load}} = \frac{se^{s\tau}((as+1)(cs-1)K_e K_{ff}K_g(s\tau_1-1)+(bs+1)(ds+1)(s\tau_2+1)(s\tau_3+1))}{(bs+1)(ds+1)(s\tau_2+1)(s\tau_3+1)(se^{s\tau}+K_0 K_e K_v)},$$

$$\tag{14.44}$$

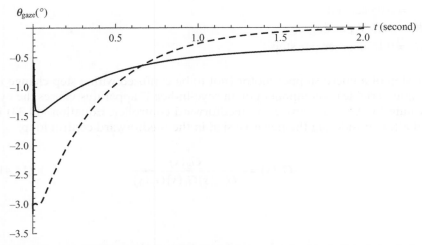

Figure 14.11. The gaze response due to step change in the head disturbance of size $-3°$. The solid and dashed lines represent the feedforward–feedback and feedback-only controllers, respectively.

and finally,

$$G_{\text{load}} = \frac{s(s(s((2.394 \times 10^{-8}\, s + 0.0000545)s + 0.00865) + 2.318) + 2.171)e^{0.05s}}{(0.0012s + 1)(0.0019s + 1)(0.0021s + 1)(5s + 1)(se^{0.05s} + 2.25)}.$$

(14.45)

Figure 14.11 shows the gaze response to a step change in θ_{head} of size -3 degrees. The controller was able to reject the effects of the disturbance. The performance of the feedback-only controller is also shown for comparison.

14.5 CONTROL OF pH

Several approaches have been described in the literature to control pH of industrial plants such as biological wastewater treatments. In general, pH is difficult to regulate because of the sigmoidal (S-shape) curve of acid–base titration processes. This example focuses on the titration between acetic acid (CH_3COOH) and sodium hydroxide ($NaOH$) in a continuous stirred-tank reactor (CSTR) (Fig. 14.12). The goal is to design a cascade control system to ensure that fluctuations in the base flow rate do not influence the system pH.

The mass balance equations for the weak acid (x_a) and strong base (x_b) concentrations are (Fig 14.12)

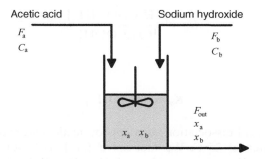

Figure 14.12. Strong base and weak acid in a CSTR.

$$V\frac{dx_a}{dt} = F_a C_a - F_{out} x_a, \tag{14.46}$$

$$V\frac{dx_b}{dt} = F_b C_b - F_{out} x_b, \tag{14.47}$$

and

$$F_{out} = F_a + F_b, \tag{14.48}$$

where V is the liquid volume in the tank. The flow rates of the process stream containing the acetic acid and the titrating stream with the sodium hydroxide are F_a and F_b, respectively. The effluent stream flow rate is F_{out}. When NaOH is mixed with CH_3COOH in the reactor, we have

$$NaOH + CH_3COOH \rightleftarrows NaCH_3COO + H_2O, \tag{14.49}$$

$$CH_3COOH \rightleftarrows H^+ + CH_3COO^-, \tag{14.50}$$

and

$$NaOH \rightleftarrows Na^+ + OH^-. \tag{14.51}$$

The electroneutrality condition leads to the following charge balance equation:

$$\left[Na^+\right] + \left[H^+\right] = \left[CH_3COO^-\right] + \left[OH^-\right], \tag{14.52}$$

where $[\cdot]$ represents a concentration. The weak acid and strong base ionic concentrations are $x_a = [CH_3COO^-] + [CH_3COOH]$ and $x_b = [Na^+]$. Two equilibrium equations are

$$K_a = \frac{[CH_3COO^-][H^+]}{[CH_3COOH]} \tag{14.53}$$

and

$$K_w = [H^+][OH^-], \tag{14.54}$$

where K_a is the acid dissociation constant for acetic acid and K_w is the ionic product for water at 25°C. We first solve for [CH_3COO^-], [CH_3COOH], and [OH^-] using Equation (14.53) and (14.54) and the relationship $x_a =$ [CH_3COO^-] + [CH_3COOH]. The resulting expressions are replaced in Equation (14.52) to yield

$$[H^+]^3 - K_a K_w - [H^+](K_w + K_a(x_a - x_b)) + [H^+]^2(K_a + x_b) = 0, \tag{14.55}$$

with $K_a = 10^{-4.75}$ and $K_w = 10^{-14}$. As an example, Figure 14.13 shows the pH–x_b relation for $x_a = 0.022$. Note that pH = $-\log[H^+]$.

To develop the pH control strategy, we first assume that the flow rate of the inlet stream (F_a) is significantly higher than that of the titrating stream (F_b) [6]:

$$V\frac{dx_a}{dt} \approx F_a(C_a - x_a) \tag{14.56}$$

and

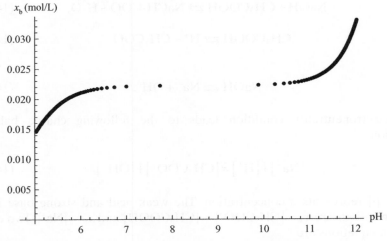

Figure 14.13. Inverse titration curve for acetic acid when $x_a = 0.022$ mol/L.

$$V \frac{dx_b}{dt} \approx F_b C_b - F_a x_b. \tag{14.57}$$

The process is further simplified by keeping x_a at its steady-state value [6]:

$$x_{a,ss} = \frac{F_{a,ss} C_a}{F_{a,ss} + F_{b,ss}}. \tag{14.58}$$

The control strategy can be stated as follows: Keep x_b (i.e., pH via Eq. 14.55) at a desired value by manipulating F_b. To simulate the control scheme shown in Figure 14.14, the following process parameters are used: $F_{a,ss} = 0.069$ m^3/s and $V = 102$ m^3 [7].

Equation (14.57) becomes

$$102 \frac{dx_b}{dt} = u - 0.069 x_b \tag{14.59}$$

or

$$G_{p1}(s) = \frac{14.49}{1478.26s + 1}, \tag{14.60}$$

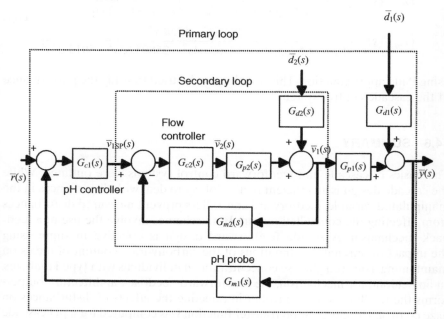

Figure 14.14. Block diagram of the pH cascade controller.

with $u = F_b C_b$ representing the manipulated variable [7]. The transfer function $G_{p1}(s)$ becomes

$$G_{p1}(s) = \frac{14.49 e^{-4s}}{1478.26 s + 1} \tag{14.61}$$

after allowing for time delays in the process. In addition, the characteristic of the NaOH flow valve is [7]

$$G_{p2}(s) = \frac{1}{2s}. \tag{14.62}$$

For the pure integrating process of the auxiliary loop, a P-only controller is adequate. Kang et al. [7] used $G_{c2}(s) = K_{c2} = 15$. The PID Cohen–Coon settings for the master controller are $K_{c1} = 34.02$, $\tau_{I1} = 9.835$, and $\tau_{D1} = 1.454$. The following functions are selected: $G_{d1}(s) = G_{d2}(s) = 1$ and $G_{m1}(s) = G_{m2}(s) = 1$. If we assume $x_{a,ss} = 0.0222$ mol/L (or 22.2 mol/m^3), x_b should be set at 0.0221 mol/L (or 22.1 mol/m^3) to have a pH equal to 7. In this case, the manipulated variable value, $u = 1.525$ mol/s, is obtained from the equation $0 = u - 0.069 x_b$. For $\bar{r}(s) = \bar{d}_1(s) = 0$, the following transfer function is derived:

$$G_{load2}(s) = \frac{G_{d2}(s) G_{p1}(s)}{1 + G_{c2}(s) G_{p2}(s) [G_{c1}(s) G_{m1}(s) G_{p1}(s) + G_{m2}(s)]}$$

$$G_{load2}(s) = \frac{28.98 s^2 - 57.96 s^3}{751.83 + 5905.59 s + 18168.64 s^2 + 25805.85 s^3 + 5913.04 s^4} \tag{14.63}$$

using Padé approximation. This function can be used to study the performance of the cascade control system.

14.6 SUMMARY

The basics of cascade and feedforward control systems were outlined. While the cascade design is an efficient methodology to deal with disturbances in the manipulated variable, feedforward controllers prevent measured disturbances from affecting the controlled output. Both schemes involve the use of a feedback mechanism. A simple feedforward design is effective in suppressing the impact of unreliable estimations of the carbohydrate content of meals on maintaining a desired glucose concentration in individuals with type I diabetes. In an oculomotor mechanism, the feedforward–feedback architecture outperforms the feedback-only controller in rejecting the effects of disturbances on gaze control. The closed-loop transfer function was computed for a pH cascade control system for further practice.

PROBLEMS

14.1. Design a cascade controller for the following process:

$$G_{p1}(s) = \frac{e^{-0.4s}}{1+0.8s}, \quad G_{p2}(s) = \frac{e^{-0.04s}}{1+0.08s},$$
$$G_{d1}(s) = G_{d2}(s) = 1, \quad \text{and} \quad G_{m1}(s) = G_{m2}(s) = 1.$$

Compare the performances of the cascade control with the single-loop control for a unit step change in d_2.

Note: Use Cohen–Coon tuning rules to design proportional (P) controllers G_{c1} and G_{c2}.

14.2. Design a cascade controller for the following process:

$$G_{p1}(s) = \frac{3.5e^{-0.2s}}{1+0.6s}, \quad G_{p2}(s) = \frac{1.2e^{-0.02s}}{1+0.05s},$$
$$G_{d1}(s) = G_{d2}(s) = 1 \quad \text{and} \quad G_{m1}(s) = G_{m2}(s) = 1.$$

Compare the performances of the cascade control with the single-loop control for a unit step change in d_2.

Note: Use Cohen–Coon tuning rules to design P controllers G_{c1} and G_{c2}.

14.3. Design a feedforward–feedback control structure for the following process (see Fig. 14.8):

$$G_p(s) = \frac{1}{0.94s+1}, \quad G_{m1}(s) = 1, \quad G_{m2}(s) = 1, \quad G_v(s) = 1,$$
$$G_d(s) = \frac{1}{0.47s+1}, \quad \text{and} \quad G_c(s) = K_c\left(1+\frac{1}{\tau_I s}\right).$$

Make sure that the closed-loop response is three times as fast as the open-loop response. Use IMC-tuning rules to calculate K_c and τ_I.

14.4. Design a feedforward–feedback control mechanism for the following process (see Fig. 14.8):

$$G_p(s) = \frac{5}{0.75s+1} \quad G_{m1}(s) = 1, \quad G_{m2}(s) = 1, \quad G_v(s) = 1,$$
$$G_d(s) = \frac{2}{0.35s+1}, \quad \text{and} \quad G_c(s) = K_c\left(1+\frac{1}{\tau_I S}\right).$$

The closed-loop response should be twice as fast as the open-loop response. Use IMC-tuning rules to compute K_c and τ_I.

14.5. Show that the inverse Laplace transform of

$$\overline{y}(s) = \frac{1,125s(460,550s+45,669)}{(25s+2)(151s+1)(313s+1)(366s+1)}$$

is

$$y(t) = 1,125\left(\frac{441,250e^{-2t/25}}{117,699,239} - \frac{6,435,469e^{-t/151}}{9,647,910} + \frac{13,833,847e^{-t/313}}{5,160,186}\right.$$
$$\left. - \frac{16,254,304e^{-t/366}}{8,056,265}\right).$$

(see Section 14.3).

14.6. In Section 14.3, the glucose response to a step change of size 1.00 mmol/min in the gut absorption rate of glucose (i.e., $\overline{d}(s)=1/s$) was plotted. Compare the dynamic behavior with the response observed when $\overline{d}(s)=1.5/s$.

Note: Plot the response in Figure 14.9, the line $y = 80$, and the new response on the same graph.

14.7. Derive the equation for the profile shown in Figure 14.11 when the feedforward–feedback controller is used (Section 14.4).

14.8. Generate the inverse titration curve for acetic acid when $x_a = 0.03$ (see Section 14.5 and Fig. 14.13).

14.9. (a) Derive the following transfer function obtained in Section 14.5:

$$G_{load2}(s) = \frac{28.98s^2 - 57.96s^3}{751.83+5,905.59s+18,168.64s^2+25,805.85s^3+5,913.04s^4}.$$

(b) Study the stability of $G_{load2}(s)$.

14.10. Use the transfer function

$$G_{load2}(s) = \frac{28.98s^2 - 57.96s^3}{751.83+5,905.59s+18,168.64s^2+25,805.85s^3+5,913.04s^4}$$

in Section 14.5 to derive $x_b(t)$ for a step change in the disturbance (d_2: fluctuation in the manipulated variable $u = F_bC_b$) of size A.

REFERENCES

1. Marlin TE. *Process Control: Designing Processes and Control Systems for Dynamic Performance*. Boston: McGraw-Hill, 2000.

2. Ogunnaike BA, Ray WH. *Process Dynamics, Modeling, and Control*. New York: Oxford University Press, 1994.

3. Marchetti G, Barolo M, Jovanovic L, Zisser H, Seborg DE. A feedforward-feedback glucose control strategy for type 1 diabetes mellitus. *Journal of Process Control* 2008; 18:149–162.

4. Viollet S, Franceschini N. A high speed gaze control system based on the vestibulo-ocular reflex. *Robotics and Autonomous Systems* 2005; 50:147–161.

5. Crane BT, Demer JL. Human gaze stabilization during natural activities: translation, rotation, magnification, and target distance effects. *Journal of Neurophysiology* 1997; 78:2129–2144.

6. Tan WW, Lu F, Loh AP, Tan KC. Modeling and control of a pilot pH plant using genetic algorithm. *Engineering Applications of Artificial Intelligence* 2005; 18:485–494.

7. Kang J, Wang M, Xiao Z. Modeling and control of pH in pulp and paper wastewater treatment process. *Journal of Water Resource and Protection* 2009; 2:122–127.

REFERENCES

CHAPTER 15

EFFECTIVE TIME CONSTANT

An important consideration in diffusive processes is the time it takes a variable of interest (e.g., concentration, flux) to reach a steady-state, or equilibrium, value. Chapter 12 introduced a first-order plus dead-time (FOPDT) model for control synthesis. Since the resulting system contains a single time constant, the strategy offers a way to measure the response speed of higher-order plants. Section 8.2 provided a variety of tools for deriving lower-order models. Such procedures can help determine a *relaxation time* constant (t_{eff}) for complex systems. However, the approaches described only apply to rational functions. This chapter focuses on a Laplace transform-based technique to estimate the t_{eff} of distributed-parameter systems. One of the advantages of this approach is that t_{eff} can be directly calculated from transcendental expressions. The method can be used to help manufacturers design devices that can release a precise dose of medication at a specific time during therapy [1]. An overview of linear ordinary differential equations (ODEs) is first provided.

15.1 LINEAR SECOND-ORDER ODEs

Homogeneous second-order ODEs are discussed. Laplace transform inversions were used to solve *initial value problems*, that is, ODEs in which the conditions are indicated at the same value of the independent variable (e.g., $t = 0$). For *boundary value problems*, the conditions are applied at different

Control of Biological and Drug-Delivery Systems for Chemical, Biomedical, and Pharmaceutical Engineering, First Edition. Laurent Simon.
© 2013 John Wiley & Sons, Inc. Published 2013 by John Wiley & Sons, Inc.

values of the independent variable (e.g., $x = 0$ and $x = 1$). Methods to solve the latter class of problems with specific engineering applications can be found in Loney [2], McQuarrie [3], Dubin [4], and Rice [5].

The form of a linear, homogeneous second-order equation with constant coefficients is

$$a_1 y'' + a_2 y' + a_3 y = 0 \quad (a_1 \neq 0), \tag{15.1}$$

where a_1, a_2, and a_3 are real numbers. Using the following definitions:

$$a = \frac{a_2}{a_1}; \quad b = \frac{a_3}{a_1}, \tag{15.2}$$

Equation (15.1) becomes

$$y'' + ay' + b = 0. \tag{15.3}$$

Let m_1 and m_2 be the roots of the *auxiliary equation*: $m^2 + am + b = 0$. Three cases are identified:

- m_1 and m_2 are real and distinct:

$$y = c_1 e^{m_1 x} + c_2 e^{m_2 x} \tag{15.4}$$

- m_1 and m_2 are real and equal:

$$y = c_1 e^{m_1 x} + c_2 x e^{m_1 x} \tag{15.5}$$

- m_1 and m_2 are complex numbers:

$$m_1 = p + qi, \quad m_2 = p - qi$$
$$y = e^{px}(c_1 \cos(qx) + c_2 \sin(qx)). \tag{15.6}$$

Example 15.1   Solve $y'' - 8y' + 15y = 0$; $y(0) = 0$; $y(1) = 2$.

Solution   The auxiliary equation is

$$m^2 - 8m + 15 = 0, \tag{15.7}$$

which gives $m_1 = 3$ and $m_2 = 5$. The general solution is

$$y(x) = c_1 e^{3x} + c_2 e^{5x}. \tag{15.8}$$

Applying the boundary conditions yields $0 = c_1 + c_2$ and $2 = c_1 e^3 + c_2 e^5$. The constants are $c_1 = -0.01558$ and $c_2 = 0.01558$. Thus, the *particular solution* is

$$y(x) = -0.01558 e^{3x} + 0.01558 e^{5x}. \tag{15.9}$$

15.2 STURM–LIOUVILLE (SL) EIGENVALUE PROBLEMS

Although SL eigenvalue systems are not the focus in this chapter, it is important to be able to identify such problems and to be aware of some properties relevant to the effective-time-constant approach. The following differential equation,

$$\frac{d}{dx}\left[p(x)\frac{dy}{dx} \right] + [q(x) + \lambda r(x)]y = 0, \tag{15.10}$$

is in an *SL* form. In this formulation, $p(x) > 0, r(x) > 0$, and $q(x)$ are continuous functions on an interval $a \le x \le b$, and λ is a constant. The SL problems are said to be *regular*, *periodic*, or *singular* depending on the boundary conditions [2]:

- In the *regular SL* problem, the boundary conditions are

$$\alpha_1 y(a) + \beta_1 y'(a) = 0 \tag{15.11}$$

and

$$\alpha_2 y(b) + \beta_2 y'(b) = 0, \tag{15.12}$$

where $\alpha_1, \beta_1, \alpha_2$, and β_2 are constants that satisfy the following conditions:

$$|\alpha_1| + |\beta_1| > 0 \tag{15.13}$$

and

$$|\alpha_2| + |\beta_2| > 0. \tag{15.14}$$

- The boundary conditions of the *periodic SL* problem are

$$y(a) = y(b) \tag{15.15}$$

and

$$y'(a) = y'(b). \tag{15.16}$$

- Any one of the following conditions applies to a *singular SL* problem:
  1. In the case of $p(a) = 0$, we only have one boundary condition:

$$\alpha_2 y(b) + \beta_2 y'(b) = 0. \tag{15.17}$$

  2. If $p(b) = 0$, the following boundary condition applies:

$$\alpha_1 y(a) + \beta_1 y'(a) = 0. \tag{15.18}$$

  3. If $p(a) = p(b) = 0$, there is no boundary condition.
  4. If the interval is semi-infinite, for example, $x \in [0, \infty)$, or infinite, the system is singular.

Both λ and the nontrivial solution $y(x) \neq 0$ should be determined when solving an SL problem.

Example 15.2   The following equation:

$$y'' + \lambda y = 0 \quad x \in [0, \pi], \tag{15.19}$$

with the conditions

$$y(0) = 0 \tag{15.20}$$

and

$$y'(\pi) = 0, \tag{15.21}$$

is an example of a *regular SL boundary value problem*. Solve the equation over the given interval.

Solution   Three cases are considered:

- $\lambda = 0$: The equation becomes $y'' = 0$ and the solution is $y(x) = k_1 + k_2 x$. Substitution of the boundary conditions leads to the trivial solution $y = 0$. As a result, $\lambda \neq 0$ because we are looking for nontrivial solutions.
- $\lambda < 0$: The equation can be written as $y'' - \alpha^2 y = 0$ using $\lambda = -\alpha^2$. The general solution is

$$y(x) = k_1 e^{\alpha x} + k_2 e^{-\alpha x}.$$

Applying the boundary conditions gives $k_1 + k_2 = 0$ and $k_1 \alpha e^{\alpha \pi} - k_2 \alpha e^{\alpha \pi} = 0$. The last expression becomes $k_1(e^{\alpha \pi} + e^{-\alpha \pi}) = 0$. Therefore, $k_1 = k_2 = 0$ and $y = 0$. Thus, $\lambda < 0$ cannot be used.

- $\lambda > 0$: The equation can be written as $y'' + \alpha^2 y = 0$ using $\lambda = \alpha^2$. The general solution is

$$y(x) = k_1 \sin(\alpha x) + k_2 \cos(\alpha x).$$

Applying the boundary conditions leads to $k_2 = 0$ and $k_1 \alpha \cos(\alpha \pi) = 0$. We know that $k_1 \neq 0$ because $k_1 = 0$ would yield the trivial solution. Then, $\cos(\alpha \pi) = 0$ or $\alpha \pi = (2n - 1)\pi/2$, which results in $\alpha = n - \frac{1}{2}$. The values $\lambda_n = \left(n - \frac{1}{2}\right)^2$ are the *eigenvalues* of the problem. The corresponding *eigenfunctions* are $\phi_n(x) = \sin\left[\left(n - \frac{1}{2}\right)x\right]$, where $n = 1, 2, 3, \ldots$.

One important property of regular and periodic SL problems is that *an infinite sequence of real eigenvalues exists*:

$$\lambda_1 < \lambda_2 < \lambda_3 < \ldots \tag{15.22}$$

and

$$\lim_{n \to \infty} \lambda_n = \infty. \tag{15.23}$$

Also, *the set of eigenfunctions is complete*, which means that any piecewise continuous function f, defined on an interval $a \leq x \leq b$, can be represented by an infinite series of the form

$$f(x) = \sum_{n=1}^{\infty} c_n \psi_n(x), \tag{15.24}$$

where ψ_n are the eigenfunctions. Equation (15.24) is known as an *eigenfunction expansion* of $f(x)$ on an interval, $[a, b]$. Details on how to calculate c_n can be found in mathematical method textbooks [2].

In Example 15.2, a function $f(x)$, defined on the interval, $[0, \pi]$, is written as

$$f(x) = \sum_{n=1}^{\infty} c_n \sin\left[\left(n - \frac{1}{2}\right)x\right].$$

Using the method of separation of variables (beyond the scope of this textbook), the SL theory is implemented to solve partial differential equation (PDE) problems, such as the heat equation and Fick's second law of diffusion. This technique shows that a function $g(x, t)$ can be expressed as

$$g(x, t) = \sum_{n=1}^{\infty} c_n(x) \phi_n(t), \tag{15.25}$$

where $g(x, t)$ could represent the concentration at position x and time t. In this case, the coefficient c_n is a function of x. We are now in a position to introduce the effective-time-constant estimation procedure.

15.3 RELAXATION TIME CONSTANT

The analysis of diffusive processes leads to auxiliary SL problems with eigenvalues (λ_n) and eigenfunctions (c_n) described by

$$g(x, t) = \sum_{n=1}^{\infty} c_n(x) \exp(-\lambda_n t), \tag{15.26}$$

where λ_n is the inverse of the characteristic time constant t_n. It can be inferred that when $\lambda_n < \lambda_m$, the inequality $t_n > t_m$ holds. In general, the system dynamics can be represented by the first two or three eigenvalues. Collins used a single *effective time constant*, or *relaxation* time, to describe how long it takes the response, $g(t)$, to reach a steady-state value, g_e [6]:

$$t_{\text{eff}} = \int_0^{\infty} t\Omega(t)dt, \tag{15.27}$$

where

$$\Omega(t) = \frac{(g_e - g(t))}{\int_0^{\infty} (g_e - g(t))dt}. \tag{15.28}$$

Equation (15.28) is equivalent to

$$t_{\text{eff}} = \frac{\lim_{s \to 0} \left(\dfrac{g_e}{s^2} + \dfrac{d\bar{g}(s)}{ds} \right)}{\lim_{s \to 0} \left(\dfrac{g_e}{s} - \bar{g}(s) \right)} \tag{15.29}$$

when $\bar{g}(s)$, the Laplace transform of $g(t)$, is used (see Simon [1]). In addition to measuring the time constant of high-order, linear lumped-parameter systems, t_{eff} can be applied to analyze PDE models. Only temporal variations are considered when studying PDE systems (e.g., drug-delivery rate or temperature at a fixed location). There is one main limitation for directly using Equation (15.29): *The system has no overshoot* (i.e., the sign of the difference $g_e - g(t)$ does not change). The finite interval $a \leq x \leq b$ requirement for regular and *periodic SL problems* still applies. The main advantages of using Equation (15.29) are the following:

- A time constant, similar to the one obtained for a first-order process, can be estimated for linear higher-order ODE and PDE systems.
- No order-reduction method is required.
- A closed-form solution, $g(t)$, is not necessary.

Example 15.3 Derive t_{eff} for a first-order process after a step change of size M:

$$\bar{Y}(s) = \frac{K}{\tau s + 1} \frac{M}{s}. \tag{15.30}$$

Solution The effective time constant is

$$t_{eff} = \frac{\lim_{s \to 0} \left(\dfrac{y_e}{s^2} + \dfrac{d\bar{Y}(s)}{ds} \right)}{\lim_{s \to 0} \left(\dfrac{y_e}{s} - \bar{Y}(s) \right)}. \tag{15.31}$$

Evaluations of the steady-state value y_e and the derivative $d\bar{Y}(s)/ds$ give

$$y_e = \lim_{s \to 0} \left[s\bar{Y}(s) \right] = \lim_{s \to 0} \left[\frac{sK}{\tau s + 1} \frac{M}{s} \right] = MK \tag{15.32}$$

and

$$\frac{d\bar{Y}(s)}{ds} = -\frac{KM(2s\tau + 1)}{s^2(s\tau + 1)^2}. \tag{15.33}$$

As a result, it can be shown that

$$\lim_{s \to 0} \left(\frac{y_e}{s^2} + \frac{d\bar{Y}(s)}{ds} \right) = KM\tau^2. \tag{15.34}$$

In addition, the numerator of t_{eff} is evaluated:

$$\lim_{s \to 0} \left(\frac{y_e}{s} - \bar{Y}(s) \right) = KM\tau, \tag{15.35}$$

which yields

$$t_{eff} = \tau \tag{15.36}$$

as we would expect for a first-order process.

15.4 IMPLEMENTATION IN *MATHEMATICA*®

The techniques involved in the computation of t_{eff} can be implemented by hand or with software packages such as *Mathematica* (Wolfram, Inc.). Several tutorials are readily available at http://www.wolfram.com/ to help learn the rudiments of *Mathematica* computing. Example 5.3 is shown in Figure 15.1 for illustration purposes.

15.5 CONTROLLED-RELEASE DEVICES

Transdermal drug-delivery devices were discussed in Sections 6.6 and 7.4. Fick's second law of diffusion results in

$$\frac{\partial C}{\partial t} = D \frac{\partial^2 C}{\partial x^2}. \tag{15.37}$$

```
(* The Laplace transform of the step response *)

Y[s_]= ( K    M )
      ( ─── ─ )
      ( τs+1 s )

  K M
─────────
s (1 + sτ)

(* Computation of the derivative *)

Yder[s_] = ∂_s (Y[s] )

   K M τ          K M
- ──────────── - ───────────
  s (1 + s τ)²   s² (1 + s τ)

(* The derivative is simplified *)

Simplify[Yder[s]]

  K M (1 + 2s τ )
- ──────────────
   s² (1 + s τ )²

(* The final value theorem is applied to calculate the steady-state value*)

ye = Limit[s Y[s], s → 0]
K M

(* The effective time constant is computed *)

            Limit[ ye + Yder[s], s → 0]
            [      s²                  ]
teff1 = FullSimplify[ ──────────────────────── ]
            [ Limit[ ye - Y[s] , s → 0]  ]
            [        s                   ]
τ
```

Figure 15.1. Estimation of the effective time constant using *Mathematica*®.

In this equation, the variable C represents the concentration in the membrane and D is the diffusion coefficient. The boundary conditions are given by

$$C(0) = KC_d; \quad C(h) = 0. \tag{15.38}$$

The flux at h, defined as the boundary between the membrane and the receiver compartment, is

$$j = -D \frac{\partial C}{\partial x}\bigg|_{x=h}. \tag{15.39}$$

Using the results of Section 6.6, the Laplace transform of the flux $(\bar{J}(s))$ takes the form

$$\bar{J}(s) = \frac{\operatorname{csch}\left(h\sqrt{\dfrac{s}{D}}\right) KC_d}{\sqrt{\dfrac{s}{D}}}. \tag{15.40}$$

The steady-state flux is given below:

$$J_{ss} = \frac{DKC_d}{h}. \tag{15.41}$$

The effective time constant is

$$t_{eff} = \frac{7h^2}{60D}. \tag{15.42}$$

If the response time is defined by $t_{Reff} = 4 \times t_{eff}$, we can easily calculate the time it takes the flux to reach 98% of its final value. This estimate agrees with experimental data [1].

15.6 SUMMARY

A method to calculate an effective time constant for complex processes was outlined. Laplace transform techniques are implemented to estimate t_{eff} for lumped- and distributed-parameter systems. SL theory, which guarantees that an infinite sequence of real eigenvalues exists, provides the basis for the technique. *Mathematica* can be used to reduce the computations involved. The approach can help determine the time it takes to reach the steady-state flux in controlled drug-delivery devices.

PROBLEMS

15.1. Solve $y'' - 5y' + 6y = 0$; $y(0) = 0$; $y(1) = 2$.

15.2. Solve $y'' + 7y' + 10y = 0$; $y(0) = 0$; $y(1) = 4$.

15.3. Find the eigenvalues and eigenfunctions of the following SL boundary value problem:

$$y'' + \lambda y = 0 \quad x \in [0, 1]$$

$$y(0) = 0$$

$$y'(1) = 0.$$

15.4. Find the eigenvalues and eigenfunctions of the following SL boundary value problem:

$$y'' + \lambda y = 0 \quad x \in [0, 1]$$

$$y'(1) = 0$$

$$2y(0) - y'(0) = 0.$$

15.5. Find the eigenvalues and eigenfunctions of the following SL boundary value problem:

$$y'' + \lambda y = 0 \quad x \in [0, \pi]$$

$$y'(0) = 0$$

$$y(\pi) = 0.$$

15.6. Using Laplace transform properties, derive the following expression:

$$t_{\text{eff}} = \frac{\lim\limits_{s \to 0} \left(\dfrac{g_e}{s^2} + \dfrac{d\overline{g}(s)}{ds} \right)}{\lim\limits_{s \to 0} \left(\dfrac{g_e}{s} - \overline{g}(s) \right)}$$

from the definition of the effective time constant:

$$t_{\text{eff}} = \int_0^\infty t\Omega(t)\,dt,$$

where

$$\Omega(t) = \frac{(g_e - g(t))}{\int_0^\infty (g_e - g(t))dt}.$$

15.7. Derive t_{eff} for a second-order process after a step change of size M:

$$\bar{Y}(s) = \frac{K}{(\tau_1 s + 1)(\tau_2 s + 1)} \frac{M}{s}.$$

15.8. Show that the effective time constant for the flux is

$$t_{\text{eff}} = \frac{7h^2}{60D}$$

for the controlled-release system described in Section 15.5:

$$\frac{\partial C}{\partial t} = D \frac{\partial^2 C}{\partial x^2}$$

with the boundary conditions

$$C(0) = KC_d; \quad C(h) = 0.$$

15.9. Given the controlled-release system described in Section 15.5, derive an expression for the effective time constant for the concentration at $x = h/2$.

15.10. Given the controlled-release system described in Section 15.5, derive an expression for the effective time constant for the flux at $x = 0$.

REFERENCES

1. Simon L. Timely drug delivery from controlled-release devices: dynamic analysis and novel design concepts. *Mathematical Biosciences* 2009; 217:151–158.
2. Loney NW. *Applied Mathematical Methods for Chemical Engineers*. Boca Raton, FL: CRC, 2007.
3. McQuarrie DA. *Mathematical Methods for Scientists and Engineers*. Sausalito, CA: University Science Books, 2003.
4. Dubin DHE. *Numerical and Analytical Methods for Scientists and Engineers Using Mathematica*. Hoboken, NJ: Wiley-Interscience, 2003.
5. Rice RG, Duong DD. *Applied Mathematics and Modeling for Chemical Engineers*. New York: Wiley, 1995.
6. Collins R. The choice of an effective time constant for diffusive processes in finite systems. *Journal of Physics D: Applied Physics.* 1980; 13:1935–1947.

CHAPTER 16

OPTIMUM CONTROL AND DESIGN

The purpose of system design is to optimize a *performance index* while satisfying specific requirements [1]. To determine an *optimum design*, the engineer first identifies *design variables*, a *cost function*, and a set of *constraints*. The best specifications for the plant are achieved through an iterative procedure. *Optimum control problems* usually consist of finding a *control law* that minimizes a criterion. For example, techniques can be implemented to tune proportional–integral–derivative (PID) controllers based on an integral of squared error. The advantage of these methods is the full incorporation of the plant dynamics without further model reduction. Such approaches are useful whether or not a feedback signal is available.

An application of optimum control theories can be found in biomedicine where treatment strategies are improved by estimating infusion rates and the number of doses necessary to keep drug concentration in the blood within a therapeutic range. The model is usually represented by ordinary differential equations (ODEs) that involve the drug levels in the compartments and a control variable. For example, the governing equations for a two-compartment model are

$$\frac{dy_1}{dt} = -k_1 y_1 + k_2 y_2 + R(t) - k_E y_1, \tag{16.1}$$

$$\frac{dy_2}{dt} = k_1 y_1 - k_2 y_2, \tag{16.2}$$

Control of Biological and Drug-Delivery Systems for Chemical, Biomedical, and Pharmaceutical Engineering, First Edition. Laurent Simon.
© 2013 John Wiley & Sons, Inc. Published 2013 by John Wiley & Sons, Inc.

where y_1 and y_2 are the amounts in compartments I and II, respectively. The mass transfer rate constants are represented by k_1 and k_2; k_E is the elimination rate constant; and $R(t)$ is the infusion rate (i.e., control variable). A typical optimal control problem determines an $R(t)$ profile that satisfies Equations (16.1) and (16.2) and the initial conditions $y_1(0) = y_2(0) = 0$ while minimizing the criterion

$$F = \int_0^\infty \left(y_1(t) - y_{1,\text{set}} \right)^2. \tag{16.3}$$

The desired plasma blood concentration is $y_{1,\text{set}}$. Several methods, including dynamic programming, have been implemented to deal with this problem and other systems arising in controlled-release technology.

16.1   ORTHOGONAL COLLOCATION TECHNIQUES

Mass transport across membranes, governed by partial differential equations (PDEs), can be formulated as an ODE system using discretization techniques, for example, finite difference and orthogonal collocation methods. The finite difference approach is based on the Taylor series expansion, where the spatial derivative of f, namely, $f'(x)$, is approximated by the ratio $[f(x + h) - f(x)]/h$. Through this procedure, an optimal control problem involving PDEs is converted into a system similar to Equations (16.1) and (16.2). One notable disadvantage of finite differences is that the procedure can lead to a large number of ODEs. With orthogonal collocation techniques, this number is reduced considerably [2]. According to this method, a variable $T(z, t)$ can be estimated by a finite series,

$$T(z, t) \approx a(t) + b(t)z + z(1 - z) \sum_{i=1}^{N} a_i(t) P_{i-1}(z), \tag{16.4}$$

after normalizing z so that $0 < z < 1$. The collocation points z_j represent the roots of $P_n(z)$, a polynomial that is orthogonal to $P_m(z)$ when $m \neq n$.

Equation (16.4) can be written as

$$T(z_j, t) \approx \sum_{i=1}^{N+2} d_i(t) z_j^{i-1}. \tag{16.5}$$

The first and second derivatives of Equation (16.5) are

$$\frac{\partial}{\partial z}T(z_j, t) = \sum_{i=1}^{N+2}(i-1)d_i(t)z_j^{i-2} \tag{16.6}$$

$$\frac{\partial^2}{\partial z^2}T(z_j, t) = \sum_{i=1}^{N+2}(i-1)(i-2)d_i(t)z_j^{i-3}. \tag{16.7}$$

Equations (16.5)–(16.7) can be written in vector forms:

$$\mathbf{T} = \mathbf{Zd}, \tag{16.8}$$

$$\frac{\partial}{\partial \zeta}\mathbf{T} = \mathbf{Qd}, \tag{16.9}$$

and

$$\frac{\partial^2}{\partial \zeta^2}\mathbf{T} = \mathbf{Rd}, \tag{16.10}$$

where $\mathbf{T} = \{T(z_j, t)\}$ and $\mathbf{d} = \{d_i(t)\}$ are represented as vectors. The matrices are $\mathbf{Z} = \{z_j^{i-1}\}$, $\mathbf{Q} = \{(i-1)z_j^{i-2}\}$, and $\mathbf{R} = \{(i-1)(i-2)z_j^{i-3}\}$. After solving for $\mathbf{d}$, Equations (16.8)–(16.10) become

$$\mathbf{d} = \mathbf{Z}^{-1}\mathbf{T}, \tag{16.11}$$

$$\frac{\partial}{\partial \zeta}\mathbf{T} = \mathbf{QZ}^{-1}\mathbf{T} = \mathbf{AT}, \tag{16.12}$$

and

$$\frac{\partial^2}{\partial \zeta^2}\mathbf{T} = \mathbf{RZ}^{-1}\mathbf{T} = \mathbf{BT}. \tag{16.13}$$

Thus, a PDE of the form

$$\frac{\partial C}{\partial t} = D\frac{\partial^2 C}{\partial z^2}$$

can be approximated by a set of ODEs with respect to time:

$$\frac{d}{dt}\mathbf{C} = D\mathbf{BC}, \tag{16.14}$$

where $\mathbf{C} = \{C(z_j, t)\}$. Using this procedure, an optimal control problem involving PDEs can be formulated as one in which state variables are represented by ODEs. Section 16.2 outlines an optimization technique that is often applied to solve these types of problems.

16.2 DYNAMIC PROGRAMMING

Dynamic programming is a method that consists of finding control and state histories of a dynamic system over a finite time period in order to minimize a performance index [3]. Such techniques have been implemented to develop the best treatment strategies. In the case of dose optimization, consider the following problem:

$$F = \min_{D_0, \dots, D_m, t_0, \dots, t_m} \int_0^{t_f} [x(t) - x_{\text{set}}]^2 \, dt, \tag{16.15}$$

where D_i is a dose administered at time t_i; $x(t)$ is the drug concentration in a particular compartment (usually the level in the blood) at time t. The goal is to maintain $x(t)$ near the target x_{set}. Equation (16.15) may also include the delivery rate and constraints such as dose size and exposure time. The state $x(t)$ may arise from processes described by PDE systems. Dynamic programming approaches are also applied to select the best dosage regimens when transdermal patches are used. In general, a software package is required to solve optimal control problems, especially for those applications involving nonlinear terms.

16.3 OPTIMAL CONTROL OF DRUG-DELIVERY RATES

This section is based on an optimal control problem that arises from the repeated application of transdermal patches [4]. To model drug transport, the system consists of the vehicle (thickness a), skin (thickness b), and the capillaries. Fick's second law is applied to the two membranes:

$$\frac{\partial C_1}{\partial t} = D_1 \frac{\partial^2 C_1}{\partial x^2}, \quad a < x < 0, \quad 0 < t \leq T \tag{16.16}$$

$$\frac{\partial C_2}{\partial t} = D_2 \frac{\partial^2 C_2}{\partial x^2}, \quad 0 < x < b, \quad 0 < t \leq T. \tag{16.17}$$

The drug diffusion coefficients in the vehicle and the skin are D_1 and D_2, respectively. Each application lasts T time units. To describe the process fully, initial and boundary conditions are necessary (see Simon [4] for further details). The drug concentration in each layer can be solved by the orthogonal collocation techniques described above. A problem of particular interest is to calculate a series of loading doses to be given at regular intervals so that the delivery rate remains close to a target value, J_{set}:

$$J(t) = -D_2 \frac{\partial C_2(b, t)}{\partial x}. \tag{16.18}$$

The optimization problem takes the following form:

$$F = \min_{D_{2o,1}\ldots D_{2o,m}} \int_0^{t_f} [J(t) - J_{set}]^2 \, dt, \tag{16.19}$$

where $D_{2o,i}$ denotes the amount of drug in the vehicle at the ith application, m stands for the total number of applications, and t_f represents the overall simulation time.

16.4   OPTIMAL DESIGN OF CONTROLLED-RELEASE DEVICES

Equations, such as Fick's second law, used to describe drug transport through a membrane do not show the effects of additives, porosity, and the type of polymer on the drug diffusivity (D). The first step toward design could be to develop an empirical equation to describe, for instance, the relationship between D and the plasticizer levels in various systems [5]. The steady-state flux can then be written in terms of N key properties:

$$J_{ss} = f(x_1, x_2, \ldots, x_N). \tag{16.20}$$

In this case, the *optimum design problem* consists of finding a set of design variables. The objective function, defined as

$$G = [J_{set} - f(x_1, x_2, \ldots, x_N)]^2, \tag{16.21}$$

is minimized:

$$\mathbf{x}_{min} = \min_{x_1, x_2, \ldots x_N} G, \tag{16.22}$$

where $\mathbf{x}_{min}$ is the optimum solution: $\mathbf{x}_{min} = [x_{1min}, x_{2min}, \ldots, x_{Nmin}]$. Constraints should be identified in the formulation process. A similar approach was taken in Simon and Fernandes [6] in order to optimize estradiol release from ethylene-vinyl acetate membrane.

16.5 IMPLEMENTATION IN *MATHEMATICA*®

The problems outlined above can be solved using several numerical packages. Two illustrations are presented here in *Mathematica*.

16.5.1 Diffusion through a Membrane

The orthogonal collocation techniques outlined in Section 16.1 are coded in *Mathematica* for drug diffusion through a membrane:

$$\frac{\partial C}{\partial t} = D \frac{\partial^2 C}{\partial x^2} \tag{16.23}$$

$$C(x, 0) = 0 \tag{16.24}$$

$$C(0, t) = KC_d = C_0 \tag{16.25}$$

$$C(h, t) = 0. \tag{16.26}$$

The steps are outlined below:

- The number of internal collocation points N1 (i.e., total collocation points: N1 + 2) using Jacobi polynomials are
  ```
  S2[n_]:=Select[Sort[Re[x/.NSolve[JacobiP[n,0,0,
  2 x-1]==0,x]]],#1≥0&];
  ROOT={1.0 10^-20,S2[N1],1.0};
  ROOT = Flatten[ROOT];
  Table[x[m]=ROOT[[m]],{m,1,N2}];
  ```
- The matrices are
  ```
  Z = Chop[Table[Table[x[j]^(i-1), {i,1,N2}], {j,1,N2}]];
  IZ = Inverse[Z];
  Q = Chop[Table[Table[(i-1)x[j]^(i-2), {i,1,N2}],
  {j,1,N2}]];
  A = Q.IZ;
  R = Chop[Table[Table[(i-1)(i-2)x[j]^(i-3),
  {i,1,N2}], {j,1,N2}]];
  B=R.IZ;
  ```
  using the following parameters: $D = 3.96 \times 10^{-8} \text{cm}^2/\text{h}$; $C_0 = 77 \, \mu\text{g/mL}$; $h = 0.001 \, \text{cm}$; $t_f = 24$ hours.
- The collocation points are
  ```
  {1.×10-20,0.025446,0.129234,0.297077,0.5,0.702923,
  0.870766,0.974554,1.}.
  ```

Similarly, the matrices **A** and **B** are

$$
\mathbf{A} = \begin{pmatrix}
-57. & 63.4982 & -8.62617 & 3.21857 & -1.82858 & 1.36027 & -1.28026 & 1.65797 & -1.00001 \\
-24.3219 & 19.1364 & 6.64761 & -2.17856 & 1.19238 & -0.873484 & 0.816192 & -1.05362 & 0.635055 \\
6.94107 & -13.9649 & 3.29473 & 5.11014 & -2.21201 & 1.49506 & -1.34856 & 1.71461 & -1.03015 \\
-3.52046 & 6.22114 & -6.94642 & 0.971748 & 4.71216 & -2.46399 & 2.0323 & -2.49434 & 1.48786 \\
2.1875 & -3.72404 & 3.28863 & -5.1537 & 3.20824 \times 10^{-7} & 5.1537 & -3.28862 & 3.72404 & -2.1875 \\
-1.48786 & 2.49434 & -2.0323 & 2.46399 & -4.71216 & -0.971747 & 6.94642 & -6.22114 & 3.52046 \\
1.03016 & -1.71461 & 1.34856 & -1.49506 & 2.21201 & -5.11014 & -3.29473 & 13.9649 & -6.94107 \\
-0.635057 & 1.05362 & -0.816194 & 0.873485 & -1.19238 & 2.17856 & -6.64761 & -19.1364 & 24.3219 \\
1. & -1.65797 & 1.28025 & -1.36027 & 1.82857 & -3.21856 & 8.62616 & -63.4982 & 57.
\end{pmatrix}
$$

and

$$
\mathbf{B} = \begin{pmatrix}
1624 & -2247.98 & 849.886 & -345.248 & 201.143 & -151.2 & 143.008 & -185.606 & 112. \\
980.781 & -1294.77 & 382.522 & -99.4202 & 50.661 & -36.0094 & 33.1692 & -42.5454 & 25.6087 \\
-61.6804 & 177.082 & -192.2 & 94.5649 & -26.5081 & 15.0638 & -12.5235 & 15.3551 & -9.15428 \\
16.8586 & -33.715 & 69.2725 & -97.0342 & 55.601 & -16.9313 & 11.0348 & -12.2114 & 7.12499 \\
-8.75 & 15.6949 & -17.7396 & 50.7947 & -80. & 50.7947 & -17.7396 & 15.6949 & -8.75 \\
7.12499 & -12.2114 & 11.0348 & -16.9313 & 55.601 & -97.0342 & 69.2725 & -33.715 & 16.8586 \\
-9.15427 & 15.3551 & -12.5235 & 15.0638 & -26.5081 & 94.5649 & -192.2 & 177.082 & -61.6803 \\
25.6086 & -42.5452 & 33.1691 & -36.0093 & 50.6609 & -99.4201 & 382.521 & -1294.77 & 980.781 \\
112. & -185.606 & 143.008 & -151.2 & 201.143 & -345.248 & 849.886 & -2247.98 & 1624.
\end{pmatrix}
$$

respectively.

- The variables and equations are defined as

```
Tv1 = Table[\!\(Tv\_i\)[t],{i,1,N2}];
eq3 =Dtv ((B.Tv1))/h2^2  ;
Tv1[t_]:=Co; TvN2[t_]:=0;
f[a_,b_]:=[a==b;
DTv=Table[D[\!\(Tv\_i\)[t],{t,1}],{i,1,N2}];
eqi5=MapThread[f,{DTv,eq3}];
eqi6=Take[eqi5,{2,N2-1}];
eqns=eqi6;
init5=Table[\!\(Tv\_i\) [0]==0,{i,1+1,N2-1}];
DQt=D[\!\(Qt\)[t],{t,1}];
FluxT=((-Dtv) Take[(A.Tv1)/h2, {N2}] [1]]);
eqns1=Join[eqns,{DQt==FluxT},init5,{Qt [0]==0}];
Tv2 = Table[\!\(Tv\_i\)[t],{i,2,N2-1}];
abc=Join[{Qt[t]},Tv2];
abc=Flatten[abc];
```

The system of ODEs (i.e., eqns1) is shown below:

$$\{Tv_2'[t] == \frac{1}{h2^2} Dtv\,(0. + 980.781\,Co - 1294.77\,Tv_2[t] + 382.522\,Tv_3[t]$$
$$- 99.4202\,Tv_4[t] + 50.661\,Tv_5[t] - 36.0094\,Tv_6[t] + 33.1692\,Tv_7[t]$$
$$- 42.5454\,Tv_8[t]),$$

$$Tv_3'[t] == \frac{1}{h2^2} Dtv\,(0. - 61.6804\,Co + 177.082\,Tv_2[t]$$
$$- 192.2\,Tv_3[t] + 94.5649\,Tv_4[t] - 26.5081\,Tv_5[t] + 15.0638\,Tv_6[t]$$
$$- 12.5235\,Tv_7[t] + 15.3551\,Tv_8[t]),$$

$$Tv_4'[t] == \frac{1}{h2^2} Dtv\,(0. + 16.8586\,Co - 33.715\,Tv_2[t] + 69.2725\,Tv_3[t]$$
$$- 97.0342\,Tv_4[t] + 55.601\,Tv_5[t] - 16.9313\,Tv_6[t] + 11.0348\,Tv_7[t]$$
$$- 12.2114\,Tv_8[t]),$$

$$Tv_5'[t] == \frac{1}{h2^2} Dtv\,(0. - 8.75\,Co + 15.6949\,Tv_2[t] - 17.7396\,Tv_3[t]$$
$$+ 50.7947\,Tv_4[t] - 80.\,Tv_5[t] + 50.7947\,Tv_6[t] - 17.7396\,Tv_7[t]$$
$$+ 15.6949\,Tv_8[t]),$$

$$Tv_6'[t] == \frac{1}{h2^2} Dtv\,(0. + 7.12499\,Co - 12.2114\,Tv_2[t] + 11.0348\,Tv_3[t]$$
$$- 16.9313\,Tv_4[t] + 55.601\,Tv_5[t] - 97.0342\,Tv_6[t] + 69.2725\,Tv_7[t]$$
$$- 33.715\,Tv_8[t]),$$

$$Tv_7'[t] == \frac{1}{h2^2} Dtv\,(0. - 9.15427\,Co + 15.3551\,Tv_2[t] - 12.5235\,Tv_3[t]$$
$$+ 15.0638\,Tv_4[t] - 26.5081\,Tv_5[t] + 94.5649\,Tv_6[t] - 192.2\,Tv_7[t]$$
$$+ 177.082\,Tv_8[t]),$$

$$Tv_8'[t] == \frac{1}{h2^2} Dtv\,(0. + 25.6086\,Co - 42.5452\,Tv_2[t] + 33.1691\,Tv_3[t]$$
$$- 36.0093\,Tv_4[t] + 50.6609\,Tv_5[t] - 99.4201\,Tv_6[t] + 382.521\,Tv_7[t]$$
$$- 1294.77\,Tv_8[t]),$$

$$Qt'[t] == -\frac{1}{h2^2} Dtv\,(0. + 1.\,Co - 1.65797\,Tv_2[t] + 1.28025\,Tv_3[t] - 1.36027\,Tv_4[t]$$
$$+ 1.82857\,Tv_5[t] - 3.21856\,Tv_6[t] + 8.62616\,Tv_7[t] - 63.4982\,Tv_8[t]),$$
$$Tv_2[0] == 0, Tv_3[0] == 0, Tv_4[0] == 0, Tv_5[0] == 0, Tv_6[0] == 0, Tv_7[0] == 0,$$
$$Tv_8[0] == 0,$$
$$Qt[0] == 0\}$$

with the following variables:

```
{Qt[t],Tv2[t],Tv3[t],Tv4[t],Tv5[t],Tv6[t],Tv7[t],
Tv8[t]}.
```

The diffusion coefficient and membrane thickness are Dtv and $h2$, respectively; Qt is the cumulative amount of drug released; Tv_i is the concentration in layer i. The function *NDSolve* can be used to solve the system of ODEs:

```
NDSolve[eqns1,abc  ,{t,0,tf}].
```

16.5.2 Multiple Dosing in Two-Compartment Models

The mass balance equations for a drug in a two-compartment model are

$$\frac{dm_1}{dt} = -k_1 m_1 + k_2 m_2 + \text{Dose}\,\delta(t) - k_E m_1 \tag{16.27}$$

and

$$\frac{dm_2}{dt} = k_1 m_1 - k_2 m_2, \tag{16.28}$$

where m_i is the amount of drug in compartment i. The initial conditions are $m_1(0) = m_2(0) = 0$. For the simulations, the following parameters are used: $k_1 = 0.068$ hour^{-1}, $k_2 = 0.109$ hour^{-1}, and $k_E = 0.00592$ hour^{-1}. Three bolus doses of sizes 1.0, 2.0, and 3.0 µg are injected at times 0, 80, and 100 hours. The simulation time (t_f) is 150 hours. These specifications are written in *Mathematica* as

```
t1s=80;
y1t1s=2;
t2s=100;
y1t2s=3;
tf = 150;
y1inis = 1.0;
k1 = 0.0682389;
k2 = 0.10944;
ke = 0.00592105;
```

The times and doses are

```
ts = {t1,t2}; dose = {y1t1, y1t2};
```

Equations (16.27) and (16.28) are written as

$$\text{func} = \{y1[t], y2[t]\} / . DSolve[\{y1'[t] == -k1y1[t] + k2y2[t] - key1[t]$$
$$+ \sum_{i=1}^{\text{Length}[ts]} (\text{dose}[[i]] DiracDelta[t - ts[[i]]]), y2'[t] == k1y1[t] - k2y2[t],$$
$$y1[0] == y1ini, y2[0] == 0.\}, \{y1, y2\}, \{t\}][[1]];$$

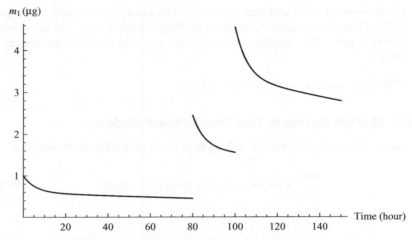

Figure 16.1. Amount of drug in the central compartment.

The two profiles are defined as

```
sol1[y1ini_?NumericQ, t1_?NumericQ, t2_?NumericQ,
y1t1_?NumericQ, y1t2_?NumericQ]=func[[1]]
```

and

```
sol2[y1ini_?NumericQ, t1_?NumericQ, t2_?NumericQ,
y1t1_?NumericQ, y1t2_?NumericQ]=func[[2]].
```

The amount of drug in the central compartment can be plotted (Fig. 16.1) using the following code:

```
Plot[sol1[y1inis, t1s, t2s, y1t1s, y1t2s],
{t,0,tf}, PlotStyle→{{Black,Thick}},LabelStyle→
{FontFamily→"Times",FontSize→14}, ImageSize→500,
AxesLabel→{"Time(h)]","m1 (μg)"}].
```

The graph (Fig. 16.2) of the amount of drug in the peripheral compartment is given by

```
Plot[sol2[y1inis, t1s, t2s, y1t1s, y1t2s],
{t,0,tf}, PlotStyle→{{Black,Thick}},LabelStyle→
{FontFamily→"Times",FontSize→14},ImageSize→500,
AxesLabel→{"Time(h)]","m2(μg)"}].
```

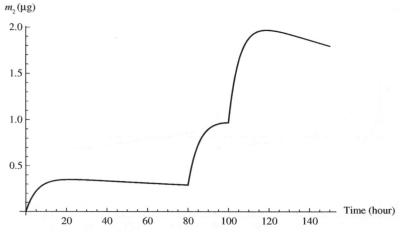

Figure 16.2. Amount of drug in the peripheral compartment.

Consider the following optimization problem:

$$F = \min_{D_0, D_1, D_2, t_1, t_2} \int_0^{t_f} [m_1(t) - m_{1\text{set}}]^2 \, dt, \tag{16.29}$$

where $m_{1\text{set}} = 2.5\,\mu g$. The objective function is first coded in *Mathematica*:

```
error[y1ini_?NumericQ, t1_?NumericQ, t2_?NumericQ,
y1t1_?NumericQ, y1t2_?NumericQ]:=NIntegrate[(y1sets-
sol1[y1ini, t1, t2, y1t1, y1t2])2, {t,0,150}].
```

As a result, Equation (16.29) becomes

```
NMinimize[{error[y1ini, t1, t2, y1t1, y1t2],0.0<
y1ini<10,0.0< t1<150,0.0< t2<150,0.0< y1t1<10,0<
y1t2<10},{y1ini, t1, t2, y1t1, y1t2}],
```

which yields

```
{10.3152,{t1→14.4035,t2→78.9673,y1ini→3.45578,y1t1
→1.20746,y1t2→0.974741}},
```

where the first number (i.e., 10.3152) is the objective value at the optimum solution.

The profiles $m_1(t)$ and $m_2(t)$ are shown in Figures 16.3 and 16.4.

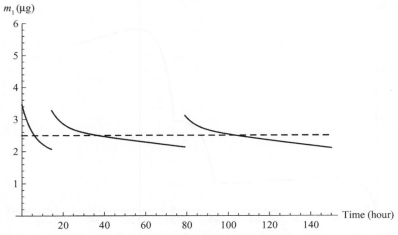

Figure 16.3. Amount of drug in the central compartment based on the following optimum dosage regimen: $D_0 = 3.45578$, $D_1 = 1.2074$, $D_2 = 0.974741$, $t_1 = 14.4035$, and $t_2 = 78.9673$.

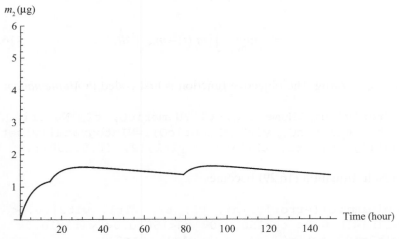

Figure 16.4. Amount of drug in the peripheral compartment after the following optimum dosage regimen: $D_0 = 3.45578$, $D_1 = 1.2074$, $D_2 = 0.974741$, $t_1 = 14.4035$, and $t_2 = 78.9673$.

16.6 SUMMARY

Optimum control and design problems encountered in drug delivery were presented. Although a complete description of the theories is beyond the scope of this book, selected examples are provided to encourage students of process control to learn about excellent opportunities and applications. With the help of numerical packages such as *Mathematica*, drug dosage strategies can be tested and the concentrations in each compartment monitored. The software can also be used to implement orthogonal collocation techniques to discretize PDE systems. This model reduction step is often necessary to tackle optimum control of distributed-parameter systems.

PROBLEMS

16.1. The discretized form of the one-dimensional Fick's second law for a planar membrane is

$$\frac{d}{dt}C(z_j, t) = \sum_{i=1}^{N+2} (i-1)(i-2)d_i(t)z_j^{i-3}$$

after using the orthogonal collocation method. Write a similar equation for the spherical coordinate.

Note: Start with $C(r_j, t) \approx \sum_{i=1}^{N+1} d_i(t)r_j^{2i-2}$.

16.2. Write the discretized form of the one-dimensional Fick's second law for a cylindrical geometry. Similar to Problem 16.1, use orthogonal collocations.

Note: Start with $C(r_j, t) \approx \sum_{i=1}^{N+1} d_i(t)r_j^{2i-2}$.

16.3. Plot the drug concentrations at the collocation points for the membrane diffusion problem in Section 16.5.1.

16.4. Plot the cumulative amount of drug released for the membrane diffusion problem in Section 16.5.1.

16.5. Redo the optimum dosage regimen problem in Section 16.5.2 using four bolus injections (i.e., calculate {y1ini, t1, t2, t3, y1t1, y1t2, y1t3}).

16.6. Generate the drug mass profiles $m_1(t)$ and $m_2(t)$ using the integral of the absolute value of the error (*IAE*) performance index:

$$IAE \approx \int_0^{t_f} |e(t)| dt.$$

Apply the specifications in Section 16.5.2.

16.7. Redo Problem 16.6 using the integral of time-weighted absolute error (*ITAE*) performance index:

$$ITAE \approx \int_0^{t_f} t|e(t)| dt.$$

16.8. Increase the diffusion coefficient in Problem 16.3 by 50% and compare the concentrations in the membrane to the profiles generated in Problem 16.3.

16.9. Study the effects of an increase in the partition coefficient (K) on the cumulative amount of drug released.

Note: Use a 50% increase in K and apply the specifications given in Section 16.5.1.

16.10. Study the effects of the membrane thickness (h) on the cumulative amount of drug released.

Note: Increase h by 10% and apply the specifications given in Section 16.5.1.

REFERENCES

1. Arora JS. *Introduction to Optimum Design*, 2nd ed. Amsterdam: Elsevier Academic Press, 2004.
2. Finlayson BA. *Nonlinear Analysis in Chemical Engineering*. London: McGraw-Hill International Book Co., 1980.
3. Bryson AE. *Dynamic Optimization*. Menlo Park, CA: Addison Wesley Longman, 1999.
4. Simon L. Repeated applications of a transdermal patch: analytical solution and optimal control of the delivery rate. *Mathematical Biosciences* 2007; 209:593–607.
5. Siepmann J, Lecomte F, Bodmeier R. Diffusion-controlled drug delivery systems: calculation of the required composition to achieve desired release profiles. *Journal of Controlled Release* 1999; 60:379–389.
6. Simon L, Fernandes M. Neural network-based prediction and optimization of estradiol release from ethylene–vinyl acetate membranes. *Computers and Chemical Engineering* 2004; 28:2407–2419.

INDEX

Control of Biological and Drug-Delivery Systems for Chemical, Biomedical, and Pharmaceutical Engineering, First Edition. Laurent Simon.
© 2013 John Wiley & Sons, Inc. Published 2013 by John Wiley & Sons, Inc.

Printed and bound by CPI Group (UK) Ltd, Croydon, CR0 4YY

16/04/2025

14658353-0001